Advances in Soft Computing 43

Editor-in-Chief: J. Kacprzyk

Advances in Soft Computing

Editor-in-Chief

Prof. Janusz Kacprzyk
Systems Research Institute
Polish Academy of Sciences
ul. Newelska 6
01-447 Warsaw
Poland
E-mail: kacprzyk@ibspan.waw.pl

Further volumes of this series can be found on our homepage: springer.com

Katarzyna M. Węgrzyn-Wolska,
Piotr S. Szczepaniak (Eds.)

Advances in Intelligent Web Mastering

Proceedings of the 5th Atlantic Web Intelligence
Conference – AWIC'2007,
Fontainebleau, France, June 25–27, 2007

 Springer

Editors

Dr. Katarzyna M. Węgrzyn-Wolska
ESIGETEL
Ecole Supérieure d'Ingénieurs en Informatique et
Genie des Telécommunications
1, rue du Port de Valvins
77215 Avon Fontainebleau
France
E-mail: kasia@wolski.net,
    katarzyna.wegrzyn@esigetel.fr

Prof. Dr. Piotr S. Szczepaniak
Institute of Computer Science
Technical University of Lodz,
ul. Wolczanska 215
93-005 Lodz
Poland
and
Systems Research Institute
Polish Academy of Sciences
ul. Newelska 6
01-447 Warsaw
Poland

Library of Congress Control Number: 2007927164

ISSN print edition: 1615-3871
ISSN electronic edition: 1860-0794
ISBN-10 3-540-72574-1 Springer Berlin Heidelberg New York
ISBN-13 978-3-540-72574-9 Springer Berlin Heidelberg New York

Springer is a part of Springer Science+Business Media

springer.com

© Springer-Verlag Berlin Heidelberg 2007
Printed in Germany

Typesetting: by the authors and SPS using a Springer LATEX macro package

Printed on acid-free paper   SPIN: 12065806   89/SPS   5 4 3 2 1 0

Preface

The World Wide Web (Web for short) provides numerous far reaching services and there is no doubt about the relevance of the Web in our daily activities at home and at work. It has become an information source of unquestionable importance and popularity. The challenge for research and technology is the enormous size and, for the most part, the lack of a well-defined structure, precision and diversity. Consequently, mastering and effective exploration of the web environment needs new ideas, adaptation of known methods and application of sophisticated and robust implementations. Among new research directions present in relation to Web applications, a remarkable place is occupied by intelligent methods defining the topic of soft computing.

The "Atlantic Web Intelligence Conferences" (AWIC) are conceived as a forum for the exchange of ideas and recently verified practical solutions in this new and exciting field. The Conference arose as an activity of the WIC-Poland and the WIC-Spain Research Centres, both belonging to the Web Intelligence Consortium - WIC (http://wi-consortium.org/). So far three AWIC conferences have been held: in Madrid, Spain (2003), in Cancun, Mexico (2004), in Łódź, Poland (2005), and in Beer-Sheva, Israel (2006).

Authors of papers contribute to diverse fields of the Web: application of artificial intelligence, design, information retrieval and interpretation, user profiling, security, engineering, etc. It presents a view on the subject matter which is wider and more complete. The contributions included in this volume were carefully selected by the reviewers. We estimate all the submitted works at their proper value and regret that not all of them can be included. The book starts with material submitted by keynote speakers and is followed by conference papers arranged in alphabetical order according to the name of the first Author.

We acknowledge the effort of the keynote speakers and we thank them for the preparation of plenary presentations. They are Professors: Ryszard Tadeusiewicz (AGH University of Science and Technology, Krakow, Poland), Jiming Liu (Hong Kong Baptist University and University of Windsor, Canada), and Ning Zhong (Maebashi Institute of Technology, Japan) - President of WIC. We are indebted to the reviewers for their reliability and hard work under considerable pressure.

Grateful appreciation is expressed to the team organizing the 5th Atlantic Web Intelligence Conference in 2007. Particular thanks are also given to Professor Janusz Kacprzyk, Editor of the Series publishing this book and for the Springer team for its practical and friendly help. The technical cooperation of the IEEE Computational Intelligence Society, is highly appreciated. We would like to extend our thanks to all involved not directly mentioned by name.

Our hope is that every reader will find in this many motivating ideas volume.

Fontainebleau, France   *Katarzyna Węgrzyn-Wolska*
June, 2007   *Piotr S. Szczepaniak*

Organization

Organizers

ESIGETEL, Avon-Fontainebleau, France
Technical University of Łódź, Poland
Systems Research Institute, PAS, Poland

Technical Co-operation

IEEE Computational Intelligence Society
Web Intelligence Consortium (WIC)

Honorary Chair

Lotfi A. Zadeh, University of California, Berkeley, USA

Chairmen

Katarzyna Węgrzyn-Wolska, ESIGETEL Avon-Fontainebleau, France
Piotr S. Szczepaniak, Technical University of Łódź, Poland

Steering Committee

Jesus Favela, CICESE, Mexico
Janusz Kacprzyk, Polish Academy of Sciences, Poland
Jiming Liu, Hong Kong Baptist University, Hong Kong, China
Masoud Nikravesh, University of California, Berkeley, USA
Javier Segovia, UPM, Madrid, Spain
Ning Zhong, Maebashi Institute of Technology, Maebashi-City, Japan

Keynotes Speakers

Jiming Liu, Hong Kong Baptist University, Hong Kong, China
Ryszard Tadeusiewicz, AGH University of Science and Technology, Poland
Ning Zhong, Maebashi Institute of Technology, Maebashi-City, Japan

Program Committee

Michel Beigbeder	ESMSE, Saint Etienne, France
Patrick Brezillon	Université Paris 6, France
Alex Büchner	University of Ulster, Northern Ireland UK
Edgar Chavez	Universidad de Michoacan, Mexico
Pedro Alexandre da Costa Sousa	Uninova, Portugal
Alfredo Cuzzocrea	University of Calabria
Gaël Dias	Beira Interior University, Portugal
Lipika Dey	Indian Institute of Technology, India
Santiago Eibe	UPM, Spain
Jesús Favela	CICESE, Mexico
Michael Hadjimichael	Naval Resarch Laboratory, USA
Enrique Herrera Viedma	Universidad de Granada Spain
Pilar Herrero	UPM, Spain
Esther Hochzstain	ORT Uruguay
Andreas Hotho	Universidad de Karsruhe, Germany
Janusz Kacprzyk	Systems Research Institute, Polish Academy of Sciences, Poland
Abraham Kandel	University of South Florida, USA
Samuel Kaski	Helsinki University of Technology, Finland
Józef Korbicz	Technical University of Zielona Góra, Poland
Jacek Koronacki	Institute of Computer Science, Polish Academy of Science, Poland
Rudolf Kruse	Otto-von-Guericke University of Magdeburg, Germany
Mark Last	Ben-Gurion University of The Negev, Israel
Pawan Lingras	Saint Marys University, Halifax, Canada
Jiming Liu	University of Windsor, Canada.
Vincenzo Loia	University of Salerno, Italy
Aurelio Lopez	INAOE, Mexico
Óscar Marbán	Facultad de Informática. UPM Spain
Oscar Mayora	ITESM-Morelos, Mexico
Ernestina Melasalvas	Technical University of Madrid, Spain
Bamshad Mobasher	DePaul University School of Computer Science, Chicago, USA
Manuel Montes-y-Gomez	INAOE, Mexico
Pierre Morizet-Mahoudeaux	University of Compiègne, France
Alex Nanolopoulos	Aristotle University, Greece
Nadia Nedjah	State University of Rio de Janeiro, Brazil
Marian Niedzwiedziński	University of Lódź, Poland
Masoud Nikravesh	University of California, Berkeley, USA
Witold Pedrycz	University of Alberta, Canada

María Pérez	Facultad de Informática. UPM Spain
Paulo Quaresma	Universidade de Évora, Portugal
Victor Robles	Facultad de Informática. UPM Spain
Danuta Rutkowska	Technical University of Częstochowa, Poland
Leszek Rutkowski	Technical University of Częstochowa, Poland
Eugenio Santos	E.U. Informática. UPM, Spain
Javier Segovia	UPM, Spain
Andrzej Skowron	University of Warsaw, Poland
Roman Słowiński	Poznań University of Technology, Poland
Myra Spiliopoulou	University of Magdeburg, Germany
Ryszard Tadeusiewicz	AGH University of Science and Technology, Poland
Andromaca Tasistro	Universidad de la republica, Uruguay
Maria Amparo Vila	University of Granada, Spain,
Anita Wasilewska	Stony Brook New York University, USA
Jan Węglarz	Poznań University of Technology, Poland
Katarzyna Węgrzyn-Wolska	ESIGETEL, Avon-Fontainebleau, France
Ronald Yager	Iona College, USA
Yiyu Yao	University of Regina, Canada
Sławomir Zadrożny	Systems Research Institute, Polish Academy of Sciences, Poland
Ning Zhong	Maebashi Institue of Technology, Japan
Wojciech Ziarko	University of Regina, Canada

Scientific Secretary

Blaise-Florentin Collin, UTEC Informatique, France

Local Organizing Committee

Catherine Bernard, ESIGETEL, France
Daphné Blanc, Euromed Marseille, France
Lamine Bougueroua, ESIGETEL, France
Laurent Daverio, École des Mines de Paris, France
Grzegorz Dziczkowski, ESIGETEL, France
Jan Wolski, AENIX, France

Contents

Part I

Keynotes

Intelligent Web Mining for Semantically Adequate Images

Ryszard Tadeusiewicz

AGH University of Science and Technology,
30 Mickiewicza Av.,
30-059 Krakow, Poland
rtad@agh.edu.pl
http://www.agh.edu.pl/english/tad/

Summary. Web mining is good established technology when we must search data in the form of plain text. For this common purpose many well designed algorithms an technologies (like ontologies) has been developed and adopted. Other situation is when the most important information, which must be searched and retrieved from multimedial databases, is presented in the form of digital image. Meanwhile we observe more and more important information which is obtained, collected, stored and retrieved in form of the many kind of images. Moreover the information is represented not only by the form of objects presented on the image, but also can be hidden in complex form in relations between objects. Therefore for achieve successful web mining for semantically adequate images we must have tools suitable for automatic understanding of image merit content. In the paper we describe the new method of automatic understanding of the images, proved in the medical images practice and ready for use for other kind of images.

Keywords: Web mining, Multimedia, Semantics of the images, Automatic understanding.

1 Introduction

Up-to-date content of the web in not only in text form, but become most and most multimedial. Moreover the merit sense of many information located in the web is expressed mainly in visual (picture) form instead of pure text. As the examples of such type visual merit content can serve medical databases with big amount of medical images (results of many modern methods of medical imaging devices) of whole human body and particular organs both healthy and showing many types of pathologies. Another interesting example can be GIS servers with geo-informatics information presented mainly in form of digital maps. Also many commercial information, like tourist information, hotels and restaurants advertisings, e-commerce offers, artistic articles presentations and so on, and so on — are based on visual information, which is not only additional information to the textual one, but often contains the most important part of the communicate from the potential viewer point of view.

K.M. Węgrzyn-Wolska and P.S. Szczepaniak (Eds.): Adv. in Intel. Web, ASC 43, pp. 3–10, 2007.

In all considered above situation we must take into account this visual part of the web content during the intelligent searching of proper source of information. Formulating this problem on most general plane we can claim, that in fact it is necessary to have tool for discovery of semantic content and merit sense also if it is hidden in the pictures presented on the selected websites. In future such tools can be very important during the web mining processes, because multimedial nature of the increasing number of websites.

It must be stressed, that the goal mentioned above needs now scientific research, because proper methods and related algorithms need to be discovered. Actual methods and tools for web mining, also if there are semantic oriented, e.g. ontologies, quite good in applications devoted to searching the merit sense hidden in text form of web content, are definitely not applicable when we must search for particular semantic content hidden on pictures, images, photos, maps and also medical imaginations. Fortunately in the domain of computer vision the new group of algorithms discovered, supporting traditional technologies of image processing, analysis, and recognition by new one: automatic understanding of the images merit sense. In the paper we try to explain, what is automatic understanding of the images and why this technology can be very fruitful in web mining when we must search the information included on web nodes only in the form of images.

2 Solutions Given by Other Authors

The problem of intelligent searching for properly selected images is known since last three or four years. Before we introduce our method, based on automatic understanding principle, we try to show, how other authors try to solve the problem under consideration. Showing such 'competitive' methods we try to explain, why such methods are not sufficient for many real problems and why the automatic understanding approach is in fact only appropriate.

Every web crawling tool offers now some image searching utilities. For example *Google's Image Search* claimed as 'the most comprehensive on the Web' is one of them. Another examples can be pointed out *Yahoo! Image Search* engine, *AltaVista*, *Picsearch* (Image Search for pictures and images), and many others. In fact image searching process offered by almost all such famous web searching programs is very primitive, because is based on annotated databases only.

If considered multimedial database is annotated (e.g. includes both images and its linguistic descriptions) the problem is very easy. If we have for every image the exact annotation of its merit content, even if it is given in arbitrary textual form — we can search very easy for proper image. This kind of searching, based of text annotations only, can be clusory. For example if we give ad the key for searching image of the tiger, we can find both the portraits of many beautiful animals and images of the flower named Bengal Tiger. Moreover this solution is acceptable for well annotated multimedial databases only, when many valuable collections of important images are not annotated at all or are annotated without deeper knowledge on the image merit content, for example when annotation is performed by webmaster and images are very complicated and needs professional knowledge (e.g. medical images).

Therefore all popular web-searching machines with option "image search" are in general not the proper solution for the whole problem formulated in this paper. Nevertheless such solutions must be mentioned above, because this way of solving by means of replacing image searching problem by easier (but not in every case equivalent) text searching problem is up-to-date most popular method, often used for practical applications.

Many authors try find another way and looking for selected properties and features, which can be computed automatically on the base of image, and may help to find answer the question, how to select and retrieve proper images form huge multimedial database (e.g. from the internet) according to the particular criteria. Most commonly considered problem is formulated in such form: how to find in database images similar — in some sense — to the images presented as the examples of interesting class of images. The methods used for solution problem formulated above are mostly based on some simple features extracted from the image. For example if we take into account color as a selected feature of the image, we can build some heuristic hypothesis for image content selection. Image with big amount of blue color pixels in upper part of the frame and green color pixels in lower part of the frame can be classified automatically to the landscape category, when another picture with big areas filled by human skin color pixels can be categorized as nudity.

Features can be off course different, for example based on image texture parameters (for classification of satellite images) or related to the particular objects, which can be easy identified on the image (for example buildings). If we can extract good set of features (or parameters) properly characterizing example image, and if we can extract the same set of parameters for every image in database — we can compare such images performing comparison of features vectors. This comparison can be based on many known distance measures — and the problems is apparently solved.

Apparently, because in fact this method can be used for the very limited number of situations only. For most examples of the searched images it is very difficult to show the features which values are sufficient for discrimination, if the image belongs to the interesting class, or not. Therefore such method can be powered using learning procedures [1]. Nevertheless approach introduced in next chapter of this paper seems to be better.

3 Solution Based on Automatic Understanding of the Image Content

All formal methods of image analysis and recognition are not sufficient for **intelligent** web mining, because it is the gap between the form of the image (which can be evaluated by means of analytical procedures and classified by means of pattern recognition algorithms) and a merit content of the image, which can be discovered only by the automatic understanding approach.

Let me use a **joke** for explanation the difference between image analysis and also image recognition - and image understanding. This joke was first presented (in more comprehensive form) in my book describing the main idea of image understanding technology [2]. Imagine, that we have big multimedial database and we try to find picture "telling the same story", as pictures given as the examples. The example images are presented on the Fig. 1.

Fig. 1. Example images

Lets think together, how to describe criteria for intelligent selection of next pictures similar (in sense of semantic content) to the shown on the Fig. 1 from a multimedial database?

Lets think together, how to describe criteria for intelligent selection of next pictures similar (in sense of semantic content) to the shown on the Fig. 1 from a multimedial database?

For solving of selected problem using classical image analysis one needs following operations (Fig. 2):

- Segmentation of all pictures and selection of important objects on all images
- Extraction and calculation of the main features of the selected objects
- Objects classification and recognition
- Selection of features and classes of recognized objects which are the same on all images under consideration
- Searching in database for images having the same objects with the same features

Similar analysis can be done for next example images (we skip here the figures, but reader can easily imagine the results). After this analysis, summarizing all information we can do such induction:

- On all images we can find two objects: Women and Vehicle
- On some images there are also object Man, but not on all - so Men can be automatically considered as not important in searching criteria

Result: computer finds and presents as desired output all images with women and vehicle (Fig. 3)

It is very easy to find out, that the method of the image selection discovered by the automate is wrong in such situation, because example of proper image from the database is shown on Fig. 4, although Fig. 4 does not contain Vehicle at all!

Fig. 2. Results of the formal analysis and classification applied to the first image under consideration

Fig. 3. Images selected by Google on the base of criteria *"women and vehicle"*

Fig. 4. Image "telling the same story" as images on Fig. 1

It was off course only joke (we apologize...), the matter although is quite serious because very often images apparently very different in fact can hide the semantically identical content - and vice versa: apparently very similar images can have dramatically different meaning.

It is evident, that when we talking about intelligent web mining for **semantically** adequate images, we must base on automatic understanding of the image content instead of using any formal criteria.

4 How Automatic Understanding of the Images Works?

When idea of automatic understanding of the images was discovered and its description was given in the book [2] we try use this new methodology for many practical applications, describer in papers[3, 4]. We showed the role of mathematical linguistic in automatic understanding approach [5, 6], and we also presented, when and why this new approach is necessary [7].

Trying to explain what automatic understanding (AU) is and how we can force the computer to understand the image content we must demonstrate the fundamental difference between a formal description of an image and the content meaning of the image, which can be discovered by an intelligent entity, capable of understanding the profound sense of the image in question. The fundamental features of automatic image understanding can be listed as follows:

- We try to imitate the natural way in which a qualified professional thinks.
- We make a linguistic description of the image content, using a special kind of an image description language. Owing to this idea we can describe every image without specifying any limited number of a priori described classes.
- The linguistic description of an image content constructed in this manner constitutes the basis for the understanding of image merit content.

The most important difference between all traditional methods of automatic image processing and the new paradigm for image understanding is that there is one directional scheme of the data flow in the traditional methods while in the new paradigm there are two-directional interactions between signals (features) extracted from the image analysis and expectations resulting from the knowledge of image content, as given by experts. In Fig. 5 we can see a traditional chart representing image processing for recognition purposes.

Classical Pattern Recognition Process

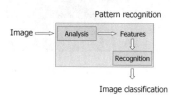

Fig. 5. Traditional method of medical image recognition

Unlike in this simple scheme representing classical recognition, in the course of image understanding we always have a two-directional flow of information (Fig. 6).

In both figures we can see that when we use the traditional pattern recognition paradigm, all processes of image analysis are based on a feed-forward scheme (one-directional flow of signals). On the contrary, when we apply automatic understanding of the image, the total input data stream (all features obtained as a result of an analysis of the image under consideration) must be compared with the stream of **demands** generated by a dedicated **source of knowledge**. The demands are always connected with a specific (selected) hypothesis of the image content semantic interpretation. As a result, we can emphasise that the

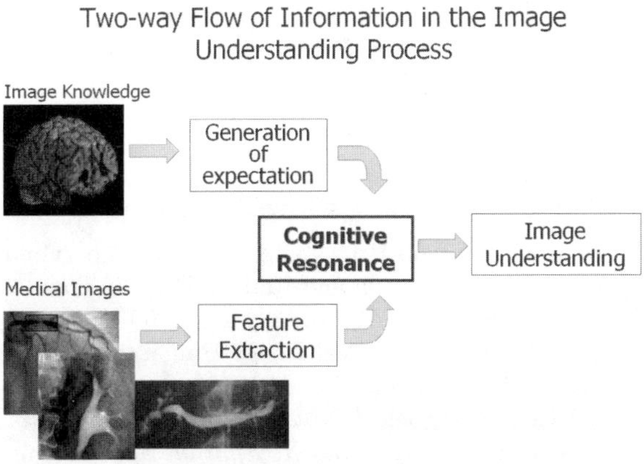

Two-way Flow of Information in the Image
Understanding Process

Fig. 6. The main paradigm of image understanding

proposed 'demands' are a kind of postulates, describing (basing on the knowledge about the image contents) the desired values of some (selected) features of the image. The selected parameters of the image under consideration must have desired values when some assumption about semantic interpretation of the image content is to be validated as true. The fact that the parameters of the input image are different **can** be interpreted as a **partial** falsification of one of possible hypotheses about the meaning of the image content, however, it still cannot be considered the final solution.

Due to this specific model of inference we name our mechanism the 'cognitive resonance'. This name is appropriate for our ideas because during the comparison process of the features calculated for the input image and the demands generated by the source of knowledge we can observe an amplification of some hypotheses (about the meaning of the image content) while other (competitive) hypotheses weaken. It is very similar to the interferential image formed during a mutual activity of two wave sources: at some points in space waves can add to one another, in other points there are in opposite phases and the final result is near zero.

Such a structure of the system for image understanding corresponds to one of the very well known models of the natural human visual perception, referred to as 'knowledge based perception'. The human eye cannot recognise an object if the brain has no template for it. This holds true even when the brain knows the object, but shown in another view, which means that other signals are coming to the visual cortex. Indeed, natural perception is not just the processing of visual signals received by eyes. It is mainly a mental cognitive process, based on hypotheses generation and its real-time verification. The verification is performed by comparing permanently the selected features of an image with expectations taken from earlier visual experience. Our method of image understanding is based on the same processes with a difference that it is performed by computers.

Acknowledgement. This work has been supported by the Ministry of Science and Higher Education, Republic of Poland, under project number N 519 007 32/0978; decision no. 0978/T02/2007/32.

References

1. Carneiro G., Chan A.B., Moreno P.J., Vasconcelos N.: Supervised Learning of Semantic Classes for Image Annotation and Retrieval, IEEE Trans. on Pattern Analysis and Machine Intelligence, vol. 29, No. 3, March 2007, pp. 394 - 410
2. Tadeusiewicz R., Ogiela M. R.: Medical Image Understanding Technology, Series: Studies in Fuzziness and Soft Computing, Vol. 156, Springer-Verlag, Berlin, 2004
3. Ogiela L., Tadeusiewicz R., Ogiela M.R.: *Cognitive Approach to Visual Data Interpretation in Medical Information and Recognition Systems*, In: Nanning Zheng, Xiaoyi Jiang, Xuguand Lan (Eds.): Advances in Machine Vision, Image Processing, and Pattern Analysis, Lecture Notes in Computer Science, vol. 4153, Springer-Verlag, Berlin - Heidelberg - New York, 2006, pp. 244-250
4. Ogiela L., Tadeusiewicz R., Ogiela M. R.: *Cognitive Computing in Intelligent Medical Pattern Recognition Systems*, In: De-Shuang Huang, Kang Li, Irwin G.W. (eds.): Intelligent Control and Automation, Lecture Notes in Control and Information Sciences, Vol. 344, Springer-Verlag, Berlin - Heidelberg - New York, 2006, pp. 851-856
5. Ogiela M.R., Tadeusiewicz R., Ogiela L.: *Graph Image Language Techniques Supporting Radiological, Hand Image Interpretations*, Computer Vision and Image Understanding, Elsevier, Vol. 103/2, 2006, pp 112-120
6. Ogiela M.R., Tadeusiewicz R., Ogiela L.: *Image Languages in Intelligent Radiological Palm Diagnostics*, Pattern Recognition, Vol. 39/11. 2006, pp. 2157-2165
7. Tadeusiewicz R., Ogiela M.R.: *Why Automatic Understanding?* In Beliczynski B. et al. (Eds.): ICANNGA 2007, Part II, Lecture Notes on Computer Science, vol. 4432, Springer-Verlag, Berlin - Heidelberg - New York, 2007, pp. 477 - 491

Towards Human-Level Web Intelligence:
A Brain Informatics Perspective

Ning Zhong

Department of Life Science and Informatics
Maebashi Institute of Technology, Japan
The International WIC Institute/BJUT, China
zhong@maebashi-it.ac.jp

Abstract. One of the fundamental goals of WI research is to understand intelligence in depth and develop wisdom Web based intelligent systems that integrate all the human-level capabilities.

In this talk, we briefly investigate several ways to develop human-level Web intelligence (WI) from a brain informatics (BI) perspective. BI can be regarded as brain sciences in WI centric IT age and emphasizes on a systematic approach for investigating human information processing mechanism.

The recently designed instrumentation (fMRI etc.) and advanced IT are causing an impending revolution in both WI and BI, making it possible for us to understand intelligence in depth and develop human-level Web intelligence.

K.M. Węgrzyn-Wolska and P.S. Szczepaniak (Eds.): Adv. in Intel. Web, ASC 43, p. 11, 2007.
springerlink.com   © Springer-Verlag Berlin Heidelberg 2007

Part II

Regular Papers

The Criteria for Effective Electronic Negotiation Systems

Oyindamola Abass[1] and Gheorghita Ghinea[2]

[1] Brunel University Uxbridge Middlesex UB8 3PH
   oyindamola.abass@brunel.ac.uk
[2] Brunel University Uxbridge Middlesex UB8 3PH
   george.ghinea@brunel.ac.uk

1 Introduction

Negotiations continue to play an important part of our lives and in particular influences can be seen in the growth in electronic commerce through various auction and online stores. Estimates of up to 7bn were predicted for Christmas 2006, a rise of about 35%-40% on the previous year [1]. This growth can only enhance the demand for electronic negotiation systems or at least highlight the lack of commercial e-negotiation systems. Such systems, though available in mostly academic circles will need to be enhanced before they can replace traditional face-to-face negotiations. As e-commerce continues to grow, we anticipate a faster uptake of negotiation systems if their design can mimic traditional systems. Our work proposes to draw the key attributes that need to be present for a successful negotiation system by examining the current state of the art in electronic negotiation frameworks. The rest of this paper is structured as follows: Section 2 elaborates on the need for electronic negotiation systems, with Section 3 examining the current literature in negotiation frameworks, whilst Section 4 presents the attributes which we found required in negotiation frameworks. Conclusions are drawn in Section 5.

2 Why E-Negotiate?

E-commerce has changed the way businesses are conducted today. Almost every medium to large-scale company has some form of web presence either via a website or an online catalogue. Factors such as lower operating cost, integration to business cycle and wider market reach with no geographical boundaries have continued to attract new and existing businesses to e-commerce. Thus, for situations requiring negotiations, one can expect more negotiations to take place in electronic rather than in traditional markets. Support for negotiations in electronic markets is therefore not only a necessity but also a critical success factor for many ecommerce market ventures. [10]. Today, electronic negotiations are being applied in various fields including electronic commerce, international relations and arbitration, contract management to name a few. Though the scope

K.M. Węgrzyn-Wolska and P.S. Szczepaniak (Eds.): Adv. in Intel. Web, ASC 43, pp. 15–20, 2007.
springerlink.com   © Springer-Verlag Berlin Heidelberg 2007

of some of these is limited, the adoption of electronic negotiations into various disciplines continues to attract researchers and practitioners from various disciplines.

3   Frameworks

Electronic negotiation systems like most other systems are built on frameworks. Most of these frameworks have focused on implementing negotiation systems whilst neglecting the modeling aspects of negotiations. Hence, unique and proprietary solutions are created repeatedly, with enormous efforts spent on integrating isolated solutions [7]. The development of generic electronic negotiation frameworks can solve this problem. Frameworks are guiding concepts or tools for modeling a class of interesting real world cases. They help us to systematically and comprehensively identify, define, and prioritize the problems in a certain domain. Various frameworks have been developed to date to support different negotiation systems. Some of these frameworks are now classified according to the applications which they support.

3.1   Applications in Auctions

An auction is a market institution with rules for resource allocation and prices on the basis of bids from the participants. Auctions typically have very small transaction costs and are used to conduct many transactions among businesses and between businesses and consumers. Negotiation frameworks to support auctions include those of Bellosta et al. [4] who put forward a multi-criteria model for electronic auctions. This reference-points based model allows the buyer agent to control the negotiation process on each attribute of the deal. In other work, Bichler et al. developed the Multidimensional Auction Platform (MAP) as a set of software modules for building multidimensional auction markets. MAP is an extensible object framework, which enables the reuse of the advanced allocation algorithms as a standard solver component in electronic markets. It provides a declarative interface and sheds developers from the complexities of a particular allocation algorithm. [5]

3.2   Applications in Multi-agent Systems

Negotiation software agents carry out negotiation activities on behalf of users. They have the potential to save the human negotiators time and find better deals in combinatorial and strategically complex settings. [7] In related work, Jennings et al. [6]developed a generic framework for classifying and viewing automated negotiations. They suggest the use of argumentation-based approaches to allow additional information to be exchanged and make it possible for agents to handle conflicting information. On the other hand, Abass and Ghinea, proposed SOLACE - a generic framework for multi-issue negotiations, which can be applied to a variety of negotiation scenarios using software agents., SOLACE supports hybrid systems in which the negotiation participants can be humans, agents or a combination of the two. [2]

3.3   Application in Electronic Markets

An electronic markets is a marketplace for negotiating the purchase and sale of goods using. Compared to traditional markets, electronic markets have fundamentally different. Strobel [10], proposed a design and application framework for electronic negotiations. Based on this framework, organizations creating an electronic market or sellers intending to offer potential buyers the option to bargain can generate, in a flexible and efficient way, customized electronic negotiation systems supporting the roles and protocols designed. Although, this classification is not exhaustive it does illustrate the breadth of approaches. In the next section, we draw out the peculiar features which the authors believe need to be present to support todays negotiation systems.

4   Criteria for Electronic Negotiation Systems

The need for general negotiation frameworks is a major requirement for todays electronic negotiation systems. Frameworks serve as a guide to developers of new systems and provide a basis for evaluation and analysis. Most of the frameworks in existence today are either too complex - [4]making difficult their implementation; too technical [12] - thus ignoring important factors such as negotiation strategies - or just consider solutions to specific problems and thus cannot be re-used in other areas. From the review of negotiation frameworks we have identified some characteristics which need to be addressed in the development of an electronic negotiation system. These characteristics have been put together in a framework - SOLACE II which is an improvement to the authors previous work on SOLACE. SOLACE II prescribes the building blocks of negotiation systems and recognizes key features of multi attributes, flexibility, negotiation strategy, platform independence, hybrid negotiation, validity and learning. These features can be applied in electronic auctions systems, multi-agent systems and electronic markets. These features are now discussed:

4.1   Multi-attribute Negotiation

Multi-attribute negotiation is increasingly becoming more ensconced in todays negotiation systems. Previously, most systems negotiated only on price, making them somewhat unrealistic because several other factors come to play in reaching agreements such as trust, security, delivery date, quality and so on. Todays negotiation systems need to cater for multiple attributes. Negotiators need to be able to bargain on several issues relevant to the negotiation making them dynamic. Participants should also be able to introduce new attributes where necessary.

4.2   Flexibility

The flexibility of negotiation frameworks can be said to be directly proportional to its usability, allowing developers to easily incorporate them into their designs. Frameworks should be flexible, specifying the basic building blocks of the electronic negotiations - protocols, participants, objects or strategies Irrespective of the specific objects or fields, the framework should be applicable to a variety of scenarios ranging from buying and selling in e-commerce to any other automated negotiation scenario. Strobel in SILKROAD [10] proposed a design and application framework for electronic negotiations. SILKROAD attempts to provide a generic framework for negotiation systems. However, implementations of systems based on this framework are very few as it appears to be too complex and is not easily adoptable.

4.3   Negotiation Strategy

Negotiation strategies drive the entire negotiation process. The negotiation strategy determines what issues will be negotiated and in what order. Such strategies could be distributive (win-lose) or integrative (win-win). Some researchers have argued against incorporating strategies into frameworks saying that negotiation strategies lead to complete automation of electronic negotiations [8],that the strategies will either be too simple (easily deciphered) or too complex to be formalised or that the strategies will not gain the trust of users. However, the importance of strategies cannot be over-emphasized. Strategies can distinguish the winners and losers in any scenario.

4.4   Platform Independence

Platform independence should be an essential feature of any framework. As well as being flexible, frameworks should not be tied to any particular development environment. For example, Bartolini et al [3]. tied their framework to the Jade multi-agent platform integrated with the Java Expert System Shell (Jess). Rule specification in the general negotiation protocol is based on Jess Assertions. This is not a truly open system and restricts developers to using this platform. Other frameworks such as SILKROAD or SOLACE on the other hand do not limit developers to any particular platform for development. The developers are given a free hand at choosing platforms that are well suited for their development scenarios.

4.5   Hybrid Negotiation

Despite all advances in automation, negotiation participants still prefer to be involved in the agreement phase of electronic transactions. In hybrid systems, structured or formalised tasks are automated, and decision support mechanisms are used to assess unstructured tasks, whereas humans interactively control the execution of the negotiation and perform the exception handling. Negotiation frameworks should therefore provide support for hybrid negotiation systems.

4.6   Time or Validity

Another important factor in negotiations is time - particularly its influence on the strategies of participants. Negotiations often breakdown if they take too long as participants may be distracted by changes in the environment or potential offers from other parties. Time could also influence the strategy of a party. In a particular negotiation, time could be an important issue if there is a deadline for reaching an agreement. The result could be a change in the strategy or sub-optimal agreements could be seen as better than no agreement. Lee [9]proposed a framework, which emphasizes the time involved in a negotiation process. Lee proposes that a time attribute be attached to each message to represent the period in which the message is valid.

4.7   Learning and Adaptation

The negotiation framework should emphasize learning and adaptation in negotiation systems. The negotiators can change tactics during the course of negotiations and possibly introduce new attributes. Learning can also be in the form of using past experience. Negotiators can use the knowledge gained in previous negotiations about the participants or scenarios to help in the negotiations. For example, if the negotiation exists in a dynamic environment and the environment changes, then the agent has to learn about the changes to be able to carry out its mission. Wong et al. [12] proposed a framework based on Case-Based Reasoning (CBR). Case-based reasoning is an approach to use past experience for choosing concrete strategy in every situation. The framework is built on top of a database with information on past negotiations. This approach, though very useful, requires a lot of information gathering and may be influenced by some historical factors present at the time the previous negotiations occurred but which may have become irrelevant in the present scenario

5   Conclusion

Approaches to electronic negotiation frameworks vary widely across board and there is no unilaterally accepted framework. The detailed evaluation carried out highlights the major factors needed for effective negotiation systems. The objective is to provide future developers of negotiation systems a guide to choosing appropriate frameworks on which to base their systems. This will reduce development time, improve consistency and in time lead to more effective and efficient systems of higher standards. The authors have incorporated the factors stated here into the development an electronic negotiation framework SOLACE II. This is an enhancement of SOLACE [2]. It is currently being used in the development of a confidence-based electronic negotiation system which incorporates the hybrid functionality discussed above. We aim to bring to the fore the need to have standards in the development of negotiation systems. With the continuous growth in electronic commerce, standardization of negotiation systems cannot be ignored for much longer and this provides the thrust of our future research efforts.

References

1. Curtis J, http://www.brandrepublic.com/bulletins/br/article/605937/what-high-street-wants-christmas accessed 22 Nov 2006
2. Abass O, Ghinea G, SOLACE: A Framework for Electronic Negotiations (2006) Intelligent Systems Special Issue 15:1–4, 15-37
3. Bartolini C, Preist C, Jennings N R (2002)A Generic Software Framework for Automated Negotiation, AAMAS'02, July 15–19,Bologna, Italy
4. Bellosta M, Brigui I, Kornman S, Vanderpooten D (2004)A Multi-Criteria Model for Electronic Auctions, ACM Symposium on Applied Computing, March 14–17, 2004, Nicosia, Cyprus.
5. Bichler M, Kalagnanam J, Lee H S, Lee J (2002) Winner Determination Algorithms for Electronic Auctions: A Framework Design in Proceedings of E-Commerce and Web Technologies: Third International Conference, EC-Web 2002, Aix-en-Provence, France, September 2–6.
6. Jennings N R, Faratin P, Lomuscio A R, Parsons S, Sierra C, Wooldridge M, (2001) Automated Negotiation: Prospects, Methods and Challenges International Journal of Group Decision and Negotiation, 10:2,199–215
7. Kersten, G E, Lo G, (2001) Negotiation Support Systems and Software Agents in E-Business Negotiations, The First International Conference on Electronic Business, Hong Kong, Dec 19–21
8. Kim J B, Segev A (2003) A Framework for Dynamic eBusiness Negotiation Processes IEEE
9. Lee K J, (2000)Time-Bounded Framework for Automated Negotiation, International Conference on Advances in Infrastructure for Electronic Business, Science, and Education on the Internet
10. Strobel, M(2001) Design of Roles and Protocols of Electronic Negotiation, Electronic Commerce Research Journal, Special Issue on Market Design, 1:3, 335–353
11. Tu M T, Seebode C, Griffel F, Lamerdorf W, (2001)DynamiCS: An Actor-based Framework for negotiating Mobile Agents.
12. Wong W Y, Zhang D M, Kara-Ali M (2000) Negotiating With Experience, AAAI2000 Knowledge- Based Electronic Markets, Technical Report WS–00:04, 85–90

Risk Mining for Strategic Decision Making

G.M. Allan[1], N.D. Allan[2], V. Kadirkamanathan[1], and P.J. Fleming[1]

[1] Rolls-Royce Supported University Technology Centre in Control and Systems Engineering, Dept of Automatic Control and Systems Engineering, The University of Sheffield, Mappin Street, Sheffield, S1 3JD, UK
jeff.allan@sheffield.ac.uk
[2] School of Management, The University of Bath, Bath, BA2 7AY, UK
mnsnda@management.bath.ac.uk

Abstract. Increasingly risk management plays a strategically important role in construction related companies. Recent high profile corporate failures and loss of shareholder value have been attributed to poor risk management. Research shows that open communication and timely information are vital for the management of strategic risks. To deliver adequate support for risk based decision making a flexible and seamlessly integrated ontology based architecture is proposed. This new architecture is constructed around two essential concepts, 1) The modeling, mining, and reuse of domain knowledge using an ontology, as well as the integration and 2) the use of semi-formal risk models for mining non-intuitive operational and process insights.

1 Introduction

STRATrisk [1], aimed at understanding and improving Board level decision making in UK construction companies and this project is used as a basis for this study. The program set out to provide a toolkit for company Boards to self assess their own understanding and to manage better the strategic risks and opportunities materially impacting on the enterprise.

Personal construct theory [2] underpins the hypothesis that experience and perception heavily influence decision making processes. Concept mapping [3],[4] was used to analyze and understand some of the interactions and emerging issues; See Fig. 1 which shows interview analysis revealing the phenomenon of "Group Think" which resulted in a hostile company takeover.

The overall aim of the STRATrisk project was to establish a rich picture of risk from the evidence and data available, whilst at the same time searching for new and original meaning. Management of risk can be seen as a specialized branch of knowledge management, and it is noted that the processes employed by STRATrisk bring together so called Soft Systems Methodologies [5] that derive tacit knowledge and contextual knowledge within work processes [6]. The soft systems process modeling and concept mapping procedures were further enhanced by the use of directed concept maps [7]. Directed maps can provide a richer view of complex environments since they are more structured.

K.M. Węgrzyn-Wolska and P.S. Szczepaniak (Eds.): Adv. in Intel. Web, ASC 43, pp. 21–28, 2007.
springerlink.com

The STRATrisk outcome was twofold, a systematic approach to the management of threats and opportunities, and an insight into issues constituting risk as an outcome of business processes. The limitation of the method is that the production of models for "what-if?" type analysis, remains elusive.

This paper proposes an extension to the Stratrisk methodologies in the following ways: 1) To create a common vocabulary for risk and an infrastructure for knowledge sharing. 2) To derive a viable architecture for knowledge management and risk mining. 3) To develop qualitative models for detailed analysis and scenario planning.

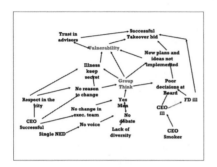

Fig. 1. Company takeover interview using concept mapping

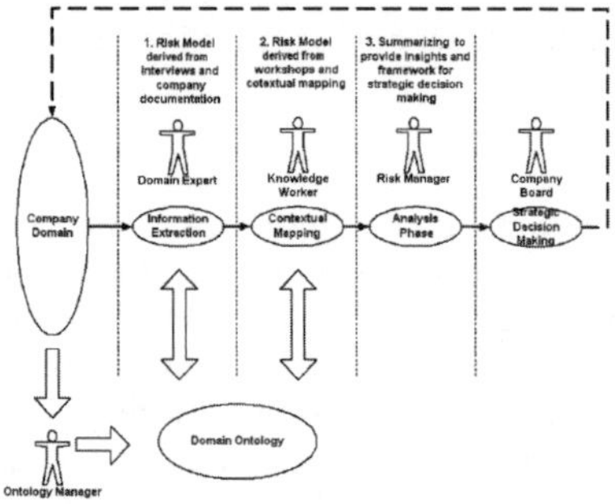

Fig. 2. Organizational Structure for Ontology Based Risk Management

2   Using Ontologies for Storing and Accessing Organisational Knowledge

2.1   Background

When considering the grand challenges for Computational Intelligence (CI) [8] a fundamental realisation is that for systems to behave with high levels of performance for complex intellectual tasks there must be an extensive knowledge domain. Indeed some authors go further and state that CI is the study of knowledge [9]. Performance and complexity management are key business issues, knowledge management thus emerges as a key motivation for organizations wanting to gain competitive advantage [10],[11],[12] Knowledge is seen as a legitimate and meaningful resource that strengthens overall management performance, and knowledge management is seen to help facilitate creating, storing, sharing, and reusing the organization's knowledge, typically using advanced technology [13],[14],[15].

Ontologies potentially enable automated knowledge sharing and reuse among both human and computer agents, because of their ability to interweave human and machine understanding through formal and real-world semantics [17].

Knowledge sharing in ontologies is not only achieved through formal and real-world semantics, but also, and this is a very important point, through the consensual engineering of ontologies themselves. Ideally, ontologies are formal vocabularies, shared by a group of individuals that are interested in a specific domain. However this vision of a shared ontology can only be achieved when using a solid methodology that guarantees the collaborative engineering process. The key advantages of ontologies are as summarized as follows: 1) They enable knowledge sharing. 2) They enable reuse of knowledge. 3) They are machine-understandable.

2.2   Ontology Based Risk Management

For organizational implementation purposes ontology driven knowledge management processes are provided by the Semantic Web. The Semantic Web is based on the idea of creating machine understandable data that can be used and exchanged.

There are a prolific number of available ontology building tools, languages, support tools, models an architectures for semantic web applications [16]. Fig. 2 suggests a generic architecture for an ontology-centric approach to risk management. An ontology for STRATrisk was built using Protoge.

Implicit information embedded in semantic web graphs, such as topography, clusters, and disconnected subgraphs is difficult to extract from text files. Visualizations of the graphs can reveal some of these features. The Protoge ontology editor, produced at Stanford University, is one of the more popular open source semantic web tools available today. It is easily extensible, and has two visualization components. OntoViz [18] is an ontology visualizer that, uses GraphViz [29] as its base. It shows classes grouped with their properties, and information about those properties, and instances grouped with lists of their properties. These groups are connected by edges indicating relationships between the objects. Jambalaya [19], another Protoge based visualization tool, displays information similarly, with instances and classes being grouped with their respective properties. Jambalaya adds additional navigation features, allowing users to look at the ontology at several levels of detail. Both of these tools are designed to allow the user to browse ontologies. In Fig. 3, we present a spring graph for part of the STRATrisk ontology. Through this and other cases it is possible to see how patterns about the underlying structure are more easily understood through the graph drawing than through text or other types of visual displays.

We argue that the general motivation for developing ontology-driven information systems is the ability to incorporate knowledge from different domains into a single framework. Using ontologies as the basis for risk simulation applications contributes to the conceptual requirement of reusability. Since the models are specified by humans to embody domain knowledge which is sometimes ambiguous, ontologies are a profound solution for reducing the semantic gap between the knowledge space and the simulation model.

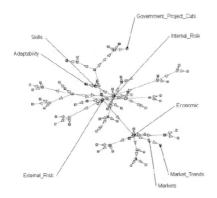

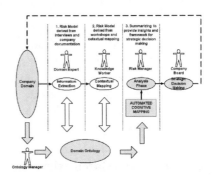

Fig. 3. Spring graph for STRATrisk ontology

Fig. 4. Improved structure for ontology based risk management

3   Modeling Strategic Risk

The basis for all simulation is the building of a simplified model which represents the reality under examination. Generally, there are three different types of models, physical, graphical, and mathematical. In this paper, combinations of mathematical and graphical models are applied. The STRATrisk project used a soft systems methodology, steering away from traditional statistical risk analysis, and employed tools enabling cognition and concept mapping. The rationale was the capture of operational complexity and human issues. The further improvement of the STRATrisk approach is the subject of further discussion here. Insights by means of simulation are vital for risk management.

3.1   Fuzzy Cognitive Maps

Cognitive maps were introduced in early studies around the 1950's [20]. An extension of cognitive maps to fuzzy cognitive maps FCMs was proposed [21]. Causal relationships and concepts in FCMs are accompanied by fuzzy sets [22]. FCMs are then able to be used as a tool for handling imprecise, or ill defined, "fuzzy" problems. This allows the intangible expression of differences in causal relationships and introduces the partial activation of concepts as opposed to binary activation in classical cognitive maps. The FCM, while still having rigor, approaches human-like thinking and communication. Additionally, verbal evaluations of concepts and relations are directly permissible [23].

The key advantages of FCMs are as follows: 1) They combine the potential for knowledge mapping of complexity, as in concept mapping, with causal relationships that then allow detailed modelling and decision support. 2) Local FCMs can be built in isolation and then later merged into holistic global models. 3) FCM's lend themselves to implementation with other CI technologies such as genetic algorithms [25] and case based reasoning [26].

Based on FCMs a new organizational risk management structure may be envisaged. See Fig. 4. A flexible platform for risk management simulation is proposed, driven by ontology, which is used for knowledge capture and also modelling purposes. The risk ontology consists of a meta risk ontology, a domain risk model, and a risk knowledge base. Manual processes, while still needed in part, can be replaced in the main by automated mining procedures, e.g. via a company intranet with specialist tools. The input of domain experts may be leveraged by the sharing of a common knowledge base.

4   Analysis of Strategic Risk Using FCM

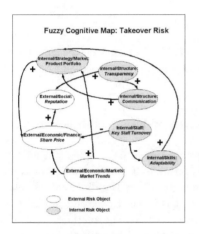

Fig. 5. FCM for Corporate Takeover Risk

Consider for illustrative purposes, a simple example and for the time being, a manual technique to create FCMs; as shown in. Fig. 5. This FCM was developed using the global STRATrisk ontology for the scenario depicted in Fig. 1. In this case a company director describes a risk that materialized in the form of a hostile takeover resulting from the weakness of "Group Think". Analysis of this relatively simple FCM can be done graphically by tracing cause and effect throughout the system or by more elaborate, but the nevertheless fairly routine, numerical analysis.

An alternative approach, and an entirely novel one, is to build FCMs from measured process data characterizing the system domain. This novel approach means deriving FCM models and node relationships purely from process data. A complex systems analyzer algorithm (CSA) [28] is implemented in proprietary software and this is used to analyze time series data representative of the system elements shown in Fig. 5. The CSA algorithm processes a set of observations conducted on the system of interest. Normally, these observations originate from repeated experiments, from process measurements, or from Monte Carlo simulations performed with a multi-disciplinary model. From this set of data, which may be defined as the Fitness Landscape of the system, the algorithm identifies the so-called attractors, i.e. locations in the landscape in the vicinity of which the system exhibits distinct and stable patterns of behaviour. These patterns, known as modes, are sets of interrelated fuzzy rules that govern the system behaviour in each mode. The relationship between system nodes is determined by perturbations. The CSA scans all the samples in the Fitness Landscape and introduces perturbations in all the directions around each sample for all the pairs of variables. The magnitude of the perturbations is equal to the width of the fuzzy intervals. In practice, the entire space is sub-divided into multi-dimensional fuzzy

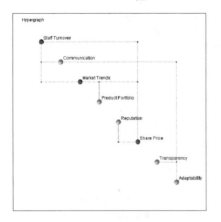
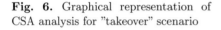
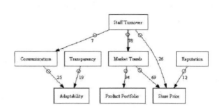

Fig. 6. Graphical representation of CSA analysis for "takeover" scenario

Fig. 7. CSA graph in conventional FCM layout

cells, and the existence, or otherwise, of reactions to the perturbations determines the rules which link the variables of a given problem, and hence creates the FCM, see Fig. 6.

Fig. 6. shows results of the CSA output in its graphical form. This is a FCM created from the time-series data provided as variables from the nodes in the system being considered - the FCM in this case is drawn on the diagonal. When Fig. 6 is exported to GraphViz [29], as in Fig. 7, it is easier to see the similarity with Fig. 5. In this case we are primarily concerned with the methodology of analysis and the opportunities presented by the novelty of the approach. Some basic analysis reveals the extent of the possibilities. Fig. 5 and Fig 7. in the "Takeover" example are seen to differ. What is being suggested here, and is of real value, is that the "actual" company data conflicts with the cognitive model that has been created and either the single interview conducted does not provide a rich enough knowledge model (which is likely), or that the metrics gathered do not characterize the system. In any event the cross-checking of models using the CSA further enhances system insights. In summary, the opportunity here is to calibrate knowledge models created from an ontological framework by objective data resources. In this approach not only can the complexity of process be captured, and insights be derived from the models, but system measures are available for comparison with other domains, companies, or industries, i.e. we can accurately compare risk models.

5 Conclusions

Risk Mangement is one particular branch of Knowledge Management where the quality of the knowledge models is essential for insight and analysis.

Ideas of the Semantic Web and the deployment of ontological frameworks within corporations are ideally suited to the gathering and mining of risk-related data and knowledge.

Cognition advances the notion of concept mapping by taking advantage of causal relationships. FCMs embrace the power of causal relationships and balance the need for "soft" processes to capture complexity as well the requirement for simulations to provide system insights. Using FCMs with the CSA we have a virtuous circle, on the one hand we have a rich knowledge model capturing all the complexities of the system process, on the other hand we have a tool that is able to asses and mine complex systems for hitherto unseen relationships, and each system enhances the potential of the other.

6   Further Work

Further work is to include expansion of STRATrisk ontology, further capture of process data for CSA analysis, and extension of the methodology for strategic risk management to other industry sectors.

References

1. N. Allan, and J. Davis, Strategic risks - thinking about them differently, Proceedings of ICE, Civil Engineering 159, November 2006, in press.
2. G. A. Kelly, The Psychology of Personal Constructs, New York, Norton, (1955).
3. C. Eden, Cognitive Mapping and Problem Structuring for Systems Dynamics Model Building, Systems Dynamics Review, 10, 2-3, 257-276, (1994).
4. A. C. McLucas, Decision Making: Risk Management, Systems Thinking and Situation Awareness, Argos Press, Canberra, Australia, (2003).
5. P. Checkland, Systems Thinking, Systems Practice, Wiley, UK, (1999).
6. A. S. Huff and M. Jenkins, (eds), Mapping Strategic Knowledge, Sage Publications, London, 2002.
7. V. Ambrosini, and C. Bowman, Mapping Successful Organizational Routines, in A. S. Huff and M. Jenkins, (eds), Mapping Strategic Knowledge, Sage Publications, London, 2002.
8. E. A. Feigenbaum, Some Challenges and Grand Challenges for Computational Intelligence, Journal of the ACM , Vol. 50, No. 1, January 2003, pp. 32-40.
9. Poole, Mackworth, Goebel, Computational Intelligence: A Logical Approach, Oxford University Press, 1998.
10. T. Davenport, and l. Prusak, Working Knowledge, Cambridge, M. A., Harvard Business School Press, 1998.
11. R. M. Grant, Prospering in dynamically competitive environments: organizational capability as knowledge integration, Organizational Science, 7(4), 375-387, (1996).
12. I. Nonake, and H. Takeuchi, The knowledge creating organization: how Japanese companies create the dynamics of innovation, New York, Oxford University Press, (1995).
13. A. Abecker, A. Bernadi, K. Hinkelmann, O. Kuhn, and M. Sintek, Toward a technology for organizational memories, IEEE Intelligent Systems, May/June, 40–48, (1998).

14. G. Heijst, R. Spek, and E. Kruizinga, Corporate memories as a tool for knowledge management, Expert Systems With Applications, 13, 41ś 54, (1997).
15. D. E. O'Leary, Using AI in knowledge management: knowledge bases and ontologies, IEEE Intelligent Systems, May/June, 34ś39, (1998).
16. J. Bruijn, Using Ontologies Enabling Knowledge Sharing and Reuse on the Semantic Web, Digital Enterprise Research Institute (DERI), Technical Report DERI-2003-10-29, October (2003).
17. D. Fensel, Ontologies: a silver bullet for Knowledge Management and Electronic Commerce, Springer-Verlag, Berlin, (2003).
18. The OntoViz Tab: Visualising Protégé Ontologies - http://protege.cim3.net/cgi-bin/wiki.pl?OntoViz.
19. M. Storey, R. Lintern, N. Ernst, and D. Perrin, Visualisation and Protégé, University of Victoria, Victoria, BC, Canada, (2004).
20. E. C. Tolman, Cognitive Maps in rats and Men, The Psychological Review, 55(4):, 189-208, (1948)
21. B. Kosko, Fuzzy Cognitive Maps, International Journal of Man-Machine Studies, 24(1): 65-75, (1986).
22. L. A. Zadeh, Fuzzy Sets, Information and Control, 8:338-351, (1965).
23. L. A. Zadeh, The concept of linguistic variable and its application to approximate reasoning I, II, III, Information Sciences, 8,9: 199-257, 301-357, 43-80, (1975).
24. D. Balder, Fuzzy Cognitive Maps and their uses as Knowledge Mapping Systems and Decision Support Systems, (2004).
25. M. S. Khan, S. Khor, A. Chong, Fuzzy cognitive maps with genetic algorithm for goal-oriented decision support, International Journal of Uncertainty, Fuzziness and Knowledge-Bsed Systems.
26. J. B. Noh, K. C. Lee, J. K. Kim, J. K. Lee, S. H. Kim, A case-based reasoning approach to cognitive map-driven tacit knowledge management, Expert Systems with Applications, 19, 249-259, (2000).
27. A. L. Rector, AIM: A personal view of where I have been and where we might be going,. Artificial Intelligence in Medicine, 23:111-127, (2001)
28. D. Lomario, G. P. De Poli, L. Fattore, J. Marczyk, A complexity-based approach to robust design and structural assessment of aero engine components, American Society of Mechanical Engineers, (2006), in press.
29. www.graphviz.org

Interrogation of Ontologies Formalized in Topic Maps with TMQL in E-Learning Context

Hakim Amrouche[1], Marie-Hélène Abel[2], Amar Balla[3], and Claude Moulin[4]

[1] National institute of computer sciences; LP 68 M, Oued-Smar, Algiers, Algeria
h_amrouche@ini.dz
[2] UMR CNRS 6599 Heudiasyc, LP 20529 University of Compiègne 60205 Compiègne
Cedex, France
Marie-Helene.Abel@hds.utc.fr
[3] National institute of computer sciences; LP 68 M, Oued-Smar, Algiers, Algeria
a_balla@ini.dz
[4] UMR CNRS 6599 Heudiasyc, LP 20529 University of Compiègne 60205 Compiègne
Cedex, France
Claude.Moulin@hds.utc.fr

Summary. In E-Learning, the selection of the pedagogical contents to teach and the organization of this content are two principal operations during the design of a course. The growth and the diversity of the informational resources used generate problems of diffusion, access, of classification and management. With the MEMORAe project, we deal with these problems and propose to answer them by using an ontology- based organisational memory. In this article, we describe how and why we exploit the formalism of Topic maps to represent the contents of a course organisational memory and the TMQL standard to interrogate it.

Keywords: E-learning, organisational memory, Topic maps, TMQL, Ontology.

1 Introduction

The growth and the diversity of the informational resources used generate diffusion/access, classification and management problems. Many works are related to the management and access problems to the online learning resources [3][4]. We can classify into two poles the produced systems: the contents presentation systems and the contents management systems.

The contents presentation systems put the emphasis on navigation, visualization and organization of information in order to facilitate their comprehension by users [5]. The learning contents management systems support the access and research of the learning data in relation with their presentation[5].

The MEMORAe(MEMoire Organisationnelle Appliquée à l'e-learning) project deals, at the same time, with the contents presentation and management. Contrarily to the traditional presentation systems, it does not use adaptive techniques for navigation but an ontology structure: a course is built around a set of concepts to apprehend. These concepts are defined by the mean of ontology and index the resources allowing their apprehension. The use of ontologies make possible, on the one hand, to define a common vocabulary; and on the other hand, to structure the learning contents of a course. In this article, we describe how and why we exploit the Topic maps formalism to represent the organisational memory contents of a course and the contribution of TMQL standard to iterogate it. The rest of the article is organized as follows: in section 2, a detailed description of the MEMORAe Project is given. In section 3, the Topic Maps formalism is specified as well as the TMQL interrogation associated standard. Thereafter, section 4, we give implementation details of the MEMORAe query mechanism using TMQL.

2   The MEMORAe Project

Learning is composed of actors (learners, teachers, etc.) and of various resources (definitions, exercises, etc), it is written in various forms (books, Web sites, etc.) like various mediums (paper, video, etc), so training is an organization. This is why within MEMORAe project [2], the resources and information are managed by means of an ontology-based learning organisational memory. The MEMORAe project deals both with the content presentation and management. The MEMORAe project does not use adaptive techniques for navigation but it uses the structure of an ontology: a training is built around a set of concepts to apprehend. Concepts, defined within ontology index the resources allowing their apprehension.

In the MEMORAe project, two types of ontologies are used: domain ontology and application ontologies. Both types of ontologies (application and domain) are jointly used to index the teaching concepts and resources. For the moment two application ontologies are defined: NF01 and B31 [2]. NF01 Ontology is designed for the training of Algorithmic course and Pascal programming. B31 Ontology is designed for the training of statistics course.

The ontologies defined in MEMORAe are formalized using Topic Maps formalisme in order to better index the informational resources and to provide a powerful means for visualization and navigation in the memory contents.

3   Topic Maps

The Topic Maps formalism [6]is a formalism for the representation and the organisation of the knowledge. Topic Maps represent knowledge by a graph made up of nodes bound by semantic relations. A node can be any object which can have a meaning in a given field. The Topics Maps model is powerful enough for navigation, visualization and information organization to facilitate the comprehension of this information by users. The Topic Maps model was standardized by ISO. The Topic Maps formalism presents three basic elements: Topic, Association and Occurrence:

- Topic: Topic is the data-processing representation of a Subject applied to a localization set (Context).
- Association: an association enables to connect two or several Topics.
- Occurrence: an occurrence (information resource) can be an article, an picture, a video, a comment, etc. It is attached to a topic.

For example, the concept ÂńÂă finite setÂăÂż is defined as follows: a finite set is a set that has a cardinal. In Topic maps, it is represented by:

- Three Topics: Finite set, Set, Cardinal.
- Two Associations: is a, has.
- One can specify that the concept of ÂńÂăfinite setÂăÂż is treated in the resource of the Web page type http://www.planete-maths.com/html /.

Note that, at the beginning, this model was not used in semantic Web. With the definition of XTM[7] standard (XML for TM), Topic Maps were introduced into the universe of the semantic Web. An XTM file has a similar syntax to an XML file.

Topics Maps can be used in various fields. Therefore, it is important to have exploitation tools of Topic Maps (creation,integration,etc), in this context, the TMQL specification was born.

3.1 TMQL

The purpose of TMQL [8] (Topic Map Query Language) is to create a standard for the interrogation of Topic maps. TMQL is not a language, but it is a specification for the interrogation languages of Topic Maps. The purpose of TMQL is to simplify the development of the Topics Map based applications.

In the TMQL recommendations, it is specified that the query must be carried out on an abstract data model, independently of the storage method (data bases or XTM). In other words, the execution of a request on a data base or on an XTM file should give the same result. The requests must also support the logical inference.

There are several implementations of the TMQL specification, such as Tolog [9], TMRQL [8], AsTMa [8]. Most of TMQL implementations remain prototypes and there is no stable implementation. Tolog is the most stable TMQL implementation, it is for that reason that we chose it to implement our interrogation prototype of Topic Maps.

4 Using Tolog to Search Concept in MEMORAe

Note that at the begining of the MEMORAe project the TMQL specification was not yet set up, so for the implementation of the MEMORAe project, a relational model of ISO Topic Maps standard was set up. Thus ontologies used are stored in a relational database. An E-MEMORAe environment was developed for the exploitation of these ontologies. E-MEMORAe allows navigation through ontologies as well as visualization, it was evaluated both on navigation and visualization sides [1]. Note that E-MEMORAe search interface is based on the execution of SQL query, since ontologies are stored in a data base. On the other hand let us note that SQL language is only intended for data base interrogation, but Topic Maps can be stored under various formats (for example: XTM file) therefore occurs the impossibility of using SQL. Besides, it is difficult with SQL, to exploit the semantic links that exist between Topics. This reduces the possibility of making deductions. It is for these reasons that we were brought to consider TMQL standard for memory interrogation with the goal of standardizing E-MEMORAe environment within information search and interrogation.

So to have more flexibility and expressivity in the query writing, we thought of integrating an interrogation environment, in MEMORAe, while working on Tolog, but the current implementation of Tolog does not allow querying Topic Maps stored in relational database. Only the interrogation of XTM files is possible. So it is necessary to extract XTM file from the data base before carrying out the query. Currently, the operation of XTM-file extraction is manual (the automatic extraction is still under development).

4.1   A Query Results

MEMORAe philosophy consists to give the most useful information for a learner. Thus the response to a request is part of the ontology that contains the relevant concept and its close neighbourhood. The close neighbour or the family of an concept is defined by the father, the brothers and first generation children of this topic. This philosophy must be respected during the use of Tolog.

In ontologies, concepts can be connected by various types of links. There are two types of significant links: the specialisation/generalisation link and the instanciation link. The specialisation/generalisation link is used to interconnect two classes in order to specify the relations (subclass, superclass). The instanciation link is used to connect a class with these instances. This link is represented by the relation instance-of.

In MEMORAe ontologies, this type of links exists, so to find the neighbourhood or the Topic family, we exploited these two types of links (or relations that express these links). Thus, we get the following definitions:

- Topic B is the parent of Topic A : if A is an instance of B or B is a superclass of A.
- Topic B is the direct child of another Topic A: if B is an instance of A or B is a subclass of A.
- Topic A is the brother of another Topic B if they have the same father.

To simplify the achievement of this operation at the time of the implementation stage, we use Tolog mechanism of the inference rules. These relations are modelled with the inference rules, as a series:

- father-of(A, B) :- { instance-of(B, A) OR superclass(B, A) } this rule describes the relation B is the father of A.

- child-of (A,B) :- { instance-of(A,B) OR subclass(A,B)} : this rule describes the relation A is the child of topic B.
- brother-of(A,B) :- { father-of(A, C) AND father-of(B, C)} : this rule describes the relation brother between A et B . The rule father-of is used to express the rule parent.

The rule that describes the relation family is defined, according to the three preceding rules, by:

family-of(A,B):-{father-of(A,B) OR brother-of(A,B) OR child-of (A,b) .}

This rule is added to the default search query. Thus, its result is added to the default search result in order to have a complete result.

4.2 Implementation

For the implementation of our prototype, we used the JAVA language with TM4J API (Topic Maps for Java) [10]. TM4J is an Open Source JAVA API that allows the integration of Tolog query. We carried out some tests on an fragment of the B31 statistics ontology. Note that the result obtained is represented in the form of an XTM file. The interpretation of an XTM file is difficult for users, it is why we used the TM4L viewer tool [11]to display the result. TM4L is an Open source tool for the visualization of Topic Maps in E-learning [11]. Our objective is to display the result directly in the E-MEMORAe environment. Figure 1 shows the result of search of the concept 'set'.

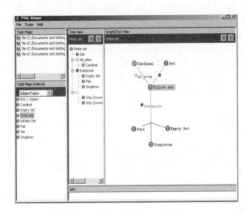

Fig. 1. Visualisation of a query results using TM4L

5 Conclusion and Further Works

Our objective with the Memorae project is to propose a new method for information organization and selection in E-learning field. Within this work, we consider that training is an organization and propose to manage, capitalize and diffuse its knowledge and resources by an organisational memory, modelled with ontologies. We chose Topic maps standard formalism to represent it.

For the memory information and interrogation search, we direct ourselves towards the use of TMQL standard for it proposes a Topic Maps powerful interface of exploitation. In order to test the power of TMQL, we developed an interrogation prototype from Tolog tool. Our tests were carried out on the B31 statistics ontology and they showed the importance of TMQL.

Our prototype should be improved and enriched in order to enable more complex queries, for example: the search for a concept in a given context, the search for a resource associated with a concept.

The choice of TMQL to interrogate our learning memory , being validated, we now work on his integration in the E-MEMORAe environment.

References

1. Benayache A, Leblanc A, Abel M-H (2006) Learning Memory, evaluation and return one experiment , In Proceedings of Workshop of Knowledge Management and Organizational Memories, Rivetted del Garda, Italy
2. Benayache A (2005) Construction of an organisational memory of formation and evaluation in a context e-learning: the MEMORAe project. Phd thesis, University of Technology of Compiegne
3. Brusilovsky P, Vassileva J (2003) Race sequencing technical for broad-scale Web-based education. International Newspaper of Continuing Engineering Education and Lifelong Learning, pp 75-94
4. Dahn I, Armbruster Mr, Furbach U, Schwabe G (2001) Slicing Books âĂŞ The Authors' Outline. In R. Bromme, E Stahl Eds.: Writing Hypertext and Learning: Conceptual and Empirical Approaches, Pergamon.
5. Outside S, Giboin A, Faron-Zucker C,Stromboni J-P (2005) Semantic Annotations To learn: Experimentation QBLS. In Acts of the Day WebLearn set of themes on the semantic Web for the e-Learning, Nice
6. IEC, " International Organization for Standardization (ISO)" (1999), International Electronical Commission (IEC), Topic Map, International Standard ISO/IEC 13250, April 19, 1999.
7. TopicMaps.org Authoring Group (2001). Xml topic maps (xtm) 1.0, topicmaps. org specification, 2001. http://www.topicmaps.org/xtm/1.0 /
8. Garshol L-M, Barta R (2003) TMQL requirements. http://www.isotopicmaps.org/ tmql/tmqlreqs.html
9. Garshol L-M (2004) Tolog language tutorial. Technical carryforward, Ontopia. http://www.ontopia.net/omnigator/docs/query/tutorial.html
10. TM4J: http://tm4j.org /
11. TM4L viewer http://compsci.wssu.edu/iis/nsdl/download.html

Using Logic Wrappers to Extract Hierarchical Data from HTML

Amelia Bădică[2], Costin Bădică[1], and Elvira Popescu[1]

[1] University of Craiova, Software Engineering Department
Bvd.Decebal 107, Craiova, RO-200440, Romania
{badica_costin,elvira_popescu}@software.ucv.ro
[2] University of Craiova, Business Information Systems Department
A.I.Cuza 13, Craiova, RO-200585, Romania
ameliabd@yahoo.com

Summary. In this note we show how logic wrappers technology can be adapted to cope with hierarchical data extraction. For this purpose we introduce *hierarchical logic wrappers* and illustrate their application by means of an intuitive example.

1 Introduction

Data published on the Web is converted from internal formats, specific to data base management systems, to suitable presentations for attracting humans. However, it is appreciated that Web data should be also reused for other purposes in tasks like searching, filtering, analysis, reasoning and integration.

Our recent work in the area of Web data extraction was focused on combining logic programming with efficient XML processing. The results were: i) definition of *logic wrappers* or L-wrappers for data extraction from the Web ([2]); ii) design of efficient algorithms for semi-automated construction of L-wrappers ([1]); iii) efficient implementation of L-wrappers using XSLT technology ([3, 6]).

So far we have only applied L-wrappers to extract relational data – i.e. sets of tuples or flat records. However, many Web pages contain hierarchically structured presentations of data for usability and readability reasons. Moreover, it is generally appreciated that hierarchies are very helpful for focusing human attention and management of complexity. Therefore, as many Web pages are developed by knowledgeable specialists in human-computer interaction design, we expect to find this approach in many designs of Web interfaces to data-intensive applications.

In this note we propose an approach for utilizing L-wrappers to extract hierarchical data. The advantage would be that extracted data will be suitably annotated to preserve its hierarchical structure, as found in the Web page. Further processing of this data would benefit from this additional meta-data to allow for more complex tasks, rather than simple searching and filtering by populating a relational database.

2 Extending L-Wrappers

L-wrappers assume a relational model by associating to each Web data source a set of distinct attributes. An L-wrapper operates on a target Web document

K.M. Węgrzyn-Wolska and P.S. Szczepaniak (Eds.): Adv. in Intel. Web, ASC 43, pp. 35–40, 2007.
springerlink.com   © Springer-Verlag Berlin Heidelberg 2007

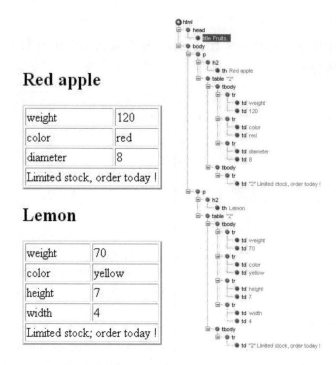

Fig. 1. A sample HTML document containing hierarchically structured data

represented as a labeled ordered tree and extracts a set of relational tuples of document nodes. See [2, 3] for details about definition, implementation and evaluation of L-wrappers.

Let us now consider a very simple HTML document that contains hierarchical data about fruits (see figure 1). A fruit has a name and a sequence of features. Additionally, a feature has a name and a value. This is captured by the schema shown in figure 2a. Note that features can be fruit-dependent; for example, while an apple has an average *diameter*, a lemon has both an average *width* and an average *height*.

Abstracting the hierarchical structure of data, we can assume that the document shown in figure 1 contains triples: *fruit-name*, *feature-name* and *feature-value*. So, an L-wrapper of arity 3 would suffice to wrap this document.

Following the hierarchical structure of this data, the design of an L-wrapper of arity 3 for this example can be done in two stages: i) derive a wrapper W_1 for binary tuples (*fruit-name,list-of-features*); ii) derive a wrapper W_2 for binary tuples (*feature-name,feature-value*). Note that wrapper W_1 is assumed to work on documents containing a list of tuples of the first type (i.e. the original target document), while the wrapper W_2 is assumed to work on document fragments containing the list of features of a given fruit (i.e. a single table from the original target document).

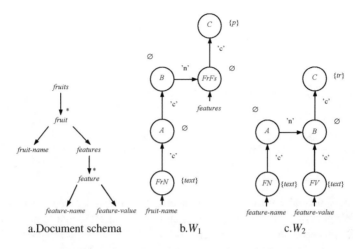

a.Document schema   b.W_1 c.W_2

Fig. 2. Wrappers for the document shown in figure 1

For example, wrappers W_1 and W_2 can be designed as in figure 2 (following the graph notation introduced in [2, 3]). Their expression in logic programming is:

```
extr_fruits(FrN,FrFs) :-
    tag(FrN,text),child(A,FrN),child(B,A),next(B,FrFs),child(C,FrFs),tag(C,p).
extr_features(FN,FV) :-
    tag(FN,text),tag(FV,text),child(A,FN),child(B,FV),next(A,B),child(C,B),tag(C,tr).
```

The combination of W_1 and W_2 into a single L-wrapper of arity 3 requires the extension of relation *extract* with a new argument of to represent the root node of the target document fragment – instead of $extract(N_1, \ldots, N_k)$ we now have $extract(R, N_1, \ldots, N_k)$, R is the new argument. It is now required that for all $1 \le i \le k$, N_i is a descendant of R in the document tree. The solution, using predicate *ancestor*(*Ancestor, Node*) and assuming that document root node has index 0, is:

```
ancestor(N,N).
ancestor(A,N) :- child(A,B),ancestor(B,N).
extr_fruits(R,FrN,FrFs) :- ancestor(R,FrN),ancestor(R,FrFs),extr_fruits(FrN,FrFs).
extr_features(R,FN,FV) :- ancestor(R,FN),ancestor(R,FV),extr_features(FN,FV).
extract(FrN,FN,FV) :- extr_fruits(0,FrN,FrFs),extr_features(FrFs,FN,FV).
```

3   Preserving the Hierarchy

While simple, the solution outlined in the previous section has the drawback that even if it was devised with the idea of hierarchy in mind, it is easy to observe that the hierarchical nature of the extracted data is lost.

Assuming a Prolog-like execution engine, we can solve the first drawback using the *findall* predicate. *findall*(*X, G, Xs*) returns the list *Xs* of all terms *X* such that goal *G* is true (it is assumed that *X* occurs in *G*). The solution and the result are shown below. We assume that i) the root node of the document has index 0 and ii) predicate *text*(*TextNode, Content*) is used to determine the content of a text node.

```
extract_all(Res) :-
  extr_fruits_all(0,Res).
extr_fruits_all(Doc,fruits(Res)) :-
  findall(
    fruits(name(FrN),FrFs),
    (extr_fruits(Doc,NFrN,NFrFs),
    text(NFrN,FrN),
    extr_features_all(NFrFs,FrFs)),
    Res).
extr_features_all(Doc,features(Res)) :-
  findall(
    feature(name(FN),value(FV)),
    (extr_features(Doc,NFN,NFV),
    text(NFN,FN),text(NFV,FV)),
    Res).
```

```
?-extract_all(Res).
Res = fruits(
  [fruit(name('Red apple'),
  features(
    [feature(name('weight'),value('120')),
    feature(name('color'),value('red')),
    feature(name('diameter'),value('8'))
    ])),
  fruit(name('Lemon'),
  features(
    [feature(name('weight'),value('70')),
    feature(name('color'),value('yellow')),
    feature(name('height'),value('7')),
    feature(name('width'),value('4'))
    ])])
```

Let us now formally introduce the concept of *hierarchical logic wrapper* or *HL-wrapper*. We generalize the data source schema from flat relational to hierarchical and we attach to this schema a set of L-wrappers.

Definition 1. *(Schema tree) Let $\mathcal{W}$ be a set denoting all vertices. A schema tree S is a directed graph defined as a quadruple $\langle A, V, L, \lambda_a \rangle$ s.t. $V \subseteq \mathcal{W}$, $A \subseteq V \times V$, $L \subseteq V$ and $\lambda_a : A \to \{'*','1'\}$. The set of schema trees is defined inductively as follows:*

i) *For all $n \geq 1$, if $u, v, w_i \in \mathcal{W}$ for all $1 \leq i \leq n$ then $S = \langle A, V, L, \lambda_a \rangle$ such that $V = \{u, v, w_1, \ldots, w_n\}$, $A = \{(u, v), (v, w_1), \ldots, (v, w_n)\}$, $L = \{w_1, \ldots, w_n\}$, $\lambda_a((u, v)) =' *'$, and $\lambda_a((v, w_1)) = \ldots = \lambda_a((v, w_n)) =' 1'$ is a schema tree.*

ii) *If $S = \langle A, V, L, \lambda_a \rangle$ is a schema tree, $n \geq 1$, $u \in L$ and $v, w_i \in \mathcal{W} \setminus V$ for all $1 \leq i \leq n$ then $S' = \langle A', V', L', \lambda'_a \rangle$ defined as $V' = V \cup \{v, w_1, \ldots, w_n\}$, $A' = A \cup \{(u, v), (v, w_1), \ldots, (v, w_n)\}$, $L' = (L \setminus \{u\}) \cup \{w_1, \ldots, w_n\}$, $\lambda'_a((u, v)) =' *'$, $\lambda'_a((v, w_1)) = \ldots = \lambda'_a((v, w_n)) =' 1'$ and $\lambda'_a(a) = \lambda_a(a)$ for all $a \in A \setminus A'$ then S' is also a schema tree.*

If Σ is a set of tag symbols denoting schema concepts and S is a schema tree then a pair consisting of a schema tree and a mapping of schema tree vertices to Σ is called a *schema*. For example, for the schema shown in figure 2a, $\Sigma = \{fruits, fruit, features, feature, feature-name, feature-value\}$ (note that labels '1' are not explicitly shown on that figure). For an L-wrapper corresponding to the relational case ([2]) if D is the set of attribute names then $\Sigma = D \cup \{result, tuple\}$. Also it is not difficult to see that in an XML setting a schema would nicely correspond to the document type definition of the output document that contains the extracted data.

Definition 2. *(HL-wrapper) A HL-wrapper consists of a schema and an assignment of L-wrappers to split vertices of the schema tree. A vertex v of the schema tree is called* split vertex *if it has exactly one incoming arc labeled '*' and $n \geq 1$ outgoing arcs labeled '1'. An L-wrapper assigned to v must have arity n to be able to extract tuples with n attributes corresponding to outgoing neighbors of v.*

An HL-wrapper for the example document considered in this paper consists of: i) schema shown in figure 2a, ii) L-wrapper W_1 assigned to the vertex labeled with symbol *fruit* and iii) L-wrapper W_2 assigned to vertex labeled with symbol *feature*.

Let us now outline a solution for HL-wrapper implementation using XSLT. For this purpose we combine the idea of the hierarchical Prolog implementation with the translation of L-wrappers to XSLT ([3]). See appendix for the resulted XSLT code.

Following [3], a single-pattern L-wrapper for which the pattern graph has n leaves, can be mapped to an $XSLT_0$ stylesheet consisting on $n + 1$ constructing rules ([5]). In our example, applying this technique to each of the wrappers W_1 and W_2 we get three rules for W_1 (start rule, rule for selecting *fruit name* and rule for selecting *features*) and three rules for W_2 (start rule, rule for selecting *feature name* and rule for selecting *feature value*). Note that additionally to this separate translation of W_1 and W_2 we need to assure that W_2 selects feature names and feature values from the document fragment corresponding to a given fruit – i.e. the document fragment corresponding to the *features* attribute of wrapper W_1). This effect can be achieved by including the body of the start rule corresponding to wrapper W_2 into the body of the rule for selecting features, in-between tags <features> and </features> (see appendix). Actually this operation corresponds to a join of wrappers W_1 and W_2 on the attribute *features* (assuming L-wrappers are extended with an argument representing the root of the document fragment to which they are applied – see section 2).

4   Related Work

Two earlier works on hierarchical data extraction from the Web are hierarchical wrapper induction algorithm Stalker [7] and visual wrapper generator Lixto [4].

Stalker ([7]) uses a hierarchical schema of the extracted data called *embedded catalog formalism* that is similar to our approach (see section 2). However, the main difference is that Stalker abstracts the document as a string rather than a tree and therefore their approach is not able to benefit from existing XML processing technologies. Extraction rules of Stalker are based on a special type of finite automata called *landmark automata*, rather than logic programming – as our L-wrappers.

Lixto [4] is a software tool that uses an internal logic based extraction language – Elog. While in Elog a document is abstracted as a tree, the differences between Elog and L-wrappers are at least two fold: i) L-wrappers are only devised for the extraction task and they use a classic logic programming approach – an L-wrapper can be executed without any modification by a Prolog engine. Elog was devised for both crawling and extraction and has a customized logic programming-like semantics, that is more difficult to understand; ii) L-wrappers are efficiently implemented by translation to XSLT, while for Elog the implementation approach is different – a custom interpreter has been devised from scratch.

5   Concluding Remarks

In this note we introduced HL-wrappers – an extension of L-wrappers to extract hierarchical data from Web pages. The main benefit is that all results and techniques already derived for L-wrappers can be applied to the hierarchical case. As future work we plan: i) to do an experimental evaluation of HL-wrappers (but, as they are based on

L-wrappers, we have reasons to believe that the results will be good); ii) to incorporate HL-wrappers into an information extraction tool.

References

1. Bădică, C., Bădică, A., and Popescu, E.: A New Path Generalization Algorithm for HTML Wrapper Induction. In: Last, M. and Szczepaniak, P.(eds.): *Proc. of the 4th Atlantic Web Intelligence Conference (AWIC 2006)*, Israel. Studies in Computational Intelligence Series, Springer Verlag, 23, 10–19, 2006.
2. Bădică, C., Bădică, A., Popescu, E.,Abraham, A.: L-wrappers: concepts, properties and construction. A declarative approach to data extraction from web sources. In: *Soft Computing - A Fusion of Foundations, Methodologies and Applications*, http://www.springerlink.com/content/3m58724n3418270g, Springer Verlag, 2006.
3. Bădică, C., Bădică, A., and Popescu, E.: Implementing Logic Wrappers Using XSLT Stylesheets. In: *International Multi-Conference on Computing in the Global Information Technology, (ICCGI'06)*, Romania. http://doi.ieeecomputersociety.org/ 10.1109/ICCGI.2006.127, IEEE Computer Society, Los Alamitos, CA, USA, 2006.
4. Baumgartner, R., Flesca, S., Gottlob, G.: The Elog Web Extraction Language. In: *Proceedings of LPAR'2001*, LNAI 2250, Springer Verlag, 548–560, 2001.
5. Bex, G.J., Maneth, S., Neven, F.: A formal model for an expressive fragment of XSLT. *Information Systems*, Elsevier Science vol.27, no.1, 21–39, 2002.
6. Clark, J.: XSLT Transformation (XSLT) Version 1.0, W3C Recommendation, 16 November 1999, http://www.w3.org/TR/xslt 1999.
7. Muslea, I., Minton, S., and Knoblock., C.: Hierarchical Wrapper Induction for Semistructured Information Sources. In: *Journal of Autonomous Agents and Multi-Agent Systems*, Springer Netherlands, vol.4, no.1-2, 93–114, 2001.

Appendix: XSLT Code of the Sample Wrapper

```xml
<?xml version="1.0" encoding="UTF-8"?>
<xsl:stylesheet xmlns:xsl="http://www.w3.org/1999/XSL/Transform" version="1.0">
    <xsl:template match="html">
        <fruits><xsl:apply-templates select="//p/*/preceding-sibling::*[1]/*/text()" mode="select-fruit-name"/></fruits>
    </xsl:template>
    <xsl:template match="node()" mode="select-fruit-name">
        <xsl:variable name="var-fruit-name" select="."/>
        <xsl:apply-templates mode="select-features" select="parent::*/parent::*/following-sibling::*[position()=1]">
            <xsl:with-param name="var-fruit-name" select="$var-fruit-name"/>
        </xsl:apply-templates>
    </xsl:template>
    <xsl:template match="node()" mode="select-features">
        <xsl:param name="var-fruit-name"/>
        <xsl:variable name="var-features" select="."/>
        <fruit><name><xsl:value-of select="normalize-space($var-fruit-name)"/></name>
        <features><xsl:apply-templates select="$var-features//tr/*/preceding-sibling::*[1]/text()"
            mode="select-feature-name"></xsl:apply-templates></features></fruit>
    </xsl:template>
    <xsl:template match="node()" mode="select-feature-name">
        <xsl:variable name="var-feature-name" select="."/>
        <xsl:apply-templates mode="select-feature-value" select="parent::*/following-sibling::*[position()=1]/text()">
            <xsl:with-param name="var-feature-name" select="$var-feature-name"/></xsl:apply-templates>
    </xsl:template>
    <xsl:template match="node()" mode="select-feature-value">
        <xsl:param name="var-feature-name"/>
        <xsl:variable name="var-feature-value" select="."/>
        <feature><name><xsl:value-of select="normalize-space($var-feature-name)"/>
            </name><value><xsl:value-of select="normalize-space($var-feature-value)"/></value></feature>
    </xsl:template>
</xsl:stylesheet>
```

SCAM: Semantic Caching Architecture for Efficient Content Matching over Data Grid

Muhammad Farhan Bashir, Raja Asad Zaheer, Zohaib Mansoor Shams, and Muhammad Abdul Qadir

Center for Distributed and Semantic Computing Mohammad Ali Jinnah University, Islamabad, Pakistan
m.farhan.bashir@gmail.com,
rajaasad@jinnah.edu.pk,
some1shams@gmail.com,
aqadir@jinnah.edu.pk

Summary. Heterogeneous, distributed multi-databases (data warehouses, grids) are growing day by day as businesses demand more economical and flexible solutions for their data storage and processing needs. One of the main issues obstructing the large scale deployment of distributed multi-databases is latencies involved in data retrieval. Such problems are usually solved with the help of persistent caches. The paper presents a semantic caching architecture which supports faster semantic matching in query processing. This paper enhances the existing architecture of semantic cache and provides the mapping of hierarchical indexing scheme over it with the help of case study in order to represent the working of architecture. This paper also provides experimental results of the query processing over updated architecture and provides the discussion over comparison of the extended architecture with the previously proposed architecture.

1 Introduction

Heterogeneous, distributed multi-databases (data warehouses, data grids) are growing day by day as businesses demand more economical and flexible solutions for their data storage and processing needs. One of the main issues obstructing the large scale deployment is latencies involved in data retrieval, is which usually solved with the help of persistent caches. Usage of cache causes another problem, that is, the cost of processing on caches might increase. The reason to this fact is that the data is not organized in such a way that supports query processing in minimum amount of time. We need to devise some indexing scheme for cached data so that the processing could become easier and cost of processing should reduce as a result. The cache architecture should provide support in order to incorporate the indexing scheme.

This paper provides the semantic caching architecture for data grid to support the notion of semantic matching over the contents of cache with the help of hierarchical semantic indexing scheme. We have enhanced the existing architecture of semantic cache [9], whose focus was to support the management issues regarding semantic caches. We have enhanced incorporated the faster matching of

K.M. Węgrzyn-Wolska and P.S. Szczepaniak (Eds.): Adv. in Intel. Web, ASC 43, pp. 41–46, 2007.
springerlink.com

semantics with the help of 4-level hierarchical indexing scheme [10]. The scheme has been explained in that paper with the help of a case study. In this paper, I have also provided the empirical results of the 4-level hierarchical indexing scheme when mapped onto this proposed architecture.

2 Background and Related Work

Researchers have performed comparisons of semantic caches with page and tuple caches. Ren [1] lists disadvantages of page and tuple caches like heavy dependence of page cache upon the network and database server, and retrieval of irrelevant data from database due to poor clustering scheme. On contrarily, semantic cache over comes these problems due to storage of semantics with data. Ren [2] is of the view that page and tuple caches are not well suited for location dependent queries. In that nick of time some researchers shifted their focus to define semantic caches unambiguously, like activities performed in [1, 2, 4, 5, 6, 7, 8]. Ahmed [9] has provided semantic caching architecture for data grid with the focus of dealing with the management related issues. With the passage of time the trend changed a bit and researchers shifted their focus on using semantic caching in other domains instead of enhancing the powers of cache. Bashir [3] has performed comprehensive survey and is of the view that the focus of researchers turned towards using it in other domains instead of improving the semantic cache itself. Bashir in another research activity [10] proposed the faster semantic indexing scheme for content matching over semantic cache. The proposed scheme was devised to work with the architecture proposed by Ahmed [9]. Initially, it was supposed that the scheme can be used with the existing architecture to support the faster semantic content matching but as the indexing scheme started getting matured, it became difficult to make both the ends meet. The reason to this fact is that the focus of existing architecture is the management related issues whereas faster query processing support was not considered as a primary issue in that architecture. So there is a need to bring some enhancements to the existing architecture.

3 Proposed Architecture

The basic purpose to enhance existing architecture is to make it capable to incorporate faster content matching. In order to meet this requirement we have partitioned the architecture into two major building blocks, Query Processing Module (*QPM*), and Cache Management Module (*CMM*) as represented in figure 1. Both of these modules work almost independently but they need to interact with each other in order to process query in quick manner. So we tried to enhance the cohesion and reduce coupling.

The basic purpose of *CMM* is to cope up with management related issues of semantic cache. Most of the sub-modules of *CMM* are inherited from existing architecture [9], but their scope and placement has been changed a bit. *CMM* is further divided into three modules, Cache Management Service (*CMS*), Local

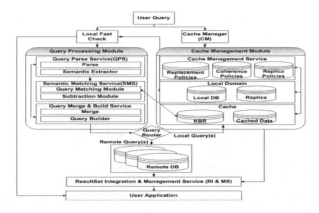

Fig. 1. Caching Architecture for Efficient Content Matching

Domain (*LD*), and Cache (*C*). Cache management is performed on the basis of policies stored in the *CMS*, includes replacement, coherence and replica policies. Cache Manager (*CM*) performs management of *LD* and *C* with the help of these policies. *LD* represents the local database and replication stored at client end, whereas the *C* contains the cashed data in cache DB and its semantics in *KBR* (*Knowledge base Repository*). The purpose of *QPM* is to deal with the issues related to faster contents matching over semantic cache. *QPM* comprises of three main components Query Parser Service (*QPS*), Semantic Matching (*SM*), and Query Merge and Build Service (*QMB*). The purpose of *QPS* is to parse the user's query and extract its semantics in order to convert it into a standard form which is then passed to *SM* where these semantics are matched with the semantics stored in the cache. *KBR* interaction, in order to perform faster semantic matching is proposed in [10]. It then performs the *Subtraction Process* in order to cut the corners of query making it asking exactly that amount of data which was required by the user. This is the time when *QMB* comes into the play where *merging and rebuilding procedure* is performed to finalize the new queries. At this stage the probe and remainder queries are ready to be posed to the cache and remote database. The details of *QBM* are not the scope of this paper. Rest of the components line *LFC* and *RI* and MS are same as described in the existing architecture [9]. We need to understand the internal structure of *KBR* presented

Table 1. Data extracted from query

Semantics	Extracted Semantics
Database Names	db1
Table Names	Emp
Column Names	n, a
Conditions	$(a > 30 and s > 20) or (s < 20 and (n = "fn" or j = "mg"))$

in figure 2 (a), because internal structure will help us understanding the way of storage. *QPS* extracts the query semantics database, table, column and conditions from the queries after parsing them and passes this data to the *SM*, where these semantics are matched against stored semantics. The storage is devised in a way that helps out in processing queries.

$QC1 - 1 : Select$**n,a**$fromdb1.emp$where$(a > 30ands > 20)or(s < 20and(n = "fn"orj = "mg"))$

The 4-level hierarchical query processing for semantic matching [10] takes full advantage of this storage scheme. Figure 2 (b) represents the internal structure of conditions table of *KBR*. Conditions are to be stored in different manner than rest of the semantics. The reason behind this fact is the same that storage should support faster matching of semantics over cache. The details of matching process are presented in the case study later in this paper.

4   Case Study

The purpose of this case study is to represent the mapping of proposed indexing scheme [10] over semantic caching architecture. *QPS* extracts the semantics (database, table and Column) from the query QC1-1 as presented in table 1, are stored in the *KBR* against some query identifier (id=1 for this scenario). We can easily identifying by looking table 1 that the conditions if stored the way they are, would not be helpful in faster query processing. There is a need to convert conditions in some standard format so that these could be stored in a way which is helpful in minimizing processing overhead. It is quite obviously known that

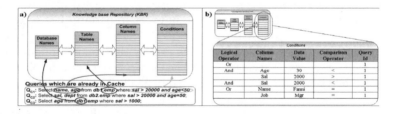

Fig. 2. a). Semantics and, b). Data Stored KBR

single condition comprises of three parts: column name, comparison operator, and data value, for example $(a < 60)$. Here column name provides the basic matching and rest two are used for the extended processing. Figure 2 (b) shows the fact that made it easy to decide Conditions table's structure. We had to face a new problem caused by the irregular shape of parenthesization of conditions. We had to regularize this structure in order to process efficiently.

$QC1 - 2 : a > 30ANDs > 20)OR(s < 20AND(n = "fn"ORj = "mg"))$

$QC1 - 3 : ORANDa30 < s20 > ANDs20 < ORn"fn" = j"mg" =$

To achieve this goal we strived for conversion into postfix notation. Following QC1-2 and QC1-3 are the infix and postfix notations against the query posed earlier in this case study. The stored data is represented in figure 2 (b). We are also using the hashing technique in order to make the matching process faster.

5   Empirical Results

We have created a running prototype of faster query matching. Figure 3 (a) provides us with the details of the implemented system. Suppose that user poses the query to the query processor, Input Validator extracts the semantics as presented in table 1, which will be passed to Query Parser. There we standardize the query to make processing easier. The process of matching user query with cached queries starts, which is discussed in details in [10]. The empirical results of QMM represent the validation and performance of our

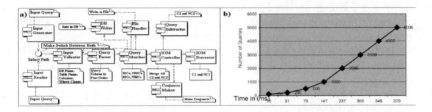

Fig. 3. a). System Design and, b). Processing Time

system, which is the theme of this paper. We can easily see from the figure 3 (b) that the number of queries as increased from 1000 to 5000 the time only increases from 147ms to 379ms only, which is very slower rate as compared to the previous ratios of 100 to 1000. The results show that after specific number of queries (1000 in this architecture), the time of processing increases with slower growth. This is one of the reasons that we consider this architecture helpful for data grids, where we have to deal with the massive number of queries in the cache.

6   Discussions and Conclusion

This section focuses on the discussion related to these similarities and difference found in both the architectures. The main difference is that the previous architecture provided solid grounds for cache management, whereas the new architecture takes care of both, the management as well as the support for faster semantic matching. Faster processing is incorporated with the help of new features.

This paper introduces the neglected area of semantic caching architecture. This paper also presents the solution to the problems in previous architecture by introducing new features. The proposed architecture incorporates support for faster contents matching in order to overcome the limitations posed by the existing architecture. The experimental results show that the matching process has become quick. The proposed architecture will help reducing query response time to the users.

References

1. Ren, Q. and Dunham M. H.: Semantic Caching and Query Processing. Southern Methodist University, TR-98-CSE-04, 1998.
2. Ren Q., Dunham M. H.: Using Semantic Caching to Manage Location Dependent Data in Mobile Computing, ACM, 2000.
3. Bashir M. F., Qadir M. A.: Survey on Efficient Semantic Content Matching over Semantic Cache, in the proceedings of 10th International IEEE Multitopic Conference INMIC 2006.
4. Godfrey P., and Gryz J.: Answering queries by semantic caches. In Proceedings of the 10th Database and Expert Systems Applications (DEXA'99), pages 485–498, August 1999.
5. Chidlovskii B., Borghoff B. M.: Semantic Caching of Web Queries, In the proceedings of VLDB Journal 9, 1 (Mar. 2000), 2-17.
6. Ishikawa Y., Kitagawa H.: A Semantic Caching Method Based on Linear Constraints, in proceedings of International Symposium on Database Applications in Non-Traditional Environments, IEEE, 1999.
7. Lee K. C. K., Leong H. V., & Si A.: Semantic Query Caching in a Mobile Environment, ACM SIGMOBILE Mobile Computing and Communications, Volume 3, Issue 2, 1999.
8. Lee D., Chu W. W.: Semantic Caching via Query Matching for Web Sources, CIKM, ACM, 1999.
9. Ahmed M. U., Zaheer R. A., and Qadir M. A..: Intelligent Cache Management for Data Grid, Australasian Workshop on Grid Computing and e-Research (AusGrid 2005).
10. Bashir M. F., Qadir M. A..: HiSIS: 4ÜLevel Hierarchical Semantic Indexing for Efficient Content Matching over Semantic Cache, in the proceedings of 10th International IEEE Multitopic Conference INMIC 2006.

Recommender System from Personal Social Networks

David Ben-Shimon[1], Alexander Tsikinovsky[1], Lior Rokach[2], Amnon Meisles[3], Guy Shani[3], and Lihi Naamani[2]

[1] Deutsche Telekom Laboratories at Ben-Gurion University
[2] Department of Information System Engineering, Ben-Gurion University
[3] Department of Computer Science, Ben-Gurion University

Summary. Recommender systems are found in many modern web sites for applications such as recommending products to customers. In this paper we propose a new method for recommender system that employs the users' social network in order to provide better recommendation for media items such as movies or TV shows. As part of this paper we develop a new paradigm for incorporating the feedback of the user's friends. A field study that was conducted on real subjects indicates the strengths and the weaknesses of the proposed method compared to other simple and classic methods. The system is envisioned to function as a service for recommending personalized media (audio, video, print) on mobile phones, online media portals, sling boxes, etc. It is currently under development within Deutsche Telekom Laboratories - Innovations of Integrated Communication projects.

1 Introduction

Systems that recommend items to users are becoming popular and can be found in many modern web sites for applications such as recommending products to customers in e-commerce sites, recommending TV programs to users of interactive TV and presenting personalized advertisements. There are two dominating approaches [8] for creating recommendation; Collaborative Filtering (CF) and Content-Based (CB) recommendations. CF considers the recommended items only by a unique identifier and recommends items that were purchased together, ignoring any attribute of the item. CB recommendations are based on an item profile which is commonly defined by the attributes of the item without considering acts of purchasing. Each of these methods has its pros and cons but it seems that a hybrid approach can overcome most of the disadvantages of the two methods. Another method for providing recommendations is based on Stereotypes [3]. Stereotypes are a way to define an abstract user that has general properties similar to a set of real users. Stereotypes are used in recommender systems for varying purposes, ranging from initial user profile creation to generating recommendations [8]. All methods use some type of a user profile or user model for recommendation. CF systems usually maintain a vector of rated items while CB systems maintain a rated set of item attributes.

K.M. Węgrzyn-Wolska and P.S. Szczepaniak (Eds.): Adv. in Intel. Web, ASC 43, pp. 47–55, 2007.
springerlink.com © Springer-Verlag Berlin Heidelberg 2007

A social network is a graph that represents connected users. The two most important characteristics of social networks are: who the people in the network are and how they are connected to each other. The usual notion of connection between people in a virtual community is related to direct social interaction [4]. This makes social networks useful for providing recommendations. Assuming that connected people have also some common interests, this paper proposes a method for recommending media items based on a personal social network of each user.

The rest of this paper is organized as follow: Section 2 surveys the current status of recommender systems and recommendations-based social network. Section 3 describes in detail the proposed method. Section 4 describes the field study and the comparative results of the proposed approach for recommending items. Section 5 presents a discussion and our conclusions.

2   Recommender Systems

With the explosion of data available online, recommender systems became very popular. While there are many types of recommender systems ranging from manually predefined un-personalized recommendations to fully automatic general purpose recommendation engines, two dominating approaches have emerged - Collaborative Filtering and Content Based recommendations.

2.1   Collaborative Filtering

Collaborative filtering stems from the idea that people looking for recommendations often ask for the advice of friends. While on the internet the population that can supply advises is very large, the problem shifts into identifying what part of this population is relevant for the current user.

CF methods identify similarity between users based on items they have rated and recommend new items that similar users have liked. CF algorithms vary by the method they use to identify similar users. Originally Nearest-Neighbor approaches based on the Pearson Correlation, computing similarity between users directly over the database of user-item ratings were implemented. Modern systems tend to learn some statistical model from the database and then use it for recommending previously rated items to a new audience. Model-based approaches usually sacrifice some accuracy in favor of a rapid recommendation generation process [5], better scaling up to modern applications. The main advantage of CF is that it is independent of the specification of the item and can therefore provide recommendations for complex items which are very different yet are often used together. The major drawback of this approach is the inability to create good recommendations for new users that have not yet rated many items, and for new items that were not rated by many users. This drawback also known as the *'cold start problem'*.

2.2   Content-Based Recommendation

The ideas of Content-Based recommendations originate in the field of information filtering, where documents are searched given some analysis of their text. Items are hence defined by a set of features or attributes. Such systems define a user using preferences over this set of features, and obtain recommendations by matching user profiles and item profiles looking for best matches.

Some researchers [8] separate methods that learn preferred attributes from rated items from methods that ask the user to specify his preferences over item attributes, but we refer to all methods that recommend based on item attribute preferences as CB recommendation. CB approaches rarely learn statistical models and usually match user profiles and item profiles directly. User and item profiles are very sensitive to profile definitions - which attributes are relevant and which attributes should be ignored. It is also difficult to create an initial profile of the user, specifying the interests and preferences of the user; Users are reluctant to provide thorough descriptions of the things they like and do not like. It is also possible that users are unaware of their preferences. For example, a user cannot know whether she likes an actor she never seen. In fact the acquisition of user preferences is usually considered a bottleneck for the practical use of these systems. Content-based recommendations may also result in very expected items and may not be able to direct the user towards items he is unaware of but may like. Nevertheless, CB systems can easily provide valid recommendations to new users, assuming that their profile is specified using a questionnaire or some other method for preferences elicitation, even if they never used the system before.

CB engines can provide recommendations for new items that were never rated before based on the item description and are therefore very useful in environments where new items are constantly added.

Hybrid approaches [10] of CF and CB can reduce the disadvantages of the methods.

2.3   Communities and Social Networks

The main idea of a Social Network (SN) is to use some relations that users sharing between them. It is the set of actors i.e. group of people, which are the nodes of the network, and ties that link the nodes by one or more relations. Social network indicates the ways in which actors are related. The tie between actors can be maintained according to either one or several relations. Moreover, the network gives not only to their actors, people that are directly connected, but also to actors of their actors, also called "friends of my friends".

The roots of link analysis predate the use of modern computers. The field of social network analysis has developed over many years as sociologists developed formal methods of studying groups of people and their relationships. The advent of computers allowed these techniques to become much more widespread and to be applied on a much larger scale in recommender systems.

Social networks can be divided into several groups in terms of different criteria:

- Dedicated - dating or business networks, networks of friends, graduates, fun clubs etc'
- Indirect - online communicators, address books, e-mails etc'
- Common activities. - co-authors of scientific papers, co-organizers of events etc'
- Local networks - people living in the neighborhood, families, employees networks etc'
- Hyperlink networks - links between home pages etc'

Recommender systems for social networks differ from the approaches which described before, since often they suggest rational human beings to other people and not just product or service. Generally the network is initiated by the users when one person initializes the relationship with another one, and the latter can respond either positively or negatively to the invitation such as in MSN messenger, ICQ etc'. After the relationship is been set up the network could be for use in some manner i.e. chatting in the MSN or ICQ examples. Such interaction is impossible with products and services. Furthermore, the relationship between people is bidirectional in opposite to the relationship between a person and product.

Most of the recommendation systems [6] which using social networks for recommendation are often need the *'trus't* parameter. That means if we are going to use the relationship between people for providing recommendation we need to know how well person A trust person B concerning the product and taste of the recommended item. So another parameter in the relations at the social network is the trust. However, there several situations where we assume the trust is high or we cannot measure i.e. ICQ or MSN messenger. There we are sure about the connection but we cannot know for sure the level of trust concerning recommendation for item of some type.

In [6] they use the trust ratings between the users in the CN as the basis for making calculations about similarity. They assume that there is correlation between trust and user similarity; that means if user A trust user B with high value, then the similarity of the preferences list of movies will also be high.

It has been shown [1] that in specific domain such as movies, users develop social connections with people who have similar preferences; however no field experiment has been provided. The results of [1] extended in [11]; their work proved that there is positive correlation between trust and user similarity in an empirical study of a real online community. There is a lot of work in the area of measuring the trust in communities and related the issue to recommender systems such as [7] and [6]. Measuring the trust is not easy task and often impossible and not accurate, thus, it is leading us to option of creating the relation on full trust and this can obtain only by explicitly order from the user i.e. in the MSN messenger and the ICQ where one cannot be related to other one unless he approved his invitation. As mentioned before we are using the suggested method for mobile application named MediaScout. In MediaScout each user can invite other users to become his friend by simply provide the mobile number; the receiver can approve or decline the invitation. Based on those direct relations the users in MediaScout can send each other recommendation about movies.

3 Personal Social Networks

3.1 Creating the Network

As described above, MediaScout is an application which provides personalized media content via mobile devices and home TV. One of the features which exist in the system is the ability to send an invitation to some other user in the system and to propose a friendship. The receiver of the invitation can accept or decline the invitation. If the second user accepts the invitation, then the two users become 'friends' of each other and can send recommendations to each other using one of the features in the application. This is a similar scenario to the one of MSN messenger and of ICQ, except for the usage goal. In the two existing applications one can chat with one's friend and in MediaScout, a pair of friends can send movie recommendations to each other. MediaScout uses a binary feedback mechanism. This means that for each media item that the user watches she can provide a feedback whether she liked it or not. This feedback helps the system to refine the profile of the user. For each user the system keeps several items that she rated positively, several items that were rated negatively and a list of friends as well.

Having this kind of information and relations about users can easily lead to the generation of a social network which involves most or all of the users of the application. Each user is a node in the network and the friendship relation is represented by edges which connect the users with a trust of 1 between each two connected nodes. For each user one can envision the first layer of friends, the users which are connected to the user by a direct link. In addition, there are friends of the friends and so on.

3.2 Constructing the Personal Network Model

Consider the network of friends, where the only absolute information we have is that the first layer of each user includes the list of the user's friends and for each user we have two lists of movies, one that she likes and one that she dislikes. We assume that there is some similarity between the preferences of a user and those of her friends and of the friends of her friends. We propose to construct recommendations that are based on the personal social network of each user. The personal social network of each user is a snapshot of the entire network which presents the relations of each user with her related friends, up to the level of 6 (friends of friends etc.). This network is constructed in the form of a social tree for each user (Ur) by using a Breadth-First Search (BFS) [9] algorithm. Figure 1 illustrates the transition between the social network into the personal network of a user. The number of the levels which we take into consideration while building the personal network will affect the recommendations later on. In the experiment at the next section, the network is constructed up to the level of six in order to cover a large range of relations among users. Denote the distance from user Ui to another user Uj by $d(Ui, Uj)$ which is computed by traversing the social tree. The tree can be viewed as a set of users described by

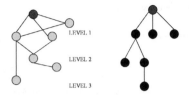

Fig. 1. The left graph describes the relations of the red user with other users in the system. The graph on the right hand side presents the resulting personal network after the run of the BFS algorithm. Note that each user appears only once in another user's personal network.

$$X(Ui, L) = \{U_j | \forall j, U_j \in \ social\ tree\ of\ user\ i\ up\ to\ the\ level\ of\ L\} \qquad (1)$$

3.3 Constructing Recommendations

Once we have the personal social network of each user, and the items that each user likes and dislikes the construction of the recommendation list proceeds as follows. The list of recommendations for each user is based on the items that her 'friends' like and dislike. If some of the users in her personal network like some media item, this item is going to appear in the list of recommendations. Items that are negatively rated by friends are not going to be recommended. Recommendations are constructed by summing the impact of ratings of the personal social network, both positive and negative. We use the notations below to define the computation of the recommendation list:

- M - Media item.
- U - User in the system.
- $R(Uj,Mi)$ - the rating of user j over media item i, The value 1 for positive rating, -1 for negative rating and 0 if user j did not rate media i.
- $d(Ui,Uj)$ - the distance of user j from user i in the social tree of user i.
- $X(Ui,L) = \{Uj \mid \forall j, Uj \in$ social tree of Ui computed up to the level of L$\}$
- K - Attenuation coefficient of the social network.

The overall rating of a media item for user i, based on her personal social network computed as in (2).

$$Rank(Mi, Ui) = \sum_{\forall Uj \in X(Ui,L)} [K^{-d(Ui,Uj)} * R(Ui, Mi)] \qquad (2)$$

The list of recommendations is simply a vector of sorted media items according to rank, where larger rank indicates higher order in the list.

The attenuation coefficient of the social network defines the impact of the distance between users, on the strength of recommendations. If $K = 1$ the impact is constant and the result is exactly equal to the popularity of the media items in the personal network of the user. If $K = n > 1$ the rating of a user at level x is equivalent to n ratings of users at level of x+1.

4 Experiments

We conducted a field experiment with real users aimed to examine whether there is any correlation between user preferences regarding movies, to the preferences of her friends? Whether providing a recommendation based on a personal social network will be more effective than recommending movies by popularity, for example.

The experiment involved 50 users, all of them from the same class at Ben Gurion University, and they were all familiar with each other. The assumption is that there are groups inside this group and the entire group of the fifty users is homogenous.

Each user needed to write down two things:

1. The list of her friends - where 'friend' is defined to be a person whose recommendation she will consider, positive or negative, concerning movies.
2. To rate 108 movies on a scale of 1 to 5, where 1 means that the user did not like the movie at all and 5 means she liked the movie very much. Note that in cases where the user was not familiar with the movies, she rated them 0.

The movies were taken from the Internet Movie Data Base (IMDB) [12]. The list of 108 movies was constructed to have both popularity and diversity in the genre attribute. This information is available on IMDB. Having the rating of each user and her list of trusted friends enabled us to compare several methods on the same data. These methods included popularity, random recommendation, CF and our proposed approach.

4.1 Evaluation Metrics

To assess the performance of recommendations lists, we sort (descending) the movies for each user according to her preferences. We construct for each user her recommendation list based on the personal social network, using $k=2$. To assess the relevance of the resulting recommendations list, we checked the location of the movies that the user liked and used the R measure (3) which is taken from [5].

$$Ra = \sum_{t \in I} \frac{1}{2^{i/a}} \qquad (3)$$

Here i is the set of locations of the movies the user liked in the recommendations list and a is the viewing half-life and in this experiment it was set to 2. This metric assumes that each successive item in the list is less likely to be viewed with an exponential decay. The grade was then divided by $Rmax$ - the maximal grade, when all the movies the user liked appear at the top of the list. The second metric we used is termed *recall*. Recall is the ratio of the number of relevant movies presented to the user to the total number of relevant movies for the user in the data. Since the users rated the items in the experiment from 1 to 5, we simply calculate the average rating for each user and considered the movies that she rated below her average as irrelevant for her and those rated above the average as relevant. Recall is quite important to measure in our case since there are a limited number of movies that we were able to present to the user as a recommendation list. We measure the recall by 10, 20, 30, and 60 based on the fact that in MediaScout users are viewing a limited number of items at once.

4.2 Comparative Results

We compare the results of the suggested method to popularity. Table 1 present the results of the popularity and the personal social network. Analysis of the results indicates that the personal social network is obtaining similar results to the popularity. In the R measure, which takes into consideration the order of the recommendation, yet all the movies which exists in the data as potential recommendations, the personal social network obtains better results. The recall measure is affected from the point of the measure. The fact that the recall at the level of sixty items (out of 108 items we got in this experiment) is close to 92% in popularity and in the social network as well is quiet impressive. It is indicating that both of the methods perform very well on this data. Generally the social network and popularity obtain similar results.

Table 1. Comparative results between popularity and the personal social network

	R measure	Recall 10	Recall 20	Recall 30	Recall 60
Popularity	111.97	0.26	0.4591	0.644	0.927
Social Network	119.734	0.2265	0.4357	0.617	0.9112

5 Conclusions

In this paper we presented a new community based recommendation method. The experimental study conducted with real subjects shows that this method can improve the recommendation performance in cases we need a long list of recommendations.

Although further experiments needed with different algorithms such as CF and different measures, we can see that the personal social network has the potential to provide very quality recommendations. One of the approaches that might be useful for improving the recommendations is to find the most appropriate attenuation factor, the K value as described in section 3, for each user. Given the list of rated item of a given user; is there any specific K which led to more quality recommendations for the user? We believe there is. It is most likely that the K will be different from user to user but we didn't manage answer the question of how to find it. Finding the personal K by some heuristic will probably improve the results of the social network in our case.

References

1. Abdul-Rahman, A. and Hailes, S. Supporting trust in virtual communities. In Proceedings of the 33rd Hawaii International Conference on System Sciences. Maui, HW, USA (2000).
2. B. Smyth and P. Cotter, Surfing the Digital Wave: Generating Personalized TV Listings Using Collaborative, Case-Based Recommendation, Lecture Notes in Computer Science, vol. 1650, pp. 561-567, (1999).
3. E. Rich. User modeling via stereotypes. Pages 329- 342, (1998).
4. Garton, L., Haythornthwaite, C., and Wellman, B. Studying Online Social Networks, in Journal of Computer-Mediated Communication, 3 (1), (1999).
5. J. S. Breese, D. Heckerman, and C. Kadie. Empirical analysis of predictive algorithms for collaborative filtering. In UAI'98, pages 43-52, (1998).
6. Jennifer Golbeck, James Hendler. 2006. FilmTrust: Movie recommendations using trust in web-based social networks. Proceedings of the IEEE Consumer Communications and Networking Conference, (2006).
7. Massa, P., B. Bhattacharjee. 2004. Using Trust in Recommender Systems: an Experimental Analysis. In Proceedings of iTrust International Conference (2004).
8. M. Montaner, B. Lpez, and J. L. De La Rosa. Taxonomy of recommender agents on the internet. Artificial Intelligence Review, 19:285-330, (2003).
9. M. E. J. Newman and M. Girvan. Finding and evaluating community structure in networks. Phys. Rev. E, 69(2):026113, Feb, (2004).
10. R. Burke. Hybrid recommender systems: Survey and experiments. User Modeling and User-Adapted Interaction, 12(4):331-370, (2002).
11. Ziegler, Cai-Nicolas, Georg Lausen. Analyzing Correlation between Trust and User Similarity in Online Communities" Proceedings of Second International Conference on Trust Management, (2004).
12. http://www.imdb.com/

Adding an Index Mechanism to an Ontology

Fathia Bettahar, Claude Moulin, and Jean-Paul A. Barthès

Compiègne University of Technology
Heudiasyc, CNRS, France
{fathia.bettahar,claude.moulin,jean-paul.barthes}@utc.fr

Summary. This paper describes the design and the implementation of an indexing process based on features added to an ontology. This process provides a way for indexing a knowledge base, i.e. attaching element identifiers to keywords or expressions found in the ontology or the knowledge base.

The features inserted in the ontology define for each element the data type properties that can be used to build and update the index. Fundamentally, an index is a hash table where each entry contains a key built using this process and where the value is the list of identifiers of ontology elements concerned by the corresponding key. We describe the indexing characteristics for the OWL language and show how they can be reduced for a simpler language.

We then give an algorithm that allows to build an index for a compliant knowledge base. We conclude by presenting an application using this indexing process. It concerns a platform dedicated to eGovernment services. Within this platform, each module requires an index to facilitate the access to information semantically attached to ontology elements.

1 Introduction

The complexity of eGovernment services and administrative processes requires to handle knowledge from multiple information sources. Keyword indexing is often not a satisfactory option because it lacks precision and does not take into account the denotation of information. Moreover, it does not allow any reasoning on knowledge structures.

Information retrieval and indexing based on ontology is proposed in academic and industrial research environments [1, 2] as a new approach to improve information discovery by adding semantic support [3]. This approach is adopted by several authors and particularly for information retrieval from the Web. Internet search engines like Google or Altavista use a central database to index information and a simple keyword based requester to reach information.

To improve the semantic of search, two major approaches are proposed by researchers. The first approach concerns an annotation techniques based on the use of ontologies [1, 7]. The annotations are used to retrieve documents. This approach is dedicated to request/answer systems like KAON [5]. The second approach is an information retrieval technique based on the use of domain ontologies like the work of Desmontils in [6]. They are dedicated for retrieving raw documents.

K.M. Węgrzyn-Wolska and P.S. Szczepaniak (Eds.): Adv. in Intel. Web, ASC 43, pp. 56–61, 2007.
springerlink.com

In both cases, tools are required to create indexes based on the vocabulary occurring in ontologies. Annotation properties could be used for that. We propose to add information inside the ontology in order to support the creation of indexes for such tools. Our approach consists in annotating the properties whose values can be used to create index entries. This approach is derived from the entry point mechanism found in MOSS, a knowledge environment developed at the University of Compiègne [9].

2 Indexing Features

2.1 General Considerations

In this section we describe the features we include in an ontology and whose semantics can be used to build the index of a knowledge base. We first give the characteristics in the OWL formalism then we present the equivalent features in a simpler formalism that we have developed. The parser generating the OWL file is in charge of the creation of the required OWL structures.

In this section, all considerations concern the OWL formalism. The name of a concept is not defined in OWL; if it were defined, it would correspond to the identifier. The main property to consider for indexing is the label (introduced by rdfs:label). This property is attached to any owl:Thing and thus we consider that each instance of owl:Thing (concept, relation and individual) can be indexed from the value of this annotation property.

Using only the label property is not generally sufficient. For example, in the domain of e-government, many elements are accessed from their acronym. In French, the acronym "RMA" can be used to access the social program whose label is "Revenu Minimal d'Activité". The values of acronym, a string data type property defined on the concept of Program, would be useful to index the instances of Program. More generally, we consider that each data type property may be a candidate property for indexing the individuals of its domain.

2.2 Principle

Any data type property, can be used for indexing. Obviously string properties are best candidates but other types of properties can also be used. If a property has for domain a union of concepts, we also consider that this property can be used for indexing for only some of these concepts. The principle is the following: when a couple - Concept (C), data type Property (p) - is considered for indexing, each instance i of the concept C can be indexed from the value of p for i.

2.3 Advantages

Our indexing proposal allows a total independence of modules compliant with an ontology. Indeed, each module is responsible of the index creation it needs, and finds in the ontology all the elements useful for this creation. It can also add any element that it requires locally. It can apply any transformation for building the strings or other elements required for the entry keys, from the labels and values found in the ontology or knowledge base.

2.4   OWL Implementation

In the OWL syntax, an annotation property is first created. It is also a functional datatype property whose range is xsd:boolean. It is considered as a restriction on a datatype property for some concepts and its domain is the class owl:Restriction:

```
<owl:FunctionalProperty rdf:ID="indexing">
<rdf:type
    rdf:resource="http://www.w3.org/.../owl#AnnotationProperty"/>
<rdf:type
    rdf:resource="http://www.w3.org/.../owl#DatatypeProperty"/>
<rdfs:range rdf:resource="http://www.w3.../XMLSchema#boolean"/>
<rdfs:domain rdf:resource="http://www.w3.org/.../owl#Restriction"/>
</owl:FunctionalProperty>
```

A concept having a datatype property intended to be used for indexing is declared as a subclass of an owl:Restriction. This restriction is annotated by the indexing property with true as the Boolean value. For example, for the following OWL code indicates that each instance of the class Z-Program must be indexed by the value of the property hasAcronym.

```
<owl:Class rdf:about="#Z-Program">
  <rdfs:subClassOf>
    <owl:Restriction>
      <owl:onProperty rdf:resource="#hasAcronym"/>
      <owl:cardinality
          rdf:datatype="&xsd;nonNegativeInteger">1</owl:cardinality>
      <indexing
      rdf:datatype="http://www.w3.../XMLSchema#boolean">true</indexing>
    </owl:Restriction>
  </rdfs:subClassOf>
</owl:Class>
```

2.5   A Simpler Syntax

The previous OWL syntax is not easy to manage because it is necessary to insert features that no OWL editor currently takes into account. In the European project, TERREGOV[1] [10], we defined a simpler frame-like formalism called Simplified Ontology Language (SOL). The parser that we developed, generates the OWL file (and other useful files) for interoperability reasons, and creates the requested OWL structures.

---

[1] The content of this paper is the sole responsibility of the authors and in no way represents the views of the European Commission or its services.

In SOL, a datatype property, that can be used for indexing, is simply declared with the (:index) feature. In the following example, the attribute acronym is intended to be used for indexing the instances of the Program concept. If the attribute acronym were defined in another concept it could be used for indexing or not, according to the presence of (:index) in its declaration.

```
(defconcept
   (:name :en "Program" :fr "Programme")
   (:att (:en "acronym" :fr "acronyme") (:unique) (:index))
   (:att (:en "documentation" :fr "documentation")) ...)
```

3   Algorithm

3.1   Description

We propose an algorithm for creating a knowledge base index from the ontology features. We consider a simple index, i.e. a hash table structure where keys are words or expressions found in element labels and as values of some data type properties and where each associated values is a list of element identifiers. We consider a knowledge base manager (KBM) for a system that keeps tracks of all concepts, relations and individuals of the knowledge base and supports SPARQL [13] queries. The top level of the algorithm is:

```
Indexing All Concepts, All Relations, All Individuals
Indexing elements from "indexing" Annotation property
```

The indexing of concepts, relations and individuals from their labels are similar. The algorithm is the following:

```
1: get the elements (concepts, relations, individuals) from KBM
     for each element el
2:     get the element labels from KBM
         for each label
           get the canonical form of the label
3:         get the list of element ids already indexed on this label
4:         add the el id to the list
           put the entry (canonic_label,list) into the index;
         endfor
     endfor
```

The previous algorithm gives the steps to create an index from the labels of the ontology elements. In Step 1 we get the elements present in the knowledge base. In step 2, we get the labels attached to an element. May exist more than one label, e.g. in different languages. In step 3, the list of ids may be empty. In step 4 we add the element id to the list. This algorithm yields a table containing labels as keys and list of ids as values.

The following algorithm gives the indexing of concepts and individuals from annotation property "indexing":

```
1: set the first query string
      (for searching concepts and relations concerned by the indexing)
2: get the query results from KBM
3: for each result (couple: concept id ; relation id)
      set the second query from concept and relation ids
      (for searching the instances of the concepts and the values
            associated through the relation)
      get the query results from KBM
4:    for each result (couple: element id, string value)
        get the string value
        get the canonical value as a string
        get the list of element ids already indexed on this label
5:      add the el id to the list
        put the entry (canonic_label,list) into the index;
      endfor
   endfor
```

The first request (written here in SPARQL) allows for searching the concepts and attributes concerned by the indexing. In this case, each concept appears as a subclass of a restriction on the relations also concerned by indexing:

```
SELECT ?cpt ?att
WHERE {?x rdf:type owl:Restriction .
        ?x owl:onProperty ?att .
        ?cpt rdfs:subClassOf ?x .
        ?x tg:indexing true }
```

For each concept and attribute the second query used in step 3 searches instances of this concept and values associated through the attribute.

3.2   Application

We consider a platform dedicated to e-Government services where a global on-tology allows the homogeneity of semantics. In this platform several modules use the ontology: the semantic registry of web services contains the semantic descriptions (written in OWL-S [12]) of services written thanks to the central ontology, the document base where documents are indexed on the concepts of the ontology. We plugged a dialog system helping a user to find concepts defini-tion, documents or services from free text queries. See [8] for more details about this dialog system.

Our dialog system receives natural language queries from civil servants, ana-lyzes them and extracts the key words or expressions that it contains. According to this analysis the module that can answer the question is selected. The dialog system creates an index based on the process described above. The words ex-tracted from queries, after a normalization process, allow to find the identifiers of the ontological elements related to the question. For example, from the identi-fiers, the document base can deliver the documents indexed on the corresponding concepts.

4 Conclusion

This paper presents a mechanism that allows to build an index from features added to an ontology. Any knowledge base, compliant with the ontology can follow the algorithm described in Section 3 to create an index from the elements labels but also from the values of some datatype properties annotated in the ontology.

This mechanism shows the relevant information that can be used as entry point for a knowledge base. It does not prevent to add any other elements that specific modules could require.

References

1. Guarino N., Masolo C., Vetere G., (1999) OntoSeek: Content-Based Access to the Web, IEEE Intelligent Systems, pp 70–80.
2. Woods, W. A. (1997) "Conceptual Indexing: a better way to organize knowledge, Technical Report SMLI TR-97-61, Sun Microsystems Laboratories, Mountain View, CA.
3. Haav H-M., Lubi T.-L. (2001) A Survey of Concept-based Information Retrieval Tools on the Web, A. Caplinkas and J. Eder (Eds), Advances in Databases and Information Systems, 5th East-European Conference ADBIS*2001, Vilnius "Technika" , 2:29–41
4. Kiryakov A., Popov B., Terziev I., Manov D., Ognyanoff D. (2005) Semantic Annotation, Indexing, and Retrieval. Elsevier's Journal of Web Semantics, Vol. 2, Issue (1).
5. Bozsak E., M. Ehrig, S. Handschuh, et al., (2002) KAON - Towards a large scale Semantic Web. DEXA 2002, Aix en Provence, France.
6. Desmontils E., Jacquin C., Simon L. (2003) Ontology Enrichment and Indexing Process, Ingénierie des Connaissances, RESEARCH REPORT 03.05, Institut de Recherche en Informatique de Nantes 2.
7. Tsinaraki C., Polydoros P., Kazasis F. and Christodoulakis S. (2005), Ontology-based Semantic Indexing for MPEG-7 and TV-Anytime Audiovisual Content. Multimedia Tools and Applications, 26(3). 299–325.
8. Moulin C., F. Bettahar, M. Sbodio, et al. (2006) Adding Support to User Interaction in Egovernment Environment. 4th Atlantic Web Intelligence Conference, AWIC'06, Beer-Sheva, Israel, 151–160.
9. Barthès J.P. (1994) Developing integrated object environments for building large knowledge-based systems, Int. J. Human-Computer Studies 41, 33–58.
10. TERREGOV project, http://www.terregov.eupm.net/
11. Web Ontology Language (OWL), http://www.w3.org/TR/owl-ref/.
12. OWL-S, OWL-based Web Service Ontology,
    http://www.daml.org/services/owl-s/
13. SPARQL Query Language for RDF, http://www.w3.org/TR/rdf-sparql-query/

Immunological Selection in Agent-Based Optimization of Neural Network Parameters

Aleksander Byrski, Marek Kisiel-Dorohinicki, and Edward Nawarecki

Institute of Computer Science
AGH University of Science and Technology,
Mickiewicz Avn. 30, 30-059 Cracow, Poland
{olekb,doroh,nawar}@agh.edu.pl

Summary. In the paper an immunological selection mechanism in the agent-based evolutionary computation is discussed. Since it allows to reduce the number of fitness assignments, it is especially useful for problems with high computational cost of individual's evaluation. A neural network architecture optimization is considered as an example of such a problem. Selected experimental results obtained for the particular system dedicated to time-series prediction conclude the work.

1 Introduction

Artificial immune systems recently began to be the subject of increased re-searchers' interest. Different immune-inspired approaches were applied to many problems, such as classification or optimization [8]. The most often used algorithms are based on clonal and negative selection processes [4].

Based on these phenomena, a specific immunological approach was proposed as a more effective alternative to the classical resource-based selection used in evolutionary multi-agent systems (EMAS) [6]. As it was shown in [2] the introduction of immune-based mechanisms may affect various aspects of the system behaviour, such as the diversity of the population and the dynamics of the whole process. The most interesting effect is the substantial reduction of the number of fitness assignments required to get the solution of comparable quality. This is of vast importance for problems with high computational cost of fitness evaluation, like hybrid soft computing systems with learning procedure associated with each assessment. An example of such approach considered in this paper is the evolution of neural network architecture [1]. The work reported focuses on the impact of the immune-based approach on the performance of EMAS applied to time-series prediction in comparison to the resource-based selection used alone.

Below, after a short presentation of the basics of evolutionary multi-agent systems and the idea of immunological selection, the description of EMAS for time-series prediction is given. Selected results obtained for the prediction of popular benchmark Mackey-Glass chaotic time series illustrate the pros and cons of the approach.

K.M. Węgrzyn-Wolska and P.S. Szczepaniak (Eds.): Adv. in Intel. Web, ASC 43, pp. 62–67, 2007.
springerlink.com ⓒ Springer-Verlag Berlin Heidelberg 2007

2   Immune-Based Selection in Evolutionary Multi-agent Systems

The term *evolutionary multi-agent system* (EMAS) covers a range of optimization techniques, which consist in the incorporation of evolutionary processes into a multi-agent system at a population level. The most distinctive for these techniques are selection mechanisms, based on the existence of non-renewable resource (called life energy). Energy is gained and lost when agents perform actions, and at the same time it determines actions an agent is able to execute [6].

In optimization problems agents representing better solutions should gain energy at worse agents expense. The realization of this resource-based evaluation of may be based on the idea of agent rendezvous: agents may evaluate their neighbours, and exchange energy. Worse agents (considering their fitness) are forced to give a fixed amount of their energy to their better neighbours. Then if low energy level increases the possibility of death and high energy level increases the possibility of reproduction, the flow of energy should cause that in successive generations survived agents would represent better approximations of the solution.

In order to speed up the process of resource-based selection, coming from the observation that "bad" phenotypes come from the "bad" genotypes, a new group of agents (acting as lymphocyte T-cells) may be introduced. They are responsible for recognizing and removing agents with genotypes similar to the patterns posessed by these lymphocytes. Of course there must exist some predefined affinity function, e.g. using real-value genotype encoding, it may be based on the percentage difference between corresponding genes.

These agents-lymphocytes are created in the system during the action of death performed by an agent, and they are granted with the pattern based on removed agent's genotype. The new lymphocytes must undergo the process of negative selection. In a specific period of time, the affinity of the immature lymphocytes patterns to the "good" agents (posessing relative high amount of energy) is tested. If it is high (lymphocytes recognize "good" agents as "non-self") they are removed from the system. If the affinity is low – probably they will be able to recognize "non-self" individuals ("bad" agents) leaving agents with high energy intact. The system working accoding to these principles will be called an *immunological* Evolutionary Multi-agent System (iEMAS).

3   Evolving Neural Agents for Time-Series Prediction

Many examples from the literature show that neural networks may be successfully used as a mechanism to model the characteristics of a signal in a system for a time-series prediction [7]. The choice of a particular architecture of the network is to a large extent determined by a particular problem. Usually the next value of the series is predicted on the basis of a fixed number of the previous ones. Thus the number of input neurons correspond to the number of values the prediction is based on, and the output neuron(s) give prediction(s) of the next-to-come value(s) of the series. According to NARX model, a multi-layered

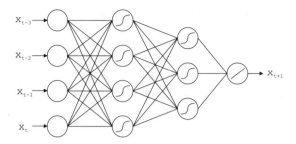

Fig. 1. Predicting multi-layered perceptron

Table 1. Optimal values of neural network parameters

Parameter	Range	EMAS		iEMAS	
Input layer neurons count	[10,50]	43	±4.5	42	±3
1st hidden layer neurons count	[10,50]	35	±4	36	±5
2nd hidden layer neurons count	[10,50]	34	±5	33	±5
1st hidden layer learning coefficient	(0.0,1.0)	0.7	±0.08	0.85	±0.1
2nd hidden layer learning coefficient	(0.0,1.0)	0.72	±0.1	0.62	±0.13
Output layer learning coefficient	(0.0,1.0)	0.65	±0.07	0.45	±0.13
1st hidden layer momentum coefficient	(0.0,1.0)	0.6	±0.07	0.55	±0.15
2nd hidden layer momentum coefficient	(0.0,1.0)	0.58	±0.12	0.50	±0.09
Output layer momentum coefficient	(0.0,1.0)	0.35	±0.14	0.5	±0.09

perceptron (MLP) may be used in this case (fig. 1), supervisory trained, using the comparison between values predicted and received as an error measure [3].

A multi-agent predicting system is a population of intelligent agents performing independent analysis of incoming data and generating predictions. Subsequent elements of the input sequence(s) are supplied to the environment, where they become available for all agents. Each agent may propose its own predictions of (a subset of) the next-to-come elements of input obtained from the posessed neural network. Of course the network is trained by the agent using the data acquired from the environment. The prediction of the whole system may be generated with the use of collective intelligence management techniques, such as voting based on PREMONN algorithm [5].

In such a system the introduction of evolutionary processes allows for searching for a neural network architecture and learning parameters most suitable for the current problem and system configuration, according to the principles of evolutionary multi-agent systems. The system should work as follows. First, the population of computational agents is initialized with random problem solutions (neural networks and learning parameters). Every agent iss able to perform predictions of subsequent time-series values, after it acquires its actual value and performes the subsequent step of the training of its neural network. Based on the value of the fitness function – MAPE error (mean absolute prediction error) the agents are able to perform the actions of rendezvous, reproduction, and death, as it was described in the previous section.

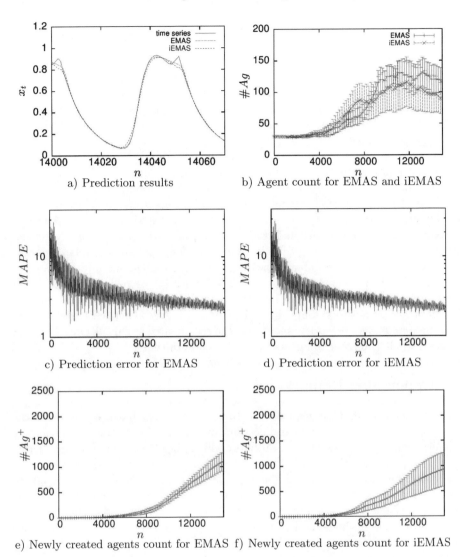

Fig. 2. Experimental results obtained for EMAS and iEMAS

4   Experimental Studies

The experiments were performed in order to show whether the introduction of
the immunological-based selection mechanism into predicting EMAS will make
the energetic selection more effective. A Mackey-Glass chaotic time series was
used as a benchmark for prediction, yet good prediction was obviously not the
main goal of the proposed approach. Rather the optimization of neural network
parameters by means of evolutionary processes was of our interest here.

Evolutionary multi-agent system consisted of three evolutionary islands, each with initial population of 10 agents. The genotype cotained the description of a feed-forward neural network (multi-layered perceptron) – the number of neurons in input and hidden layers and learning coefficients. The networks were taught by the means of backpropagation algorithm with momentum. Before maturing, every lymphocyte undergone the negative selection process for 150 steps.

Each plot of the figure 2 shows an average (and the value of standard deviation) of the prediction error obtained from the 10 runs of optimization with the same parameters. It seems that the system was able to approximate correctly the time series, as it may be observed in figure 2(a). The count of agents in every iteration is presented in figure 2(b), after initial raise, it tends to stabilize, which is important because of the dynamically changing number of agents in the population.

The figures 2(c) and 2(d) show the prediction error MAPE (Mean Absolute Prediction Error) for EMAS and iEMAS. Comparing the plots, it seems that introduction of the immunological selection allows for obtaining a similar accuracy of prediction, but the number of newly introduced individuals into the population is lower in case of iEMAS (see fig. 2(e) and 2(f)) what indicates the lower computation cost. In the table 1 the obtained results are presented, describing optimal neural network found in EMAS and iEMAS, along with the starting parameters. It may be clearly seen, that the results are similar, although iEMAS needed smaller number of agents in order to obtain them.

5   Concluding Remarks

In the paper an immune-inspired selection for the agent-based evolution of the architecture of a predicting neural network was considered. By means of negative selection mechanism, agents that probably will not be able to construct an efficient neural network, may be removed before they complete the process of training. Experimental results confirm that the approach allows for improper individuals to be removed from the system faster than using classical energetic selection. Thus solutions of comparable quality are obtained in less computational effort.

Future research should consider a wider range of problems to justify the above conclusion. A precise study on the influence of various parameters on the system effectiveness is also indispensable.

References

1. A. Byrski and M. Kisiel-Dorohinicki. Immune-based optimization of predicting neural networks. In Vaidy S. Sunderam et. al., editor, *Proc. of the Computational Science Ű ICCS 2005: 5th International Conference, Atlanta, GA, USA*, Lecture Notes in Computer Science, LNCS 3516, pages 703–710. Springer Verlag, 2005.
2. A. Byrski and M. Kisiel-Dorohinicki. Immunological selection mechanism in agent-based evolutionary computation. In M. Kłopotek and et al., editors, *Intelligent Information Processing and Web Mining : proceedings of the international IIS: IIPWM '05 conference: Gdańsk, Poland, June 13Ű16, 2005*, pages 411–415. Springer Verlag, 2005.

3. G. de A. Barreto and A. F. R. Araújo. A self-organizing NARX network and its application to prediction of chaotic time series. In *Proc. of the IEEE-INNS Int. Joint Conf. on Neural Networks, (IJCNN'01)*, volume 3, pages 2144–2149, Washington, D.C., 2001.

4. W.H. Johnson, L.E. DeLanney, and T.A. Cole. *Essentials of Biology*. New York, Holt, Rinehart and Winston, 1969.

5. A. Kehagias and V. Petridis. Predictive modular neural networks for time series classification. *Neural Networks*, 10:31–49, 1996.

6. M. Kisiel-Dorohinicki. Agent-oriented model of simulated evolution. In William I. Grosky and Frantisek Plasil, editors, *SofSem 2002: Theory and Practice of Informatics*, volume 2540 of *LNCS*. Springer-Verlag, 2002.

7. T. Masters. *Neural, Novel and Hybrid Algorithms for Time Series Prediction*. John Wiley and Sons, 1995.

8. S. Wierzchoń. Function optimization by the immune metaphor. *Task Quaterly*, 6(3):1–16, 2002.

A Hybrid Approach to Product Information Extraction on the Web

Stefan Clepce, Sebastian Schmidt, and Herbert Stoyan

Department of Computer Science 8 (Artificial Intelligence),
Friedrich-Alexander University of Erlangen-Nuremberg,
Am Weichselgarten 9, 91058 Erlangen, Germany
stefan.clepce@informatik.stud.uni-erlangen.de,
{sebastian.schmidt, stoyan}@informatik.uni-erlangen.de

Summary. In this paper we present a hybrid approach to product information extraction on the web that combines the main advantages of different web-scale algorithms. Like redundancy-based systems, it utilizes the redundancy of the web and uses a search engine to autonomously search for relevant web pages. Additionally it extracts whole product specifications instead of simple facts or binary relations. Like structure-based systems, it employs heuristic assumptions about the general structure of lists and tables, but similarly to ontology-based systems uses domain-specific knowledge to guide the search and the extraction. The difference to those is that our ontology is a feature template that is much simpler to generate and maintain. Also, experimental results are provided to demonstrate the effectivity of our approach.

Keywords: template filling, web-scale information extraction.

1 Introduction and Related Work

When someone wants to buy a technical product like a digital camera, the decision is based on its appearance, price and some important technical features like optical resolution, zoom factor, weight and so on. In this paper we propose a way to automatically extract such technical features of products from the world wide web while focusing on semistructured data as found in tables in many product description pages on e-commerce web sites.

The previous approaches to information extraction of semistructured data on the web can be divided into four categories.

Wrapper based methods use simple extraction procedures (wrappers) for a specific set of pages – usually similar pages from a single web site like an online catalog. Wrappers can be generated semi-automatically using machine learning algorithms for wrapper induction [5] or fully automatically [1] with loss of adaptability.

In contrast to this, we follow a web-scale approach that regards the whole web as one source of data and combines the advantages of *redundancy based* [2, 4],

K.M. Węgrzyn-Wolska and P.S. Szczepaniak (Eds.): Adv. in Intel. Web, ASC 43, pp. 68–73, 2007.
springerlink.com

structure based [6, 9] and *ontology based methods* [3]. Like redundancy-based systems, ours utilizes the redundancy of the web and uses a search engine to autonomously search for relevant web pages and to validate a part of the extraction results. Additionally it extracts whole product specifications instead of simple facts or binary relations. Like structure-based systems, it employs heuristic assumptions about the general structure of lists and tables, but similarly to ontology based methods it uses domain-specific knowledge to guide the search and the extraction. In contrast to ontology based systems we use a simple feature template that can be learned automatically (as proposed in [7]) instead of a complex ontology.

2 Implementation

2.1 System Overview

Figure 1 gives an overview on our implementation. The system works in several consecutive steps which are implemented as distinct modules. The input is a domain description that basically consists of a product domain like "digital camera" and a template for the relevant features of products in this domain. The features are defined by their name (e.g. "LCD size"), their type (bool, number, range or text), alternative keywords under which the feature occurs in web pages (e.g. "LCD" for "LCD size") and a measuring unit for numerical features (like "pixels" or "inches").

First, a list of manufacturer names for the given domain is gathered, since it is essential for some of the subsequent extraction steps (Sec. 2.2).

Next, the system generates search engine queries to find URLs of potentially relevant pages, i.e. pages that list the technical specifications of one product. Additional search terms like manufacturer names, single feature names ("resolution", "weight", etc.) or known online catalogs ("dooyoo.de", etc.) lead to additional, more specific results.

Afterwards, the system downloads the found pages and tries to identify one product in each page by its name (Sec. 2.2) and its features (Sec. 2.3).

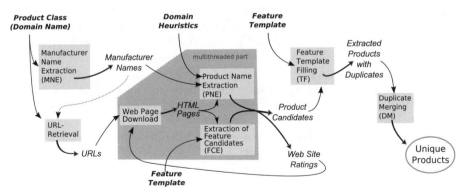

Fig. 1. System Overview

The extraction result is a table of products with their features, which initially contains some duplicate products that have been found on different pages. All duplicates for one product are merged in order to combine feature values that are filled in one instance but not in others. Duplicates are mainly identified by the product name.

2.2 Product Name Identification

First, potential *manufacturer names* are gathered using different search engines: often they can be found following the phrase "manufacturer name: " and in drop-down menus that contain the phrase "choose manufacturer". The extracted name candidates are validated by using two search engine queries: [*candidate* ■man-ufacturer digital camera■] [1] and [■*candidate* digital camera■]. If both queries return more hits than a specific (experimentally determined) threshold, the name is accepted as valid.

Later, the manufacturer names are used to identify the *product name* (manu-facturer name + model, as in "Olympus C-5050 Zoom") in a product description page. We try to identify one of the known manufacturers in the page title or first heading. Then up to four subsequent words are considered as a potential model name. The validity of the product name is checked by counting search engine hits with decreasing number of words (e.g. Hits for [■Olympus C-5050 at Amazon■], then Hits for [■Olympus C-5050 at■] etc.). A phrase with much more hits than the previous (longer) one is a valid name. If no valid name can be found the page is discarded.

2.3 Extraction of Features

The feature extraction is divided into two steps to make the system more modular and clearly structured.

The first step – *the extraction of feature candidates* – is linked to the page download in order to rate the viewed sites and skip pages from irrelevant ones. Each HTML page is transformed into a DOM tree using the open source li-brary htmlparser [8]. Any table or list that contains at least two feature key-words is extracted: each row becomes a feature candidate for the product in the current page. In table rows, the feature name is the text encountered in the leftmost nonempty column, and the value is the text in the second nonempty column (example HTML: <tr><td>Max resolution</td><td></td><td>1280 x 960</td></tr>).

In list elements, the feature name is the text in front of the first appearance of a colon (":") and the value is the text behind it (example HTML: <li>Max resolution: 1280 x 960</li>).

Later, the *template filling*, the second step that runs offline, fills the features that are defined in the template with the appropriate values from the feature candidates of one product. The filling algorithm works as follows:

---

[1] [■A B■] denotes a search query for the exact phrase "A B", whereas [A B] looks for pages where both "A" and "B" must occur, but may be separated.

```
1. for each feature in the template:
2.    for each feature candidate:
3.        if the candidate name is equal to one of the feature keywords:
4.            try to fill the feature with the current candidate's value
5.            else (or if the filling did not work): try next candidate
```

The actual template filling (line 4 in the above pseudocode) depends on the feature type. The filling of bool and text features is trivial. Numeric values, on the other hand, are filled as follows:

```
1. filter out numbers and number ranges with units from the value-text
2. if one of the filtered number-unit-pairs' unit matches the template's unit:
3.    parse the number/range as new value and normalize it.
```

For step 1 we have developed a number-unit-filter based on regular expressions. That is necessary because sometimes there are more features encoded into one candidate (e.g.: "`Sensor: 1/1,8" CCD-Chip with 5,000,000 pixels`" for "sensor size" *and* "sensor resolution"). For step 2 a number parser that takes into account representation problems like non-integer values (e.g. 1.2 or 1,2 in German) versus dots or commas as thousand-delimiters (6,000 or 6.000 in German) as well as combined numbers like "1280 x 960".

3 Evaluation

We evaluated our system by searching for digital cameras in German web pages. The main evaluation measures were the number of extracted products and the filling rate of the resulting table. From 2966 URLs provided by our base search we extracted 1159 unique products (before merging: 1690 products) with a filling rate of 73.1% (filling rate before merging: 66,8%).

The page download, including the extraction of product names and feature candidates, took approximately 20 minutes for 3000 pages and 2-3 hours for 20.000 pages (while many pages from irrelevant servers were discarded) on a 1GHz Intel Centrino with 1GB RAM and 2000 MBit/s DSL. The other steps take significantly less time.

In addition, the different extraction steps have been evaluated separately. Table 1 shows the precision of each step. The recall could only be measured for the template filling since we do not know the total number of relevant products, nor the number of relevant web pages.

The *manufacturer name extraction* was evaluated with three different domains. For digital cameras we ran one extraction with experimentally determined thresholds for the validation queries and one experiment with both thresholds

Table 1. Precision of the different extraction steps

MNE	MNE (thresh. 1)	URLs	PNE	FCE (full pages)	FCE (candidates)	TF
87	83.5	70.4	90	72	88.9	98.1

= 1. The first test returned 47 valid manufacturer names with some duplicates (like "HP" and "Hewlett Packard"), so we actually found 41 manufacturers. This is a precision of 87% with no error[2]. These names were also used in the evaluation of the other modules. The second test returned four invalid names (manually verified) and 63 valid variations of 55 manufacturers. So here the error rate is 6% and the precision is 82.1%.

A test search for "DVD-Player" with the same thresholds as the first test returned 80 names, of which the first 50 ones were valid *and* relevant (i.e. important manufacturers). A search for "Handy" (German for "mobile phone") with threshold 1 returned 127 hits with many duplicates and zero error rate within the first 60 names.

The *URL retrieval* with the simplest search yielded 2966 URLs from 35 web sites. 2088 of these URLs leaded to pages with relevant information, i.e. at least 10 features *and* a product name. Thus, the precision was 70.4%.

The *product name extraction* module failed in all 288 pages from one site (out of 2966) because of the used heuristic. Of the remaining pages we examined 134 taken from 12 sites (5% sample): all were extracted or discarded correctly. This leads to a total precision of over 90%.

The *extraction of feature candidates* was evaluated with the same 2966 pages. In 799 (all from one site) only ca. 60% of the features were found due to the limitations of our heuristics. From the rest we examined 110 pages (5%): 95 were extracted completely, 13 were discarded correctly and 2 were erroneously discarded. Thus, the candidate extraction recall wrt. full pages is 72.4% and it is aproximately 88% wrt. number of features candidates.

To evaluate the *feature template filling* we measured the filling degree of the resulting products table, which is just the evaluation of the complete extraction. Additionally, we also evaluated this module separately: From 2088 product candidates (a product name and at least 25 feature candidates) we could extract 1690 products (i.e. more than 10 features could be filled) yielding a recall of 80.9% according to the number of products. Then we analyzed the quality of the single features on a sample of 105 products (5%): a feature filling is seen as a failure if the feature is defined in the template and is present in the candidate list but its value is not filled. On the feature level we measured a precision of 98% and a recall of 89.8%. The recall was higher (over 95%) when we did not consider failures that occured due to a missing keyword in the template (e.g. "LCD displ." not recognized as "LCD display").

4 Conclusion

The use of heuristics for extracting manufacturer names, product names and feature candidates leads to incomplete results. The negative impact of this fact is lessened by extracting data from different sources, i.e. by exploiting the redundancy of the web. Future work should concentrate on adding more sophisticated

---

[2] Since different names for one manufacturer occur on the web, we need all variations to identify different instances, so the variations are useful and not errors.

structure based algorithms for the extraction of feature lists (as proposed in [1] or in [6]) to ensure the adaptability to other domains.

More fundamental limitations are the current focus on German pages only, and that our present implementation of the product merging does not work for product types with a wide variety of configurations under one product name. Cars are such a type, where for example "Honda Civic" denotes a variety of car body types, motors, etc. Here we would need to rely much stronger on the found features for merging products than we do currently.

It is remarkable though that our approach with shallow methods and heavy use of search engines is able to autonomously populate a product database with detailed information on over 1000 products within 30 minutes while automatically focusing on the most popular products. Thus, the results are comparable to ontology based methods as in [3], while our ontology (the template) is much easier to generate and to maintain.

References

1. S T Ahmed, S Vadrevu, and H Davulcu. Datarover: An automated system for extracting product information from online catalogs. In *AWIC 2006*, volume 23 of *Studies in Computational Intelligence*, pages 1–10. Springer, 2006.
2. S Brin. Extracting patterns and relations from the world wide web. In *WebDB 98*, volume 1590 of *Lecture Notes in Computer Science*, pages 172–183. Springer, 1998.
3. D W Embley, D M Campbell, Y S Jiang, S W Liddle, D W Lonsdale, Y-K Ng, and R D Smith. Conceptual-model-based data extraction from multiple-record web pages. *Data Knowl. Eng.*, 31(3):227–251, 1999.
4. O Etzioni, M J Cafarella, D Downey, A-M Popescu, T Shaked, S Soderland, D S Weld, and A Yates. Methods for domain-independent information extraction from the web: An experimental comparison. In *16. IAAI/19. AAAI*, pages 391–398. AAAI Press/The MIT Press, 2004.
5. N Kushmerick and B Thomas. Adaptive information extraction: Core technologies for information agents. In *AgentLink 2003*, volume 2586 of *Lecture Notes in Computer Science*, pages 79–103. Springer, 2003.
6. B Liu and Y Zhai. Net - a system for extracting web data from flat and nested data records. In *WISE 2005*, volume 3806 of *Lecture Notes in Computer Science*, pages 487–495. Springer, 2005.
7. S Schmidt and H Stoyan. Web-based extraction of technical features of products. In *GI Jahrestagung (1)*, volume 67 of *LNI*, pages 246–250. GI, 2005.
8. D Udani. The html parser, 2006.
9. Y Yang and W-S Luk. A framework for web table mining. In *WIDM 2002*, pages 36–42. ACM, 2002.

A Rough-Fuzzy Ontology Generation Framework and Its Application to Bio-medical Text Processing

Lipika Dey[1], Muhammad Abulaish[2], Rohit Goyal[1], and Kumar Shubham[1]

[1] Indian Institute of Technology Delhi, Hauz Khas, New Delhi-16, India
lipikadey@gmail.com
[2] Jamia Millia Islamia, jamia Nagar, New Delhi-25, India
abulaish@ieee.org

Summary. In this paper, we present a rough-fuzzy ontology generation framework which supports rough concept descriptions at multiple levels of granularity, using fuzzy property descriptors. The rough approximations for concept descriptions and valid fuzzy property descriptors can be obtained through text-mining.

Keywords: Rough-fuzzy ontology structure, ontology-based text processing, natural language processing.

1 Introduction

Ontology facilitates domain knowledge representation in a structured and machine-interpretable form and allows information sharing across applications [1]. In spite of the successful use of ontology in building various applications, a domain-agnostic assessment of ontology-based systems reveals the following real-life problems:

- Ontology is generally designed to be a pre-defined structure with crisp concept descriptions and inter-concept relations, with well-defined semantics. However, crisp definitions cannot capture the semantics of real-life knowledge completely, which often requires conceptualization to capture inherent vagueness. For example, a disease as a concept can be characterized by accompanying information about symptoms, signs, medication, nature of disease etc. But each instance of disease can only be identified with a "possible" set of symptoms or a set of medications that are "usually" prescribed. A symptom is observed with varying degrees of intensity in different patients or different diseases, and have to be suitably represented. For example, after consulting several sources of information, it is observed that the "most common" symptom of *breast cancer* is the presence of a tumor, whose size and location is variable and determines the strength of the disease, and other "less common" symptoms include *painful breasts, skin disfiguring, nipple discharge etc.* It is extremely important to capture the

K.M. Węgrzyn-Wolska and P.S. Szczepaniak (Eds.): Adv. in Intel. Web, ASC 43, pp. 74–79, 2007.
springerlink.com

essence of this variability in the concept space since this characterization can influence the recording of data in studies conducted for future research. The two data sets on *breast cancer* stored in the machine learning repository (http://www.ics.uci.edu/ mlearn/MLRepository.html) stores patient data using two different sets of attributes. While both sets store size of tumor, none of them store information about key aspects of breast cancer like presence of *Estrogen and Progeterone receptors, ploidy of cells etc.* These attributes are considered to be of very high importance in deciding the medication and its effectiveness in tackling *breast cancer.* In other words, it can be stated that the idea of using *significance of attributes* is essential at the point of conceptualization itself.

- Acquisition of relevant knowledge for a domain and structuring it are both non-trivial tasks and require substantial involvement of domain experts. *Knowledge acquisition through text mining can provide a viable way to create and enhance concept ontologies.*
- Locating relevant concepts, their descriptors and reasoning about their presence or absence within text automatically is a complex problem and involves amalgamation of uncertainty-based reasoning techniques along with Natural Language Processing (NLP).

In this paper we propose a *rough-fuzzy ontology framework* that extends the regular ontology structure, with concepts of *rough approximation* for concept descriptions and associates fuzzy membership functions to property descriptors to allow variable-precision descriptions. We have shown how text mining aided by NLP tools can be used to extract concept descriptors, possibly non-crisp, from text to generate a rough-fuzzy ontology structure.

2   Related Works

Design and implementation of fuzzy ontology structures and ontology-based text information processing systems have received quite a lot of attention. Widyantoro and Yen [7] have shown how co-occurrence of ontology concepts in manually annotated documents can be used to build a fuzzy concept hierarchy. Wallace and Avrithis [6] have extended the idea of ontology-based knowledge representation to include fuzzy degrees of membership for a set of inter-concept relations defined in an ontology. Quan et al. [5] have proposed an automatic fuzzy ontology generation framework - FOGA by incorporating fuzzy logic into formal concept analysis to handle uncertainty information for conceptual clustering and concept hierarchy generation. Parry [4] proposed a fuzzy ontology structure for storing the Medical Subject Heading (MeSH) ontology, which has 21836 terms, of which 10072 are overloaded. In [4] it is proposed that while stating a query, the user can associate a fuzzy membership value to each query term, which determines its relative weight while retrieving relevant documents. Lee et al. [2] have proposed a fuzzy ontology structure as an extension to the domain ontology with crisp concepts for a Chinese news summarization application, which learns a domain ontology by analyzing news events marked by domain experts.

3 Proposed Rough-Fuzzy Ontology Model

An ontology is a collection of concepts which are described using a set of property descriptor triplets ⟨*has_property, value, constraints*⟩, where *has_property* is a binary relation defined between two concepts. The set of properties associated with each concept is pre-defined and deterministic. We propose that a concept can be described at multiple levels of granularity using *approximation sets* of descriptors, where each concept descriptor can be represented as a *fuzzy property descriptor*. A *fuzzy property descriptor* is characterized by the fact that the property value is accompanied by a qualifier, and both value and qualifier are defined as *fuzzy sets*. This framework allows defining the *property-value* of a concept with differing degrees of fuzziness, without actually changing the concept description paradigm. Further, using rough-set theory, each concept is represented by two approximations. The *lower* approximation $P_L(C)$ consists of a set of property descriptors that are definitely observable in the concept. The *upper-approximation* $P^U(C)$ on the other hand contains property descriptors that are possibly associated with the concept but may not be observed.

Definition 1 (Fuzzy property descriptor). A *fuzzy property descriptor* p_F is defined as a quintuple of the form ⟨$C_0, \Re, v_F, q_F, f$⟩, where C_0 is a valid ontology concept, $\Re$ is a structural or semantic relation defined for the underlying ontology, v_F represents fuzzy attribute value and could be either *fuzzy numbers* or *fuzzy quantifiers*, q_F models linguistic qualifiers and are implemented as *hedges*, which can alter the strength of an attribute value and f is the set of restriction facets applicable on relation $\Re$ in the context of concept C_0 and the value v_F. The restriction facets consist of *type* (f_t), *cardinality* (f_c), and *range* (f_r).

Linguistic qualifiers are particularly useful for developing a variable precision concept description for text processing applications, since these qualifiers are responsible for altering the property value of a concept within text documents. Fuzzy numeric values can either reflect varying precision for a property value, or can be easily adapted to reflect strength of association of a property descriptor to the concept. The choice of *fuzzy numbers* or *fuzzy quantifiers* for values is dictated by the nature of the underlying attribute and also its restriction facets.

Definition 2 (Rough Fuzzy Ontology). A rough-fuzzy ontology, O_F^R is a non-crisp domain ontology represented as ⟨$C(F(C), P_L(C), P^U(C)), \Re_a, \Re_s$⟩ where

- C is a set of *roughly defined* concepts defined for the domain. Domain Objects encountered in any related application are expected to be instances of ontology concepts.
- F(C) denotes the set of fuzzy property descriptors associated with concept C. $F(C) = \{p_F | C_o = C \wedge \Re \in \Re_a \cup \Re_s\}$, where $\Re_a$ and $\Re_s$ denote structural and semantic relations.

- $P_L(C)$ and $P^U(C)$ denote the lower approximation and upper approximation of concept descriptions for C and are defined as follows:

  $P_L(C) = \{p_F(C, \Re, v_f, q_f, f) |$ for all known instances I of C there exists a valid tuple $p_F(I, \Re, v_F, q_F, f)$ that describes I appropriately$\}$. $P_L(C)$ denotes the set of properties that have been identified as essential for an object to qualify as an instance of C and are adjudged *mandatory* properties for an object to be judged as an instance of C.

  $P^U(C) = \{p_F(C, \Re, v_f, q_f, f) |$ at least η known instances of C have been found to have this property, where η is a pre-defined threshold$\}$. $P^U(C)$ denotes a generic set of properties from which some, though not all, are likely to be observed in an instance of C. $P^U(C)$ denotes the optional set of properties to be associated to definition of concept C.

- $\Re_a \subseteq C \times C$ is a set of structural relations - *is-a, part-of, kind-of*, defined between concepts in O_F^R. The cardinality of these relations could be *one-to-one, one-to-many and many-to-many*.

- $\Re_s \subseteq C \times C$ is a set of semantic relations defined between concept-pairs in O_F^R. These relations are denoted by $\rightarrow_{name}$. $\leftarrow_{name}$ denotes the conceptual inverse of the semantic relation and has a user-defined semantics associated with it.

The extraction principles for concept descriptors is presented in section 4. Using the descriptors extracted from the medical text documents, a partial description of some of the chronic diseases is obtained as follows:

P_L *(Heart disease and stroke)*= $\{\langle is\_causes\_by,$ *High blood pressure/ Cigarette smoking/ Atrial fibrillation/ High blood cholestrol, Boolean, 1*$\rangle\}$

P^U *(Heart disease and stroke)* = $\{\langle is\_causes\_by,$ *geographic locations/ Alcohol abuse/ Drug abuse/ Socioeconomic factors, often/ usually/ yes/ no, 0..2*$\rangle\}$

P_L *(Cancer)* = $\{\langle has\_symptoms,$ *weight loss/Fatique, Boolean, 1*$\rangle\}$

P^U *(Cancer)* $\{\langle has\_symptoms,$ *Fever or night sweats/ Pain/ Skin changes, often/ usually/ null, 1..n*$\rangle\}$

P_L *(Tuberculosis)*= $\{\langle has\_early\_symptoms,$ *Fever/ Chills/ Sweating/ Weight Loss/ Weakness/, Boolean, 1*$\rangle\}$

P^U *(Tuberculosis)* = $\{\langle has\_early\_symptoms,$ *Persistent cough/ Chest pain/ Breathing difficulty bar, usually, 1..n*$\rangle\}$

P_L *(Hypoglycemia)* = $\{\langle syndrome\_of,$ *abnormally low blood sugar,1*$\rangle\}$

4   Rough-Fuzzy Ontology Creation Using Text Mining

We now present a complete reasoning framework to implement rough-fuzzy ontology-based text-processing systems. Given a set of domain documents, relevant entities and their descriptors are located in these text documents using NLP techniques. A feasible set of relations among these entities are also extracted using a rule-based system. Relation extraction involves traversing the dependency graph output by the Parser (Satnford Parser) to identify domain-specific fuzzy relationships. Dependencies output by the Parser are analyzed to identify subject, object, modifier and various other relationships among elements in a sentence. Rule 1 presents a sample rule used for identifying possible relations in the domain, assuming entities are identifiable as noun phrases. WordNet [3] is used for establishing synonym relations.

Rule 1. If there exist two dependencies involving two different entities but the same verb satisfying the following condition [Dep(Entity1, verb + preposition) & Dep(verb + preposition, Entity2)], where Dep is a known dependency, then the verb+preposition composition is identified as a valid relation between the two entities. It can be characterized as an instance of a binary relation represented by Entity1 → verb ← Entity2.

Rule 2. If (Entity1 → verb ← Entity2) is identified as a relation, then if there exist auxiliary dependencies of the verb or there are adverb modifiers associated to this verb, then auxiliary dependency or the modifier decides whether the verb will be used for a mandatory or an optional property descriptor among the entities. If the auxiliary verb is a member of the set {*can be, may be, might be, likely to be, etc.*} or the modifier is a member of the set {*usually, sometimes, often, etc.*} then the relation is a member of the upper approximation, i.e. $P^U(C)$. If the verb is definitive then the relation is added to the $P_L(C)$. Simple numerical assessment may also be used.

Both property values and qualifiers are modeled as fuzzy membership functions. Qualifier sets are modeled as ordered, graded sets, and similarity between two qualifiers q_i and q_j is a function of their distance $d(q_i, q_j)$ in the graded set and represented as $f(q_i, q_j)$. Dilution and intensification of qualifiers and values are implemented using a sign function. An element t_i is a dilution with respect to element t_j in the graded set if $i < j$, while t_j is an intensifier with respect to t_i. This is expressed using a sign function namely $Sgn(t_i, t_j) = +1$, and $Sgn(t_j, t_i) = -1$ respectively.

Let μ denotes the fuzzy membership function associated to the value set. The similarity between two qualified values, for example "severe headache" and "mild headache", is expressed as $\lambda_{(q_i, v_i)}(q_j, v_j)$ and is defined as a composite fuzzy membership function of μ and f, as shown in equation 1.

$$\lambda_{(q_i, v_i)}(q_j, v_j) = \begin{cases} \mu(v_i, v_j)^{\frac{1}{d(q_i, q_j)+1}} & if\, sgn(q_i, q_j) \times sgn(v_i, v_j) = -1 \\ \mu(v_i, v_j)^{d(q_i, q_j)+1} & if\, sgn(q_i, q_j) \times sgn(v_i, v_j) = +1 \\ \mu(v_i, v_j) \times f(q_i, q_j) & if\, sgn(q_i, q_j) \times sgn(v_i, v_j) = 0 \end{cases} \quad (1)$$

5   Concept Matching over Rough-Fuzzy Ontology

The system can be used to identify a possible disease, when queried by a set of symptoms. A query string q consists of a m ⟨qualifier, value⟩ pairs, and concept C, is denoted by $M(q, C)$ and is computed as follows: $M(q, C) = (\alpha \times B_L(C) + (1-\alpha) \times B^U(C))^\gamma$ where $B_L(C) = \frac{1}{|P_L(C)|} \times \Sigma_0 \Sigma_j (\mu_{(q_0, q_j)}(v_0, v_j))^{1/\beta}$ is the lower similarity match denoting the degree of match in definite properties of a concept, where (q_0, v_0) is a ⟨qualifier,value⟩ pair, possibly NULL, present in query string, and (q_j, v_j) is part of a property descriptor in ontology. The upper similarity match $B^{\overline{U}}(C)$, is similarly defined. β is a function of proportion of descriptor string match, which is one - if relation name, value and qualifier all match exactly. For example, given a query string of symptoms "fever, muscular rigidity, cognitive dysfunction", the suspected disease is *Neuroleptic Malignant Syndrome*. Since the lower approximation of the disease contains the property "recent treatment with neuroleptics within past 1-4 weeks", this is verified through additional querying.

6 Conclusion and Future Work

In this paper we have proposed a rough-fuzzy ontology generation framework which supports concept descriptions at multiple levels of granularity using non-crisp property descriptors. Presently, a complete rough-fuzzy ontology based text information processing system is on the way for medical domain.

References

1. Fensel D, Horrocks I, Harmelen F van, McGuinness DL, Patel-Schneider P (2001) OIL: Ontology Infrastructure to Enable the Semantic Web. IEEE Intelligent Systems 16(2), 38-45
2. Chang-shing L, Zhi-wei J, Lin-kai H (2005) A fuzzy ontology and its application to news summarization. IEEE Transactions on Systems, Man, and Cybernetics - part B: Cybernetics, 35(5): 859
3. Miller G (1997) Wordnet: An online lexical database. International Journal of Lexicography 3(4)
4. Parry D (2004) A fuzzy ontology for medical document retrieval. ACSW frontiers: 121-126
5. Quan TT, Hui SC, Cao TH (2004) FOGA: A Fuzzy Ontology Generation Framework for Scholarly Semantic Web. In: Proceedings of the 2004 Knowledge Discovery and Ontologies Workshop (KDO'04), Pisa, Italy
6. Wallace M, Avrithis Y (2004) Fuzzy Relational Knowledge Representation and Context in the Service of Semantic Information Retrieval. In: Proceedings of the IEEE International Conference on Fuzzy Systems (FUZZ-IEEE), Budapest, Hungary
7. Widyantoro DH, Yen J (2001) A Fuzzy Ontology-based Abstract Search Engine and its User Studies. In: Proceedings of the 10th IEEE International Conference on Fuzzy Systems, Melbourne, Australia: 1291-1294

Accessing the Web on Handheld Devices for Visually Impaired People

Gaël Dias and Bruno Conde

Centre for Human Language Technology and Bioinformatics, University of Beira Interior, 6200-001 Covilhä, Portugal
{ddg,bruno}@hultig.di.ubi.pt
http://hultig.di.ubi.pt

Summary. In this paper, we propose an automatic summarization system to ease web browsing for visually impaired people on handheld devices. In particular, we propose a new architecture for summarizing Semantic Textual Units [2] based on efficient algorithms for linguistic treatment [3][6] which allow real-time processing and deeper linguistic analysis of web pages, thus allowing quality content visualization. Moreover, we present a text-to-speech interface to ease the understanding of web pages content. To our knowledge, this is the first attempt to use both statistical and linguistic techniques for text summarization for browsing on mobile devices.

1 Introduction

Visually impaired people are info-excluded due to the overwhelming task they face to read information on the web. Unlike fully capacitated people, blind people can not read information by just scanning it quickly i.e. they can not read texts transversally. As a consequence, they have to come through all sentences of web pages to under-stand if a document is interesting or not.

To solve this problem, we propose an automatic summarization server-based ar-chitecture for web browsing on handheld devices. In particular, we introduce five different efficient methods for summarizing subparts of web pages in real-time. Two main approaches have already been proposed in the literature. First, some method-ologies such as [2][14] use simple but fast summarization techniques to produce results in real-time. However, they show low quality contents for vi-sualization as they do not linguistically process the web pages. Second, some works apply linguistic processing and rely on ad hoc heuristics [7] to produce compressed contents but can not be used in a real-time environment. Moreover, they do not use statistical evi-dence which is a key factor for high quality summarization. As a consequence, we propose a new architecture, called XSMobile, for summarizing Semantic Textual Units [2] based on efficient algorithms for linguistic treatment [3][6] that allow real-time processing and deeper linguistic analysis of web pages, thus producing quality content visualization as illustrated in Figure 1.

K.M. Węgrzyn-Wolska and P.S. Szczepaniak (Eds.): Adv. in Intel. Web, ASC 43, pp. 80–86, 2007.
springerlink.com   © Springer-Verlag Berlin Heidelberg 2007

Fig. 1. Screenshot of the XSMobile architecture

2   Text Unit Identification

One main problem to tackle is to define what to consider as a relevant text in a web page. Indeed, web pages often do not contain a coherent narrative structure [1].

For that purpose, [15] propose a C5.0 classifier to differentiate narrative paragraphs from non narrative ones. However, 34 features need to be calculated for each paragraph which turns this solution impractical for real-time applications. In the context of automatic construction of corpora from the web, [5] propose to use a lan-guage model based on Hidden Markov Models using the SRILM toolkit [12]. This technique is certainly the most reliable one as it is based on the essence of the lan-guage but still needs to be tested in terms of processing time. Finally, [2] propose Semantic Textual Unit (STU) identification. In summary, STUs are page fragments marked with HTML markups which specifically identify pieces of text following the W3 consortium specifications. However, not all web pages respect the specifications and as a consequence text material may be lost. In this case, unmarked strings are considered STUs if they contain at least two sentences.

3   Linguistic Processing

On the one hand, single nouns and single verbs usually convey most of the informa-tion in written texts. On the other hand, compound nouns (e.g. hot dog) and phrasal verbs (e.g. take off) are also frequently used in everyday language, usually to pre-cisely express ideas and concepts that cannot be compressed into a single word. As a consequence, identifying these lexical items is likely to contribute to the performance of the extractive summarization process.

Subsequently, each STU in the web page is first morpho-syntactically tagged with the efficient TnT tagger[1] [3]. Then, multiword units are extracted from each STU based on an efficient implementation of the SENTA[2] multiword unit extractor [6] which shows time complexity O(N log N) where N is the number of words to proc-ess. Then, multiword units which respect the following regular expression are se-lected for quality content visualization:

[Noun Noun* | Adjective Noun* | Noun Preposition Noun | Verb Adverb].

This technique is usual in the field of Terminology [14]. A good example can be seen in Figure 1 where the multiword unit "Web Services" is detected, where exist-ing solutions [2][7][14] would at most consider both words "Web" and "Services" separately. Finally, we remove all stop words present in the STU. This process al-lows faster processing of the summarizing techniques as the Zipf's Law shows that stop words represent 1% of all the words in texts but cover 50% of its surface.

4   Summarization Techniques

Once all STUs have been linguistically processed, the next step of the extractive summarization architecture is to extract the most important sentences of each STU. In order to make this selection, each sentence in a STU is assigned a significance weight. The sentences with higher significance become the summary candidate sen-tences. Then, the compression rate defines the number of sentences to be visualized.

Simple tf.idf: This methodology is mainly used in Information Retrieval [13]. The sentence significance weight is the sum of the weights of its constituents divided by the length of the sentence.

A well-known measure for assigning weights to words is the *tf.idf* score [11]. The *tf.idf* score is defined in Equation 1 where w is a word, *stu* a STU, *tf(w, stu)* the number of occurrences of *w* in *stu*, | stu | the number of words in the *stu* and *df(w)* the number of documents where *w* occurs.

$$tf.idf(w, stu) = \frac{tf(w, stu)}{|stu|} \times \log_2 \frac{N}{df(w)} \tag{1}$$

In our case, a dictionary of *idf* values is processed for each website where XSMo-bile is installed based on the collection of texts present in it. The process is web-based. All texts in the collection of the website are first linguistically processed as explained in Section 3. Then, the *n* most frequent words of the collection are ex-tracted to produce query samples sent to the web search engine GoogleŹ. For each query, the first 10 most relevant urls are gathered given rise to 10*n urls. Then, a web spider processes each url as deeply as possible in the hypertext structure and extracts all texts related to the initial query. Finally, after automatically gathering huge quantities of texts to approximate as best as

---

[1] http://www.coli.uni-saarland.de/˜thorsten/tnt/

[2] http://senta.di.ubi.pt/

possible the ideal *idf* values of the words, a XML dictionary of <*word, idf*> entries is produced. So, the sentence significance weight, weight1(S, stu), is defined straightforwardly in Equation 2 where |S| stands for the number of words in S and wi is a word in S.

$$weight_1(S, stu) = \frac{\sum_{i=1..|S|} tf.idf(w_i, stu)}{|S|} \tag{2}$$

Enhanced tf.idf: In the field of Relevant Feedback, [13] propose a new score for sentence weighting that proves to perform better than the simple *tf.idf*. In particular, they propose a new weighting formula for word relevance, W(.,.). It is defined in Equation 3 where argmax w(*tf(w,stu)*) corresponds to the word with the highest fre-quency in the STU.

$$W(w, stu) = \left(0.5 + \left(0.5 \times \frac{tf(w, stu)}{argmax_w(tf(w, stu))}\right)\right) \times \log_2 \frac{N}{df(w)} \tag{3}$$

Based on this weighting factor, [13] define a new sentence significance factor weight 2(*S,stu*) which takes into account the normalization of the sentence length. The subjacent idea is to give more weight to sentences which are more content-bearing and central to the topic of the STU as shown in Equation 4 where arg-max(|S|) is the length of the longest sentence in the STU.

$$weight_2(S, stu) = \frac{\sum_{i=1..|S|} W(w_i, stu) \times |S|}{(argmax_s(|S|))} \tag{4}$$

The rw.idf: Recently, [10] have proposed the TextRank algorithm. The basic idea of the algorithm is the same as the PageRank algorithm proposed by [4] i.e. the higher the number of votes that are cast for a vertex, the higher the importance of a vertex. Moreover, the importance of the vertex casting the vote determines how important the vote itself is. The score of a vertex Vi is defined as in Equation 5 where $In(V i)$ is the set of vertices that point to it, $Out(V j)$ is the set of vertices that the vertex Vj points to and d is a dumping factor[3].

$$S(V_i) = (1 - d) + d \times \sum_{j \in In(V_i)} \frac{1}{|Out(V_j)|} S(V_j) \tag{5}$$

In our case, each STU is represented as an un-weighted oriented graph being each word connected to its successor following sequential order in the text as in Figure 2.

After the graph is constructed, the score associated with each vertex is set to an initial value of 1, and the ranking algorithm is run on the graph for several iterations until it converges. So, each word is then weighted as in Equation 6 and the sentence significance weight, weight3 *(S, stu)*, is defined straightforwardly in Equation 7 where |S| stands for the number of words in S and wi is a word in S.

---

[3] d was set to 0.85 as referred in [4].

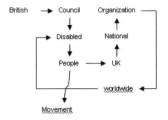

Fig. 2. Graph representation of the text: "The British Council of Disabled People is the UK's National Organization of the worldwide Disabled People's Movement"

$$rw.idf(w, stu) = S(w) \times \log_2 \frac{N}{df(w)} \tag{6}$$

$$weight_3(S, stu) = \frac{\sum_{i=1..|S|} rw.idf(w_i, stu)}{|S|} \tag{7}$$

Cluster Methodologies: Luhn suggested in [9] that sentences in which the greatest number of frequently occurring distinct words are found in greatest physical prox-imity to each other, are likely to be important in describing the content of the docu-ment in which they occur[4]. The procedure proposed by [2], when applied to sentence S, works as follows. First, they mark all the significant words in S. A word is signifi-cant if its *tf.idf* is higher than a certain threshold T. Second, they find all clusters in S such that a cluster is a sequence of consecutive words in the sentence for which the following is true: (i) the sequence starts and ends with a significant word and (ii) fewer than D insignificant words must separate any two neighboring significant words within the sequence. Then, a weight is assigned to each cluster. This weight is the sum of the weights of all significant words within a cluster divided by the total number of words within the cluster. Finally, as a sentence may have multiple clus-ters, the maximum weighted cluster is taken as the sentence weight.

5   Text-to-Speech Interface

The Text-to-Speech module is a crucial issue for accessibility of Visually Impaired People to web page contents. For this purpose, we have integrated the Microsoft Speech Server into our architecture using the SALT markup language following the architecture proposed in Figure 3.

However, in future work, we will integrate a Speech-to-Speech module on the proper device in order to avoid the overload of the Microsoft Speech Server which has shown limitations for high amounts of requests.

---

[4] [2] based their sentence ranking module on this paradigm.

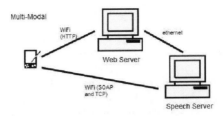

Fig. 3. Text-to-Speech Interface

6 Conclusions

In this paper, we proposed an automatic summarization system to help web browsing for visually impaired people on handheld devices. Unlike previous works [2][7][14], it is based on efficient algorithms [3][6] for linguistic treatment that allow real-time processing and deeper linguistic analysis for quality content visualization. The first results are every encouraging in terms of (1) quality of the content of the summaries, especially with the rw.idf, (2) processing time although the architecture is not still distributed over different processing units and (3) user interaction satisfaction. How-ever, improvements must be taken into account. In particular, current work involves the integration of a Speech-to-Speech control interface which may provide a solution capable to compete with Braille PDAs that are expensive and difficult to use.

References

1. Berger, A., and Mittal,V.: Ocelot: a System for Summarizing Web Pages. In Proc. of SIGIR. (2000).
2. Buyukkokten, O., Garcia-Molina, H. and Paepcke, A: Seeing the Whole in Parts: Text Summarization for Web Browsing on Handheld Devices. In Proc. of the 10th International World Wide Web Conference. (2000).
3. Brants, T.: TnT - a Statistical Part-of-Speech Tagger. In Proc. of the 6th Applied NLP Conference. (2000).
4. Brin, S., and Page, L.: The anatomy of a large-scale hypertextual Web search engine. Com-puter Networks and ISDN Systems. 30(1-7). (1998).
5. Dolan, W. B., Quirk, C., and Brockett, C.: Unsupervised Construction of Large Paraphrase Corpora: Exploiting Massively Parallel News Sources. In Proc. of COLING. (2004).
6. Gil, A., and Dias, G.: Using Masks, Suffix Array-based Data Structures and Multidimen-sional Arrays to Compute Positional Ngram Statistics from Corpora. In Proc. of the Work-shop on Multiword Expressions of the 41st ACL. (2003).
7. Gomes, P. Tostão, S., Gonçalves, D. and Jorge, J: Web-Clipping: Compression Heuristics for Displaying Text on a PDA. In Proc. of Workshop on HCI with Mobile Devices. (2001).
8. Justeson, J. and Katz, S.: Technical Terminology: some Linguistic Properties and an Algo-rithm for Identification in text. Natural Language Engineering. (1). (1995).

9. Luhn, H.P.: The automatic creation of literature abstracts. IBM Journal of Research and Development. (1958).
10. Mihalcea R., and Tarau, P.: TextRank: Bringing Order into Texts. In Proc. of the Confer-ence on Empirical Methods in Natural Language Processing. (2004).
11. Salton, G., Yang, C.S., and Yu, C.T.: A Theory of Term Importance in Automatic Text Analysis. Amer. Soc. of Inf. Science. (26)1. (1975).
12. Stolcke, A.: SRILM – An Extensible Language Modeling Toolkit. In Proc. of International Conference on Spoken Language Processing. (2002).
13. Vechtomova, O., and Karamuftuoglu, M.: Comparison of Two Interactive Search Refine-ment Techniques. In Proc. of HLT-NAACL. (2004).
14. Yang, C., and Wang, F.L.: Fractal Summarization for Mobile Devices to Access Large Documents on the Web. In Proc. of the International World Wide Web Conference. (2003).
15. Zhang, Y., Zincir-Heywood, N., and Milios, E.: Summarizing Web Sites Automatically. In Proc. Conference of Canadian Society for Computational Studies of Intelligence. (2003).

RRSS - Rating Reviews Support System Purpose Built for Movies Recommendation

Grzegorz Dziczkowski[1,2] and Katarzyna Wegrzyn-Wolska[1]

[1] Ecole Superieur d'Ingenieurs en Informatique et Genie des Telecommunicatiom (ESIGETEL) 1,Rue de Port de Valvins 77-215 Avon-Fontainebleau Cedex
[2] Ecole des Mines de Paris 35, rue Saint-Honore 77305 Fontainebleau
{grzegorz.dziczkowski,katarzyna.wolska}@esigetel.fr

Summary. This paper describes the part of a recommendation system designed for the recognition of film reviews (RRSS). Such a system allows the automatic collection, evaluation and rating of reviews and opinions of the movies. First the system searches and retrieves texts supposed to be movie reviews from the Internet. Subsequently the system carries out an evaluation and rating of the movie reviews. Finally, the system automatically associates a digital assessment with each review. The goal of the system is to give the score of reviews associated with the user who wrote them. All of this data is the input to the cognitive engine. Data from our base allows the making of correspondences, which are required for cognitive algorithms to improve, advanced recommending functionalities for e-business and e-purchase websites. In this paper we will describe the different methods on automatically identifying opinions using natural language knowledge and techniques of classification.

1 Introduction and Issue

With the growth of the Web, e-commerce has become very popular. A lot of websites offer online-sales. To increase their sales, online shops include the special recommended systems (RS) to suggest products to the clients. While people like to check out the recommendations of other users before creating their own opinion, those predictions become very useful for the customers. RS allow customers to make the choice without any personal knowledge of alternatives. Algorithms for suggestion are based on the experience and the opinion of other users. It is helpful to find recommendations from people who are familiar with the same problem, who have made their choice in the past, whose perspective is valued, or who are recognized experts [12]. RS also provide correspondences between users, who have a similar profile. A new user has to create his own profile. The RS will suggest a new precise choice based on the similar taste of other users. The efficacity of such system depends on data quality and quantity. This is why RS need huge databases of user profiles: the more profiles it gets, the beter the algorithms are. RS proposes the choice to the user, which is based on correspondences between the users' opinions. Our system (RRSS) furnishes the users' profiles, which are necessary for algorithms of cognitive engines. This result

K.M. Węgrzyn-Wolska and P.S. Szczepaniak (Eds.): Adv. in Intel. Web, ASC 43, pp. 87–93, 2007.
springerlink.com © Springer-Verlag Berlin Heidelberg 2007

cannot depend on commercial reasons, because it could make people distrustful. RCSS consist of two principal modules:

- extraction and filtering opinions from the text, which consists in the identification of quite precise information in natural language and its representation in a structured form [8].
- assigning a mark only to subjective sentences, which express or describe opinions, evaluations, or emotions [10][15].

The relative failure of the generic systems is well-known today. Many researchers try to describe natural languages in the same way as formal languages. Maurice Gross undertook with his team (LADL; French Laboratory for Linguistics and Information Retrieval) the exhaustive examination of simple sentences of French [5], in order to have reliable and quantified data predicted to rigorous scientific treatments. To exploit the linguistic knowledge, the LADL developed a special application named Unitex [9]. This is an enhancement environment used to build formalized descriptions for broad coverage of natural languages and apply them to substantial texts. Unitex processes the texts of several mega-bytes to morpho-syntactic indexation in real time, to search for set phrases or semi-fixed phrases, to produce agreements and statistical evaluation of the results. The linguistic resources used to achieve the information retrieval and extraction are as follows: dictionaries, networks of recursive transitions (local grammar) and lexicon-grammar tables.

Another way to analyse an opinion automatically from the text is to use statistical classifiers. All of the analyzed objects are assigned to the previously prepared classes. Statistical methods suppose that descriptions of the same class respect a specific structure of the class. For classification of huge corpora we often use the special learning methods based on tested instances (examples). Problems consist in constituting a representative corpus of the evaluated field, and finding the rules or constituting an operational model of this corpus. The model created allows the system to predict the behaviour for new candidates. At present, classification of opinions as subjective/objective or positive/negative is a very interesting challenge for research: Turney, Littman [13], Dave, Lawrance [3], Pang, Lee [7]. Classifiers assign the new objects for analysis to correspondend to previously prepared classes. The classifiers performance depends on the model for each base learning class.

2 Marking a Review

RRSS has modular architecture. The principle tasks are: collecting the reviews from Internet; checking if the text found is a review; assigning a mark to the review and presentation of the results. This paper focuses only on the review marking module. Generally the mark assignment process distinguishes the linguistic and probabilistic parts. In our approach the linguistic part is responsible for pre-processing of the text, creating a learning base and finding behaviour of identical mark groups [paragraph 2.1]. The probabilistic part classifies reviews to the mark [paragraph 2.3]. Our algorithm follows the next steps:

- learning base creation,
- vector representation,
- classification.

The process of mark assignment to the review consist of two main steps: first estimation of a mark based on the behaviour of the same review mark groups and the final assignment of the mark [fig 1]:

- gathering the reviews according to their mark,
- finding the behaviour of each group of mark,
- for a new review the first estimation of the mark directly from the characteristic of the group behaviour ,
- creation of a learning base for Bayes classifiers,
- assignment of a final mark to the review.

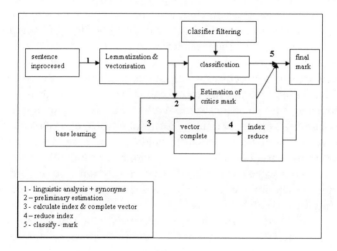

Fig. 1. Mark assignment process

2.1   Learning Base

To perform the review assessment we need a group of characteristics already evaluated - a learning base. Different websites publish film reviews with its mark assigned (e.g. IMDB, Amazon). We used this data (reviews, users, marks) to create our learning base. We use the scale of marking from 1 to 5. We regrouped all the reviews according to their mark. This way, we obtained 5 different groups of film reviews: a group according to reviews with score 1, 2..5. Then, we tried to determine the characteristics for each group. We supposed that delimited parameters characterize behaviour of a group. These characteristics are for example: a typical word, typical expression, a size of a sentence, the frequency of characteristic word repetition, the number of punctuation marks (!, ;), ?) and so on. For group categorizing, we used a linguistic analyser Unitex, to lemmatize the words, to assign semantic classes to the words, to add synonyms [4] and to detect negation. For this task we used a linguistic processing, which requires lexicons and specialized grammar.

The development of such resources is a long and tiresome task, which generally requires an expertise in the field and knowledge in data-processing linguistics; techniques of filtering, categorization of documents and extraction of information. The linguistic processing needs a good text comprehension. It means transduction, which transforms a linear structure into a conceptual structure, i.e. text (the linear structure) is transformed into an intermediate logico-conceptual representation, which is then used to make conclusions. The semantic analysis aims at producing a structure representing, as accurately as possible, a unit of the sentence, with its meanings and its complexity [1][11][6]. Semantico-conceptual structures can be more or less broad, rich and complex and more or less ambiguous [4].

To determine the behaviour of a group we parse the large corpus of reviews, which were assigned with the same mark to find the characteristic. Our linguistic resources are the dictionaries and local grammars. The electronic dictionaries describe the simple words and the complex words of a language associating them with a lemma made up of a series of grammatical, semantical and inflexional codes. Grammars are representations of linguistic phenomena by recursive transitions (RTN). Generally a grammar represents sequences of words and produces linguistic information such as for example information on the syntactic structure. The local grammars, represented in the forms of graphs, describe elements which concern the same syntactic or semantic field. On fig 2 we show an example of local grammars used to determine the behaviour of groups. We assigned the semantic classes to our word corpus. To do this we used subjective word dictionary - General Inquirer Dictionary [1]. Then we parsed the corpus using local grammars to obtain statistical results.

Finally, we obtained a series of characteristics, which precisely determine a group. The characteristics are different for all of the study groups and generally they describe the statistical scores of typical words, their synonyms and they take into account negations. The results shoved strong differences between the characteristics of those groups. The creation of the group behaviour allows the determining of to which group a new review belongs. We used characteristics of groups for a preliminary estimation of the mark [fig 1; action 2]. This estimation helps us in the selection of classifier, which will process reviews.

2.2 Vector Representation

The vector representation of a corpus requires an initial linguistic pre-processing to eliminate all of the "empty" words not taking an active part in the meaning of the document. A first step is to build the index of the text learning base. Then, each text is represented by its coordinates in this index. Introducing the classes to initial index filters reduces the dimension of the vector representation used by the classifier. A linguistic filter is applied in order to eliminate

---

[1] http://www.wjh.harvard.edu/inquirer/spreadsheet_guide.htm

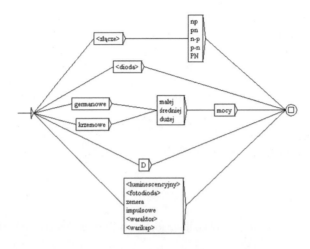

Fig. 2. Example of local grammar

some types of words considered useless for the categorization. All language sub-tleties contained in the text in order to analyze are necessary to guarantee a good performance of categorization. We added the synonyms by using the semantic classes. Then, we built a vector representation of all the text in the learning base corpus. The dimensions of the vector correspond to the complete index. The vector components are frequencies of the index terms in the document [fig 1; action 3].

Finally, the dimension of learning base space vector is enough to proceed the classification. Very often, the vector selected from the classifier includes many components with the value equal to zero. Those values do not have any incidence on the classification process. Thus, it is possible to reduce the size of the index to improve the performance of the classifier [fig 1; action 4]. Several methods were proposed to carry out a selection of the words representative of the field [14]. We chose the method of mutual information measurement proposed by Cover [2], which is especially well adapted for application in natural language processing.

Definition 1. *For the group of documents under consideration, the average mutual information I is the difference between the entropy of variable C and its conditional entropy relative with the word m_t.*

$$I(C, M_t) = \sum_{c \in C} \sum_{m_t \in M_t} P(c|m_t) \times \log \left(\frac{P(c, m_t)}{P(c) \times P(c|m_t)} \right) \qquad (1)$$

where : P(c) is the number of documents of the class C divided by the total number of documents.

P(m$_t$)is the number of documents containing word m$_t$ divided by the total number of documents.

P(c, m$_t$) is the number of documents of the class C and containing word m$_t$ divided by the total number of documents.

C is the random variable associated with all the classes (c),

M$_t$ is the random variable, representing existence of the word m$_t$ in a document.

This technique allows the calculation a reduced index dimension used by the classifiers. This method largely decreases the size of the index of a classifier. We select the words of which the mutual information is higher then a given threshold (si). The reduced index of each classifier define a new vector space dedicated to the classifier.

2.3   Bayes Classifier

The way of carrying out the classification is to find characteristics for each class and to associate a function of belonging. Among the methods using this process we can quote the decision trees, the Bayes classifiers, the method of SVM, etc. For our first approach we have used the Bayes classifier, which is a categorizer of the probabilistic type founded on the theorem of Bayes [13]. In our approach, we have presented five different classifiers, each classifier corresponds to a group of marks. The description of review behaviours, which belongs to different groups of marks, were done manually. The opinions are analysed sentence by sentence. Each classifier gives a mark (from 1 for 5) to the sentence. The classifier privileges the same mark, which was received in preliminary estimation process. For example, a classifier that corresponds to mark 1 will privilege assigning a mark 1[fig 1; action 5]. At the end of our process, we obtain the mark for all the sentences of the reviews processed. A final mark assignment of the reviews is the value of the arithmetic mean of all sentences treated. Our algorithm of rating the opinion is composed of two steps: first, the initial estimation of mark by the behaviour classification of the groups and finally the assignment of a mark by using the appropriated classifier allocated by an initial mark. By using this architecture we hope to improve the F-scores of systems, which directly use classifiers.

3   Conclusions

The objective of our work is to build a system for collecting, evaluating and ranking movies reviews. RRSS the Rating Reviews Support System is the proposal for the system, which carries out a collection and marking of reviews. This paper presents only the evaluating and ranking part of the system. RRSS will be a support to RS. The goal of our work is to automate the whole system particularly to improve the estimation of individual user's reviews. The system allows an automatic assignment of a mark; however to increase the research on other fields it will be necessary to create a linguistic base and a new analyzis of the different elements of the group's behaviour.

References

1. Altai, H., The core language Engine. MIT Press (ACL-MIT Press Series in Natural language Processing), Cambridge, 1992.
2. COVER Cover, Thomas. Elements of Information Theory. John Wiley 1991.
3. Dave, K., Lawrance, S., Pennock, D.M.2003. Mining the Peanut Gallery: Opinion Extraction with HMM Structures Learned by Stochastic Optimization. In AAAI-2000.
4. Dziczkowski, G., Wegrzyn-Wolska, K., Graph based system purpose - built for automatic retrieval and extraction of the electronics data. To appear in proceeding of Euro-IMSA, Mars 2007.
5. Gross, M., The construction of local grammars. In Finite-State Language Processing, Cambridge, MIT Press, pp 329-354, 1997.
6. Kamp, H., Evenements representations discursives et reference temporelle. Langages, nb 64, 1981, pp.34-64.
7. Pang, B., Lee, L., 2004. A Sentimental Education: Sentiment Analysis Using Subjectivity Summarization Based on Minimum Cuts. In ACL-04.
8. Panzienza, M.T., Information extraction (a multidisciplinary approach to an emerging information technology). Springer Verlag (Lecture Notes in Computer Science), Heidelberg, 1997.
9. Paumier S., De La reconnaissance de formes linquistique a l'analyse syntaxique. These, Marne-la-Valee, 2003.
10. Riloff, E., Wiebe, J., Philips, W., Exploiting Subjectivity Classification to Improve Information Extraction.
11. Sowa, J., Conceptual Structures. Information processing in Mind and Machine. Addi son Wesley Publishing CO., Reading, 1984.
12. Tarveen, L., Hill, W. (2001). Beyond recommender systems: helping people help each other. In HCI in the millennium , J. Caroll, ed., Addison-Wesley, pp 1-21, 2001.
13. Turney, P., Littman, M., 2003. Measuring Praise and Criticism: Inference of Semantic Orientation from Association. ACM Transaction on Information Systems (TOIS) 21(4):315-346.
14. Wang, Y., Hodges, J., Tang, B. Classification of Web Documents using a Naive Bayes Method. IEEE, pp 560-564. 2003.
15. Wiebe, J., Wilson, T., Bruce, R., Bell, M., Martin, M., 2004. Learning Subjective Language. Computational Linguistics 30(3):277-308.

Towards User Context Enhance Search Engine Logs Mining

S. Eibe[1,*], M. Valencia[1], E. Menasalvas[1], J. Segovia[1,**], and P. Sousa[2]

[1] Facultad de Informatica, Universidad Politecnica de Madrid, Spain
   seibe@fi.upm.es, mvalencia@zipi.fi.upm.es,
   emenasalvas@fi.upm.es, fsegovia@fi.upm.es
[2] Universidad Nova de Lisboa, Portugal
   pas@uninova.pt

Summary. Making search engines responsive to human needs requires understanding user intention when submitting a query. Intention, context and situation are intimately connected [8]. Thus context modelling is paramount when mining search engine logs but the process should be sistematized and standarized for the future generation of autonomous data mining components. In this paper, we propose to integrate context as metadata represented in PMML. The complete knowledge discovery process would benefit from the metadata and in particular final and intermediate evaluations can use this information for interpretation. The paper also presents results of integration of GUMO ontology to conceptualize user context to improve interpretation of query mining on the weblogs of the site search engine.

1 Introduction

The information explosion on the Internet has placed high demands on search engines [12] that struggle with vague queries, impatient users and an enormous and rapidly expanding collection of documents. The performance of search engines crucially depends on their ability to capture the meaning of a query most likely intended by the user.

Recently,[2] search engines are being modified with the aim of improving search results by focusing on the users, rather than on their submitted queries. This is a challenging activity as, knowing the user intention, requires a considerable amount of skill in order to satisfy non-trivial information needs.

In [3] context is defined as "any information that can be used to characterise the situation of entities". The authors in [1] wonders about the possibility of implicit factors being input to the system as some aspects of context, for example, the mood of the user cannot be directly sensed.

Context and situation of the user are intimately connected. According to [8] a situation consists of two parts: the user-related factors that are intrinsically tied to a user as abilities, goals or personal traits and the environments in which the user perceives and acts that can be described by factors that are independent of an individual user. In [6] apart from these two parts, the author distinguishes the system's factors added to be either modelled as context or resource.

---

* Project partially financed by Project CCG06-UPM/ESP-0259.
** Project partially financed by Project TIN2004-05873.

In [9] authors present an approach that uses past behaviours of users to re-rank future searchs in a way to recognise the implicit preferences of communities of searchers. They present this approach to deal, according to the authors, with the limitations of approaches only based on information retrieval perspective limited as they do not consider the past search behaviour of users.

User context in a certain way is also dealt in [11] where authors present the problem of mapping a search engine query to those nodes of a given subject taxonomy that characterise its most likely meanings.

Context for recommendation is presented in [13] where search engine user's sequential search behaviour is analysed and results are integrated with a content based similarity method to deal with the shortness problem of queries.

According to [10] making the web truly responsive to human needs means understanding the user needs. This author also analyses some of the problems that arise when using metadata to add information that search engines can use to more accurately "understand" the pages and retrieve and/or manipulate them for the user. The author mentions the problem of interpreting multiple domains, contents and purposes that can lead to different applications interpreting the same metadata in different ways [10].

In this paper, we present an approach in which context factors are used in order to improve query clustering. For this purpose a standard user modelling ontology [7] is integrated and used to define metadata in PMML [4].

Standardising the process to integrate user context is twofold: on the one hand context factors are taken into account so intention of the user is modelled and consequently results of the search engine could be improved. On the other search engine query represent a data stream where autonomous mining components are required. Once the deriving process is ready, each time a mining process over the search engine logs is needed the standard method can be used to derive and enrich attributes to be mined with context information that will be stored as metadata.

The rest of the paper is organised as follows. In section 2 the conceptualisation approach is described, firstly GUMO ontology is briefly analysed and later the method to conceptualise context is outlined. In section 3 the instantiation of the proposed method to a real web site that uses GSA [5] as search engine is shown. In the case study two ontologies are used: the GUMO ontology for user context modelling and a particular taxonomy of concepts to model the particular domain of the problem. To conclude 4 shows the advantages of the proposed method as a method to capture metadata towards systematisation of the mining process.

2   Our Approach to Analyze User Context

The Oxford English Dictionary defines *context* as "parts that precede or follow a passage or word and fix its meaning...; ambient conditions". In particular for the case of search engines, understanding these "parts that precede or follow" is of outermost importance when trying to define the intention of the user when submitting a query to the engine. Two problems arise: on the one hand, retrieving context on the other, representing it. Assuming that some factors have been captured (is not the focus of this paper) the main problem is to represent these factors so that no conflict arises when processing the information they represent. GUMO [7] is a general user model ontology for the uniform interpretation of distributed user models that contributes to simplify the exchanging of user model metadata between different user-adaptive systems.

```
<xs:element name="DerivedField">
  <xs:complexType>
  (...)
      <xs:attribute name="fieldType" type="FIELDTYPE" />
  </xs:complexType>
</xs:element>

<xs:simpleType name="FIELDTYPE">
  <xs:restriction base="xs:string">
    <xs:enumeration value="general" />
    <xs:enumeration value="user_context_contact_info" />
    <xs:enumeration value="user_context_demographics" />
    <xs:enumeration value="user_context_temporal" />

  (...)
  </xs:restriction>
</xs:simpleType>
```

Fig. 1. PMML extension mechanism: fieldType generation

What we propose is to map the information that can be measured, user context in particular, with the concepts in the ontology and store this information as metadata in PMML (see figure 1) in the system prior to any data mining analysis. Once the data mining process starts, data will be labelled according to the mapping in the data understading phase and later phases will benefit from this labelling. As an example, imagine that we have identification of navigators so the navigator name, age, address and telephone are available for analysis. Prior to analysis these attributes will be labelled as **contact information** or **demographics** according to mapping in table 1. In the same way, other input attributes will be labelled as **behavioural** or **temporal**.

Table 1. Contact Information and Demographics in GUMO

GUMO's Dimension	Attributtes	Input Attribute
Contact Information	Name, Mail, Address, Telephone, ...	Nome, Mail Direçao
Demographics	Gender, Age, ...	Idade
Temporal	hour, day, week	request_hour, request_day

Therefore, we propose a two step process:

1. Using a standard ontology to define context. For the case of user context we propose to integrate GUMO ontology [7]. In general, we propose for this step to use any ontology that describe in a uniform way the domain of the data and process to be analysed.
2. Define the mapping between the attributes either in the source data (the weblog in the case of a search engine) or any other derived attribute and the concepts in the corresponding ontology.

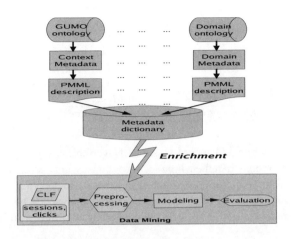

Fig. 2. Mining with contexts

We propose to use PMML to describe both the derivation process as well as the mapping. A possible representation of the metadata in PMML is shown in figure 1 while the process can be depicted in diagram 2.

Observe that once the context attribute derivation process has been defined each time a log of a search engine has to be mined the standard process can be applied in a autonomous way. The advantages of the proposed process are as follows:

- Attributes and their mapping with the ontology can be defined altough in some environments the source data is not available.
- More than one attribute can be defined for the same semantics or concept in the ontology. Thus, different input attributes depending on the situation will be mapped, derived and used.
- The attributes can be redefined or customised according to different domain to avoid misunderstanding problems.
- The metadata defined and stored can be used by software components along the knowledge discovery process to interpret and evaluate results.

This approach aids to the systematic development of the future generation of semi-autonomous mining components as it is the basis for:

- Labelling attributes of data in semi-automatic way in data understanding phase.
- Labelling results of mining process so interpretation and evaluation of results potentially improves.

3   Case Study

The proposed method has been applied in a site working as estate-agent. GSA [5] is used as search engine for user to search for properties either to rent, buy or sell.

For the case-study the search log containing 69760 records corresponding to activity in the site. Although the GSA does not generate session identifiers, the site has been

Table 2. Attributes in the search box

Domain's Category	Input Attribute
Location	Lisboa, Porto, Madeira
Action	Arrendamento, Venda, Outro
Price	p1, p2, p3
Kind of property	Terreno, Moradia

configured so to generate and record sessions. Nevertheless some preprocessing was required in order to reconstruct the whole set of session arising a total of 43914 sessions.

These data were used as sampling data so to generate metadata and automatise the method. In a second step data from the logs are being analysed in a semi-autonomous fashion. The source data used in the data understanding step were as follows:

1. Input data: log in CLF enriched with session id. Additionally, we have some info from questionnaires filled by the web users
2. Domain information. Information related to properties (type, size, price, . . .), as well as information about the possible actions of user was retrieved. All this information was available in portuguese. From this information a taxonomy was derived to enrich queries. The taxonomy reflects information to categorised actions one can do (sell, buy, rent, . . .), things one can ask for (flats, apartment, ..), location (cities, sea, mountain, . . .), prices, sizes, . . .
3. GUMO ontology

With all this information first of all, the mapping of the ontology with the attributes was made. An example of the mapping can be seen in tables 2 and 1, although the information in table 1 is in portuguese, the method presented is language independent. Later, logs of following weeks were analysed using the mapping to label input data and query clustering results. After validating the process a component has been deployed in order to support user in clustering web logs data.

4   Conclusion

The performance of a search engine heavily depends on the ability to interpret the user intention when submitting a query. Intention, situation and context are intimately connected [8]. In this paper, we have proposed a method to conceptualise context prior to mining the web logs. Context information stored as metadata in PMML can be later used to interpret and evaluate results. In order to avoid conflict problems we propose to map attributes (either from the source data, or derived attributes) with an accepted (business understanding phase) and standard ontologies. Once the mapping is defined the mining process benefits from the description of attributes. In the paper, we have shown the application of the proposed approach to the case of a site using GSA as search engine. Two ontologies have been used in this case: on the one hand the commonly accepted GUMO ontology to map user context attributes and on the other a taxonomy of domain concepts to map information of the query search box. The method is a first step towards autonomy in search engines as all the attributes from the web log in CLF format have been mapped together with the possible information in the query search box. This labelling process makes it possible to label results and to generate mining components that do not depend on the user to choose the appropriate attributes. Nevertheless, the challenge of automatically mapping attributes in standard and commonly accepted ontologies remains. Consequently, we are currently working on the generalisation of such process for the case of search engines in a established domain.

References

1. Sarab Annand. Putting the user in context, 2006. ECML PKDD 2006 Workshop on Ubiquitous Knowledge Discovery for users (UKDU'06). Berlin.
2. Ricardo A. Baeza-Yates. Web mining in search engines. In *ACSC*, pages 3–4, 2004.
3. Anind K. Dey. Understanding and using context. *Personal and Ubiquitous Computing*, 5(1):4–7, 2001.
4. DMG. PMML Version 3.0, (2006). Online at http://www.dmg.org/pmml-v3-0.html.
5. Google. Google search appliance frequently asked questions, accessed 2006. Available online at http://www.google.com/appliance/faq.html.
6. Dominik Heckmann. Distributed user modeling for situated interaction. In *GI Jahrestagung (1)*, pages 266–270, 2005.
7. Dominik Heckmann, Tim Schwartz, Boris Brandherm, Michael Schmitz, and Margeritta von Wilamowitz-Moellendorff. Gumo - the general user model ontology. In *User Modeling*, pages 428–432, 2005.
8. C. Kray. Situated interaction on spatial topics., 2003. Volume 274 Dissertations in Artificial Intelligence-Infix November.
9. Barry Smyth, Evelyn Balfe, Jill Freyne, Peter Briggs, Maurice Coyle, and Oisin Boydell. Exploiting query repetition and regularity in an adaptive community-based web search engine. *User Modeling and User-Adapted Interaction*, 14(5):383–423, 2005.
10. Tiffany Ya Tang and Gordon I. McCalla. Student modeling for a web-based learning environment: A data mining approach. In *AAAI/IAAI*, pages 967–968, 2002.
11. David Vogel, Steffen Bickel, Peter Haider, Rolf Schimpfky, Peter Siemen, Steve Bridges, and Tobias Scheffer. Classifying search engine queries using the web as background knowledge. *SIGKDD Explor. Newsl.*, 7(2):117–122, 2005.
12. Ji-Rong Wen, Jian-Yun Nie, and HongJiang Zhang. Clustering user queries of a search engine. In *WWW*, pages 162–168, 2001.
13. Zhiyong Zhang and Olfa Nasraoui. Mining search engine query logs for query recommendation. In *WWW '06: Proceedings of the 15th international conference on World Wide Web*, pages 1039–1040, New York, NY, USA, 2006. ACM Press.

Semantic Annotation of Web Resources Using IdentityRank and Wikipedia

Norberto Fernández, José M. Blázquez, Luis Sánchez, and Vicente Luque

Telematic Engineering Department. Carlos III University of Madrid
{berto,jmb,luiss,vlc}@it.uc3m.es

Summary. In this paper we introduce the IdentityRank algorithm developed to address the problem of named entity disambiguation. It is used for semantic annotation of Web resources taking Wikipedia as knowledge source.

1 Introduction

In order to make the Semantic Web [1] vision become a reality, the semantics of the data needs to be described in a computer understandable manner. This process is known in the literature as *semantic annotation*.

In [2] we introduced a system that exploited user queries to generate annotations and used the information generated and maintained by Wikipedia[1] editors as knowledge source for the annotation process.

As is indicated in [2], our system had some limitations, for instance, it is a manual system so it requires user collaboration to gather metadata. More automatized semantic annotation approaches seem more appropriate for the annotation of high volumes of information due to scalability reasons.

In order to deal with these limitations, we have extended our initial proposal by including an information extraction tool, ANNIE[2], into our system. With such tool, we process the textual contents of the Web resources to be annotated, extracting occurrences of named entities: persons, locations and organizations. Once we have these entities, in order to generate semantic annotations, we need to link each entity (e.g. the person *Alonso*) with its Wikipedia page. As there are usually several instances that can be associated to a certain entity[3], we need to disambiguate that entity selecting the Wikipedia page that best represents it in the context of the document being annotated. To address this need, we have developed an algorithm for named entity disambiguation based on Google's PageRank [3] that we name IdentityRank (a.k.a. IdRank).

---

[1] http://www.wikipedia.org

[2] http://gate.ac.uk/ie/annie.html

[3] For instance for the person *Alonso* we have among others Fernando Alonso, a Formula 1 driver, *http://en.wikipedia.org/wiki/Fernando_Alonso*, and José Antonio Alonso a Spanish Minister. *http://en.wikipedia.org/wiki/Jos%C3%A9_Antonio_Alonso_Su%C3%A1rez*

K.M. Węgrzyn-Wolska and P.S. Szczepaniak (Eds.): Adv. in Intel. Web, ASC 43, pp. 100–105, 2007.
springerlink.com

The rest of this paper describes the IdRank algorithm and is organized as follows: section 2 describes the IdRank algorithm and shows some results of an initial evaluation, section 3 elaborates on related work and finally, section 4 with concluding remarks and future lines ends this paper.

2   IdRank

In this section we will describe in detail the IdRank algorithm and the results of an initial evaluation of that algorithm. Due to the lack of space we will not describe here the PageRank algorithm. The interested reader can find a comprehensive description of that algorithm in [3].

2.1   IdRank Process

The manual annotation process described in [2] required from the user the annotation or disambiguation of the terms in his/her query using concepts represented by Wikipedia pages and the usage of relevance feedback to indicate that a certain Web resource was relevant for that query. By doing so a new annotation was generated linking the Web resource with the person who has generated the annotation and with the concepts in the annotated query.

The system has been now extended, so when a user provides a manual annotation as described above, an automatic process starts. This process consist in downloading the Web resource which is going to be annotated and automatically extracting from its contents the entities mentioned there using ANNIE. For each entity, the entity text and the entity type (*person, location, organization*) are provided. Additionally, the links to Wikipedia pages in the contents are also extracted, because they can be considered as annotations introduced by the page author at authoring time.

Now IdRank can run, using the information already available: the manually-generated annotation, the links to Wikipedia pages mentioned by the Web resource and the entities. The IdRank process consist of the following steps:

Candidate finding. The system finds the URLs of the Wikipedia pages which are candidates to represent each of the input entities. In order to do so, the system uses Yahoo APIs[4] to query Yahoo with a site restriction *wikipedia.org* as many times as different entities (different pair entity text/entity type) need to be disambiguated. The resultant set of Wikipedia URLs is modified by adding the Wikipedia URLs extracted from the Web resource content and the ones used in the manual annotation. In the case of URLs obtained from queries, we store also the position of each URL in the original Yahoo result set for later usage.

Duplicate removal. The algorithm processes the Wikipedia URL set to filter duplicates. One of the difficulties of this filtering process is the fact that there are several Wikipedia URLs representing the same concept (pages in different

---

[4] http://developer.yahoo.com/search/

languages, redirections). Due to this, the filtering process requires to download and process the candidate Wikipedia pages extracting the language links from that pages and detecting HTTP redirections when downloading a certain page. Once that we know the different Wikipedia URLs that can represent the same concept, we can assign a unique identifier to the concept (a URI) and store the mapping between that URI and the original Wikipedia URLs. So given the original Wikipedia URL set we obtain a set of unique URIs in which each URI, each concept, appears only once. In this page-processing step, we also extract the links between Wikipedia pages, which will be used in next step.

Ranking computation. A semantic network is built with the URIs that result from the duplicate removal process. In such network, nodes are concepts represented by URIs. There can be two kinds of links between that nodes:

1. A bidirectional *anchor link* between node u and node v appears if there is an HTML link between any of the Wikipedia pages that represent the concept u and any of the Wikipedia pages that represent the concept v or vice versa.
2. A bidirectional *cooccurrence link* between nodes u and v appears if there are former manual annotations defined by this or other users which use the concept u and the concept v in annotating the same Web resource (exploits the information about cooccurrence of concepts in Web resources).

We will give weights to these links. The anchor links are handled in the same way as in original PageRank, that is, each node gives the same weight to all of its forward links. The weight of the cooccurrence links, not included in PageRank, is computed using the cooccurrence frequency of the linked concepts. Mathematically, this can be expressed as:

$$\alpha_{uv} = \frac{f_{uv}}{\displaystyle\sum_{k \epsilon C_v} f_{kv}} \tag{1}$$

Where f_{uv} is the cooccurrence frequency of concepts u and v, that is, the number of Web resources annotated both with u and v divided by the number of Web resources annotated with v. C_v is the set of concepts in the semantic network that cooccur with v in at least the annotations of one Web resource apart from the one being analyzed.

Apart from link information, the original PageRank algorithm included a vector E used for ranking personalization giving more weight to certain nodes in the network. In IdRank, the values of this vector are computed taking into account the usage in the recent past of the concept u in the annotations of the same user who is defining the current annotation. In that sense the algorithm learns from past user annotations. In practice, the value of the u component of the vector E, $E(u)$, is directly proportional to the number of times the concept u has been used in the last M annotations performed by the user, being M a parameter of the algorithm.

Taking into account all these contributions we obtain the following equation, adaptation of the original PageRank equation in [3]:

$$R(u) = k_A[\sum_{\forall v \in S} (\beta_1 \frac{1}{N_v} + \beta_2 \alpha_{uv})R(v)] + k_E E(u, M) \tag{2}$$

Where $R(u)$ is the ranking of the node u, S is the set of nodes in the semantic network, $\beta_1 = 1/2$ if there is an anchor link between v and u or 0 otherwise, $\beta_2 = 1/2$ if there is a cooccurrence link between u and v ($u \neq v$) or 0 otherwise and N_v is the number of anchor links of v. In order to control the influence on the final results of each of the components of the algorithm we use two constants k_A and k_E such as $k_A + k_E = 1$.

We solve this set of equations for each value of u using appropriate numerical methods, as the one described in [3], obtaining as result a weight for each of the candidate concepts in the semantic network. Then we translate back the URIs of the concepts to Wikipedia page URLs using the table generated in the duplicate removal step. Each URL, associated to a certain URI, is assigned the same weight as the algorithm gives to the URI. For each of the original entities the algorithm assigns as Wikipedia representation the candidate whose URL has highest weight. If a certain entity has more than one candidate with maximal weight, the algorithm uses the original Yahoo ranking to decide.

2.2 Evaluation

We have carried out a basic experiment to test the behavior of the IdRank algorithm. In that experiment we use a corpus of ten documents that were obtained by querying a repository of news items looking for *Alonso* and selecting randomly some documents. The entities in these documents were automatically detected, but, in order to avoid the noise introduced in the evaluation of the disambiguation algorithm by the errors in the entity extraction process, the entities were reviewed by two human users. At the end we got 118 entities, 65 of them unique.

For each entity, we looked for the entity text in Yahoo with a literal query (among quotes) and a restriction *site:wikipedia.org* in order to find its candidates. We limited to ten the number of results returned by the search engine and filtered special pages of Wikipedia (like user pages and talk pages) from the result set. Additionally, we have manually reviewed the candidates information in order to check whether ten results per entity were enough for the process, and we got that only in 7 cases (4 different entities) there was not any Wikipedia page in the result set that could be used to represent the real meaning of the entity.

We compared our algorithm with two other naive algorithms: one simply assigns to each entity the first result obtained from Yahoo when looking for the entity text in Wikipedia. The other one simply computes the Levenshtein distance between the entity text and the Wikipedia page title using the SimMetrics library[5] and assigns to each entity the Wikipedia page whose title is more similar

---

[5] http://sourceforge.net/projects/simmetrics/

to the entity text. Additionally we tested our algorithm in two modes: working with user history (past annotations) and without user history (that is, only using the information on anchor links in the disambiguation process). We build the history by randomly selecting two entities in each document and manually annotating them.

The parameters of the algorithm were: $k_A = 0.7$, $k_E = 0.3$ and $M = \infty$, that is, we use all the annotations in the history as context for the disambiguation process. We ran the different algorithms over the corpus and then manually checked the correctness of the assignments entity-Wikipedia page.

The results of these experiments are shown in table 1 where the number of right assignments entity/page are shown. *First* represents the first result algorithm, *Sim* the text similarity algorithm, *Links* IdRank using only the anchor links information and *All* IdRank taking all the information into account.

Table 1. Evaluation results

First	Sim	Links	All
78	60	89	98

3 Related Work

There are several approaches in the state of the art dealing with named entity disambiguation. These different approaches can be characterized according to a number of criteria. One of these possible criteria is the *context* used to disambiguate the entity. Some approaches use the complete document [5] as context. Others use a number words before and after the entity like [10, 9]. Although some approaches use both common words and named entities as context [10], others suggest that better results can be obtained using as context only other named entities [9]. Another criteria is the use of *knowledge sources* like lexical databases, ontologies, etc. There are approaches that make use of such knowledge sources [4, 8] and approaches that try to cluster the named entities without any reference to an available list of possible instances [10, 9]. Finally we can further calssify the approaches with respect to the *disambiguation algorithms* used: statistical procedures [7, 10, 9], morphosyntactic analysis [9, 5], algorithms exploiting the information and structure provided by an ontology [8], etc.

The usage of a semantic network ranking algorithm, which also takes into account the temporal component of users' interests are the main differences of our approach compared with the ones in the state of the art.

4 Conclusions and Future Lines

In this paper we introduced the IdRank algorithm developed to address the problem of named entity disambiguation. It is used for semantic annotation of Web resources taking Wikipedia as knowledge source. Though some initial results on evaluation are reported, more intensive tests need to be run, for instance in order to measure the influence of the parameters of the algorithm in the final results.

Acknowledgements

This work has been partially funded by the Spanish Ministry of Education and Science under contract ITACA (TSI2006-13409-C02-01).

References

1. Berners-Lee T, Hendler J, Lassila O (2001) The Semantic Web: A new form of Web content that is meaningful to computers will unleash a revolution of new possibilities. Scientific American, May 2001.
2. Fernández N, Blázquez JM, Sánchez L, Luque V (2006) Exploiting Wikipedia in Integrating Semantic Annotation with Information Retrieval. In 4th Atlantic Web Intelligence Conference, AWIC 2006. Israel, June 2006.
3. Page L, Brin S, Motwani R, Winograd T (1999) The PageRank Citation Ranking: Bringing Order to the Web. Stanford Technical Report available online at: http://dbpubs.stanford.edu/pub/1999-66, 1999
4. Aswani N, Bontcheva K, Cunnigham H (2006) Mining Information for Instance Unification. In 5th International Semantic Web Conference. Ed. Springer, LNCS 4273, pp. 329-342. USA. November 2006.
5. Bagga A, Baldwin B (1998) Entity-Based Cross-Document Coreferencing Using the Vector Space Model. In 17th International Conference on Computational Linguistics. Canada. August 1998.
6. Ginter F, Boberg J, Ärvinen J, Salakoski T (2004) New Techniques for Disambiguation in Natural Language and their Applications to Biological Text. Journal of Machine Learning Research, 5: 605-621, 2004.
7. Han H, Giles L, Zha H, Li C, Tsioutsiouliklis K (2004) Two Supervised Learning Approaches for Name Disambiguation in Author Citations. In Joint ACM/IEEE Conference on Digital Libraries. USA. June 2004.
8. Hassell J, Aleman-Meza B, Arpinar IB (2006) Ontology-Driven Automatic Entity Disambiguation in Unstructured Text. In 5th International Semantic Web Conference. Ed. Springer, LNCS 4273, pp. 44-57. USA. November 2006.
9. Mann GS, Yarowski D (2003) Unsupervised Personal Name Disambiguation. In 7th Conference on Natural Language Learning. Canada. June 2003.
10. Pedersen T, Purandare A, Kulkarni A (2005) Name Discrimination by Clustering Similar Contexts. In 6th International Conference on Computational Linguistics and Intelligent Text Processing. Ed. Springer, LNCS 3406. Mexico. February 2005.

Learning Fuzzy Cognitive Maps from the Web for the Stock Market Decision Support System

Wojciech Froelich[1] and Alicja Wakulicz-Deja[2]

[1] Institute of Computer Science, Silesian University, Sosnowiec, Poland
`froelich@konto.pl`
[2] Institute of Computer Science, Silesian University, Sosnowiec, Poland
`wakulicz@us.edu.pl`

1 Motivations

In this paper we would like to propose a new hybrid scheme for learning approximate concepts and causal relations among them using information available from the Web. For this purpose we are applying fuzzy cognitive maps (FCMs) as a knowledge representation method and an analytical tool. Fuzzy cognitive maps are a decision-support tool, analytical technique, and a qualitative knowledge representation method with large potential for real world applications. FCMs are able to express the behavior of a system through the description of cause and effect relationships among concepts. FCMs can be represented as directed graphs consisting of concepts (nodes) and cause and effect relationships (branches) among them. The concepts represent states that are observable within the domain. The directions of branches indicate the causal dependency between source and target concepts. In spite of a quite simple construction and relatively easy interpretation, which can play a key role while constructing decision support systems, it's expected that FCMs can express complex behaviors of dynamic systems. The basic formalism of FCMs is presented in section 2. Obviously, there are also drawbacks of FCMs, that have been mentioned, e.g., in [4]. Also, it can be mentioned that, among the many extensions to FCMs, there is still lack of common formalism, which causes some difficulties when comparing one with another.

In spite of many disappointing factors such as random walk theory, a problem with reliable verification, and other known difficulties, we have selected stock market modeling and financial time series prediction as the targeted application areas for the proposed theoretical approach. The problem of time series prediction has been extensively investigated applying statistical models. Another well-known approach to modeling stock market behavior is to apply machine-learning techniques. The literature on both of these approaches is very extensive and far beyond the scope of this paper. In spite of a large number of research efforts, the problem continues to be an interesting challenge. The review of applicable methods can be found in [1]. Among many possible solutions for this kind of application, there is the possibility of applying approximate reasoning

K.M. Węgrzyn-Wolska and P.S. Szczepaniak (Eds.): Adv. in Intel. Web, ASC 43, pp. 106–111, 2007.
springerlink.com

methods [9], particularly the theory of fuzzy sets [13] and fuzzy cognitive maps [5, 6]. The application of FCMs for the problem of financial time series prediction has been investigated in [6].

From the point of view of the application targeted in our research, most of the existing works deal with relatively small numbers (5-10) of concepts. However, in many practical applications such as stock market decision support systems, it's possible to recognize thousands of concepts. In addition, most of the concepts are approximate, or they depend on parameters (e.g. time dependant) which are initially unknown. Managing and learning a large number of parameterized concepts and the relationships among the concepts are challenges that are addressed in this paper. Another objective of this study is to evaluate the usefulness of the learned concepts, considering their mutual relationships that are learned at the same time.

2 Fuzzy Cognitive Maps

The notion of a concept in the context of information systems has been defined in [8] and investigated in generalized approximation spaces [9]. The idea of cognitive maps (CMs) has been introduced in psychology [12] and used primarily as a modeling tool in the social sciences [2]. FCMs introduced by Kosko [5] combine in their representation schemes some of the achievements from fuzzy logic, neural networks, and statistical (Bayesian networks) theories. There are many extensions to the generic FCM model. A review of them can be found in the literature [7, 10, 11]. The dynamic cognitive networks (DCN) [6] introduce, among other ideas, the possibility of representing nonlinearities and temporal phenomena in data. The proposal of high order FCMs [11] overcomes the limitation of generic FCMs in modeling high order dynamics by adding memory to the concepts' nodes. For the purposes of our research, the state of a generic, fuzzy, cognitive map at a discrete time step is defined by Eq. (1). The definition intends to involve only the objects that will change over time, and, if the time is considered, the respective objects will be complemented by letter t .

$$FCM = \langle C, A, W \rangle, \tag{1}$$

where:

C is the finite set of approximate concepts, such that every concept $c \in C$ can be mapped to a real number by the respective function from $a \in A$.

A is the set of functions, and every $a_i : C \to [0, 1]$ defuzzyfies a concept c_i and can be interpreted as the degree of the concept's so-called activation level. For simplicity a_i will be understood as the value of $a_i(c_i)$.

W is the set of weights such that w_{ij} estimates a function $\mu(c_i, c_j)$ in time step t that represents an a priori, unknown causal relationship between any two concepts $\mu(c_i, c_j) \in [-1, 1]$, where: $c_i, c_j \in C$. Intuitively, the negative value of w_{ij} indicates that the activation of c_i causes the deactivation of c_j.

Basically, FCMs can operate in two modes: the exploitation mode and the learning mode. While in the exploitation mode, the activation level of every concept is obtained on the basis of the transformation function stated in Eq.2.

$$a_i(t+1) = a_i(t) + f(\sum_{j=1,j\neq i}^{n} w_{ij}a_{ij}),$$ (2)

where f is the threshold function that serves to reduce unbounded values to a strict range. It can be defined as a sigmoid function $f = \frac{1}{1+e^{-\lambda x}}$, in which the parameter λ determines the shape of the function and is usually set to 5 [11]. A discrete time simulation is performed by iteratively applying Eq. 2. During the iterative process, the state vector $\bar{a}$ may converge to the: a) fixed-point attractor ($\bar{a}$ remains unchanged), b) limit cycle ($\bar{a}$ repeats), c) chaotic attractor ($\bar{a}$ changes and repeating states never occur). The inference in FCMs can be expressed by the matrix multiplication $\bar{a} = \bar{A} \times \bar{W}$ with matrices $\bar{A}, \bar{W}$ built on the basis of the corresponding sets.

As mentioned before, the second mode of FCM operation is its learning, a focus of our research. The algorithms proposed in the literature target different application requirements and try to overcome some limitations of FCMs. The learning methods of FCMs can be of two basic types, i.e. evolutionary [10] or adaptive [4], which are mainly based on, but not restricted to, the application of Hebbian rule: $w_{ij}(t+1) = w_{ij}(t) + \alpha(t)[\nabla a_i \nabla a_j - w_{ij}(t)]$, $\alpha(t) = 0.1[1 - \frac{t}{1.1N}]$, where N is the parameter. The advantage of good global exploration of evolutionary algorithms has been exploited in [10] . The promising results applying local adaptive learning have been reported by many researchers e.g. [7]. A new hybrid algorithm [7] combines the preliminary stage of adaptive learning with the evolutionary fine-tuning of weights afterwards. The proposal of a balanced, differential-learning algorithm [4] addresses the problem of too-strong, accumulative, many-to-one inductive bias while applying generic Hebbian learning. Most of the existing approaches assume that the set of concepts is fixed by the domain expert by means of the cardinality of C and the identification of particular concepts. It has been also proposed [3] to describe the concepts by a set of fuzzy rules. Thus, the process of concept identification can be automated by matching observations to the set of fuzzy membership functions incorporated in the concept description.

3   A New Scheme for Learning Concepts and Causal Relations

Before presenting our new scheme for joint learning of concepts and causal realtions among the concepts, let us introduce a basic notation. As mentioned earlier, the concepts are indexed in C by the letter i. Assume that $C_k \subset C$ is a subset of concepts and assume $\forall_k card(C_k) = const$ over time and $card(C) \gg card(C_k)$. Finally, if c_i is a concept, then $c_i \in C_k \subset C$. For the concepts c_i from any C_k, it is possible to construct a vector $\bar{i}$ of their respective indices in C. Let $a_i^p : C \rightarrow [-1,1]$ be a parameterized function, where p denotes the parameter of a_i. In particular case of temporal concept, the parameter p can be related e.g. to time. Obviously, the number of parameters can be increased if attention is paid to the size of the space P. Let $\bar{p} \in P$ be a vector of parameters in the corresponding parameters' space. Discretization of the parameters' values in the a priori determined intervals should be made: $p_i \in < p_i^d, p_i^g >, p_i, p_i^d, p_i^g \in \mathbf{N}$. The discretization for the set of wieghts W should also be performed. Assume that all random generators work with uniform probability distributions over their domains and that the time constants $t_1, t_2, t_3, t_{finish}$ are the parameters. A general scheme of the proposed learning algorithm is presented below:

1. Let $t = 0$, generate randomly a subset C_k from C,
2. Construct a vector $\bar{i}$
3. For all $c_i \in C_k$, randomly generate the values of $\bar{p}$
4. Construct a population of genotypes $\bar{g}$ such that every: $\bar{g} = < \bar{i}, \bar{p} >$
5. Randomly generate $card(C_k)^2$ weights $w_{ij} \in W$
6. Construct (or complement if necessary) the set of phenotypes such that (according to E.q.1): $fcm = < C_k, A, W >$
7. For every fcm, run a learning process with the discretization of time (simulated or real time) $t \in < t_1, t_2 >$ (apply the incremental, adaptive-learning algorithm e.g. Hebbian-like)
8. Switch every fcm to the exploitation mode for the time interval $t \in < t_2, t_3 >$ and evaluate each on the basis of the number of the correct predictions
9. Perform the selection of the fcm individuals for the reproduction pool
10. Complement the set of genotypes (applying evolutionary operators)
11. If $t < t_{finish}$, go back to step 6
12. Finish

The idea of the proposed algorithm is to bind both types of learning, evolutionary searching in the concepts' space, with the process of adaptive training of causal relationships among them. Thus, dependency exists between the evaluation of the concept alone and its effectiveness while constructing relationships with other concepts. This seems to be essential for the targeted application. It should also be noted that our algorithm involves the possibility of incremental learning. The set of concepts C can be constructed partially by experts or complemented dynamically by applying diverse machine-learning methods, even those that work on the basis of real-time source data. Thus, the large number of possible concepts can be verified for the ability to provide predicting capabilities.

4   Collecting Data from the Web for the Stock Market Decision Support System

The concepts considered in our research can be completely different types. On the basis of heterogeneous off-line and real-time data sources from the Internet, we are trying to investigate the possibilities of constructing parametrized concepts and then identifying dependencies among them that could reflect the behavior of the stock market. Different kinds of source data, both quantitative and qualitative, have been considered:

1. static data repository that doesn't change during the learning process (These data are extracted from web pages, xml data sources, databases.)
2. data streams (These are mainly so-called 'tick-by-tick' data, read continously from the stock market servers during trading sessions.)

The algorithm proposed in this paper is planned as a part of the decision support system that is still under development and demands extensive testing. In its preliminary version we have tested among others, the possibility to predict share prizes on the basis of the parameterized temporal concepts. The results are partial and therefore cannot be treated as the final achievement of the proposed algorithm. The main role of the presented partial results is to illustrate the potential of the proposed algorithm for the targeted application area.

The part of one of the discovered fcm is reflected in the table 1. Assume, we consider two simple concepts: $Volume$ and $SharePrice$, the p denotes the time parameter for the concept $Volume(t_0-p)$, where t_0 is the reference time point. In the second column of the table, $w_{(Volume, SharePrice)}$ denotes the learned weight of the causal branches $Volume(t_0 - p) \rightarrow SharePrice(t_0)$.

Table 1. Causal relationship between share price and volume movement

p	$w_{(Volume, SharePrice)}$
1	-0,152438
2	-0,121069
3	0,4430142
4	0,4420352

5   Conclusions

In this paper, we have proposed a new approach to train fuzzy cognitive maps. The process of selection and evaluation of concepts has been bound with the adaptation stage of learning causal relationships among the concepts. The same evaluation (fitness) function for evolutionary and adaptive learning has been assumed with the intention of achieving the synergetic effect of the knowledge optimization. We have also proposed to apply the presented solution to the automation of learning FCMs on the basis of the large amount of heterogeneous source data available from the Web.

References

1. A. Abraham, N. Philip, and P. Saratchandran. Modeling chaotic behavior of stock indices using intelligent paradigms, 2003.
2. Robert Axelrod. *Structure of Decision–The Cognitive Maps of Political Elites.* Princeton University Press, 1976.

3. J. Carvalho and J. Tom. Rule based fuzzy cognitive maps – qualitative systems dynamics.
4. Alberto Vazquez Huerga. A balanced differential learning algorithm in fuzzy cognitive maps. *In Proceedings of the 16th International Workshop on Qualitative Reasoning*, 2002.
5. Bart Kosko. Fuzzy cognitive maps. *International Journal of Man-Machine Studies*, 24(1):65–75, 1986.
6. Dimitris E. Koulouriotis, Ioannis E. Diakoulakis, Dimitris M. Emiris, and Constantin D. Zopounidis. Development of dynamic cognitive networks as complex systems approximators: validation in financial time series. *Appl. Soft Comput*, 5(2):157–179, 2005.
7. Elpiniki Papageorgiou and Peter P. Groumpos. A new hybrid method using evolutionary algorithms to train fuzzy cognitive maps. *Appl. Soft Comput*, 5(4):409–431, 2005.
8. Z. Pawlak. *Rough Sets: Theoretical Aspects of Reasoning about Data.* Kluwer Academic Publishers, Dordrecht, 1991.
9. Andrzej Skowron, Roman W. Swiniarski, and Piotr Synak. Approximation spaces and information granulation. In *Rough Sets and Current Trends in Computing*, pages 116–126, 2004.
10. Wojciech Stach, Lukasz A. Kurgan, Witold Pedrycz, and Marek Reformat. Genetic learning of fuzzy cognitive maps. *Fuzzy Sets and Systems*, 153(3):371–401, 2005.
11. Wojciech Stach, Lukasz A. Kurgan, Witold Pedrycz, and Marek Reformat. Higher-order fuzzy cognitive maps. *In Proceedings of NAFIPS2006 (International Conference of the North American Fuzzy Information Processing Society)*, 2006.
12. Edward C. Tolman. Cognitive maps in rats and men. *The Psychological Review*, 55(4):189–208, July 1948.
13. L. A. Zadeh. Fuzzy sets. *Information and Control*, vol. 8: 338–353, 1965, http://www-bisc.cs.berkeley.edu/Zadeh-1965.pdf.

Agent-Based Adaptive Learning Provisioning in a Virtual Organization[*]

Maria Ganzha[2], Marcin Paprzycki[2], Elvira Popescu[1] , Costin Bădică[1], and Maciej Gawinecki[2]

[1] Software Engineering Department, University of Craiova, Craiova, Romania
{badica_costin,popescu_elvira}@software.ucv.ro
[2] Systems Research Institute, Polish Academy of Science, Warsaw, Poland
{maria.ganzha,paprzyck,gawinec}@ibspan.waw.pl

Summary. In this note we consider design of a learning provisioning subsystem for an agent-based virtual organization. Flexible delivery of learning content is based on matching of ontologically demarcated user profiles, domain specific knowledge and learning modules.

1 Introduction

Let us start from a sample scenario and consider an organization in which teams of researchers are engaged in R&D projects. Let us assume that teams and/or their members are geographically distributed (though this is not necessary as the proposed approach will work in either case). Team work requires collaboration between members and such collaboration should be supported by an appropriate technology.

It is obvious that support of collaborative research has to go beyond, even most sophisticated forms of, document versioning and flow of resources in the hierarchical structure of the organization. What needs to be taken into account is: (1) representation of domain specific knowledge – to provide context for management of resources pertinent to the project (e.g. establishing a specific "place" of a resource within the domain knowledge allows for resource indexing and clustering); (2) representation of structure and flow of interactions in the project – to route resources based on project needs and responsibilities of team members; (3) representation of user profiles (situated within the domain knowledge and the structure of the project) – specifies team member *interests*, *needs* and *skills* (e.g. specifies what to do with new/incoming resources); (4) adaptability of the system – as the time passes domain of interest to the project may expand, contract or shift; functional interrelationships between team members can change; their interests, needs and skills may evolve.

It is relatively easy to see that these four points can be generalized beyond the collaborative work scenario that we started with. Let us assume that we extend the second point by utilizing a notion of a virtual organization and within such an organization defining roles and interactions. Therefore, it is important to keep in mind, that the collaborative work scenario is used only as an example, while the overarching application

---

[*] This work was partially sponsored by the KIST-SRI PAS "Agent Technology for Adaptive Information Provisioning" grant.

K.M. Węgrzyn-Wolska and P.S. Szczepaniak (Eds.): Adv. in Intel. Web, ASC 43, pp. 112–117, 2007.
springerlink.com   © Springer-Verlag Berlin Heidelberg 2007

is adaptive personalized information provisioning in a virtual organization. In this note we focus on the way in which in such an organization we can design an adaptive learning content delivery subsystem.

To this purpose, in the following section we overview our approach to modeling virtual organizations, with focus on need of worker education. The next three sections illustrate the use of ontologies, agent systems and resource matching respectively. Finally, we draw some conclusions, outlining future research directions.

2   Overview of the General Approach

Knowledge management is at the core of our approach and nowadays it is very often claimed that the best technique for knowledge representation is ontological demarcation. In this context, representation and management of knowledge flow can be achieved as a result of a two-step process. First, roles of participants are specified, and second, the real-world organization is represented as a virtual agent-based system. For human resources, agent roles are combined with domain ontologies, while for other resources only domain ontologies are used. In both cases an overlay model allows specification of profiles of individual resources (see [3], [12], [16], [13], [4]).

Let us add that ontological representation of resource profiles naturally supports various forms of automatic reasoning (e.g. resource matching, query rewriting etc.). Furthermore, ontologies, overlay-based profiles and agent systems are naturally adaptable. Ontologies can be easily modified, adaptation of overlay-based resource profiles involves changes in weight of individual features, while changes in virtual organization are easily transformed into changes in agent interactions ([5]).

We can now summarize the fundamental features of our approach to building an environment for supporting context aware personalized resource provisioning:

1. *Domain knowledge* will be represented in terms of ontologies.
2. *Organizational structure* will be decomposed into interacting agents.
3. *Overlay model* will be used to represent resource profiles.
4. *Resource matching* will utilize reasoning involving resource profiles.
5. *System adaptability* will be obtained through: a.adapting structure of the agent system; b. adapting resource profiles.
6. *Human resources adaptability* will also be achieved by learning.

Let us now observe that in such a system we can immediately conceptualize *learning*. Imagine that a given worker has been assigned a task with a given profile. Additionally, the worker currently has a set of skills that are represented within her profile. If reasoning signals a mismatch of the human profile with the task profile then an e-learning task may be triggered. Based on the mismatch, an initial goal of the learning task is formulated and the worker (now learner) is enrolled to an e-learning process. At the end of the learning process, assuming that during learning the learner has also been appropriately evaluated, her profile is updated accordingly.

3   Ontologies

It is a well known fact that relations described in ontologies allow the discovery of knowledge which has not been captured explicitly. For instance, such reasoning could be used for: (1) inferring interest in concepts along relation of domain ontology, e.g. from interest in extreme programming and UML specification, it can be inferred that a given team member is interested in software engineering in general; (2) classification of keywords found in resources with respect to the definition of classes specified in the domain ontology.

To be able to properly support the learning process, we will have to extend the utilization of ontologies in the system.

For non-human resources we will use standard ontologies for educational material (LOM – IEEE Learning Object Metadata [6], IMS MD – IMS Learning Resource Metadata [8]). This will allow us to introduce relations between concepts such as *prerequisites*; e.g. each non-human resource will be linked to a list of concepts that are intended to be mastered after studying that resource content. Second, the user profile will be extended to include such learning-related features as: performance level and learning preferences (this extension could be based on PAPI – IEEE Personal and Private Information [7], or IMS LIP – IMS Learner Information Profile [9]). Finally, relations describing interactions between the human and non-human resources (type of interaction, start and ending time, etc.) will be included.

Note that a lot of the information about learners can be inferred from their interactions with resources (consulting a certain educational material or asking a peer for help on a specific problem increases the likelihood that the learner acquired the corresponding knowledge; a somewhat more accurate indication of the learner knowledge level is the outcome of his interaction with an assessment type resource). All these interactions will lead to adjustments of weight in the overlay represented profile.

4   Agent System

Following [1], we have envisioned the following basic types of agents in the system: i) *Personal Agent* (known also as *Interface Agent*); ii) *Task Agent*; and iii) *Middle-Agent*. The latter two categories belong to *infrastructure agents*.

The most basic agent in the system will be a *Personal Agent* (*PA*) representing each worker in the organization (regardless of her/his position). *PA*'s responsibilities include human user assistance and management of user profile. On the other hand, *TAs* are specialized for dedicated tasks like connecting to a database or provision of resources. Finally, *MAs* are specialized in intermediating between requesters and providers (usually users and/or their *PAs* and appropriate *TAs* acting as resource managers) and their responsibilities include tasks like matchmaking and/or brokering.

A number of additional infrastructure agents will be created, e.g. meeting scheduling, resource management (searching, indexing and clustering), resource profile updating, etc. While the initial set of infrastructure agents will be identified during the requirement specification phase, we may decide later to incorporate additional agents. Here, note that one of important advantages of agent-based software engineering is that adding functions to the system is relatively easy as it involves adding agents and defining their interactions with agents already existing in the system.

Finally, in the context of this note, an important feature of the agent-based approach is that it can be easily integrated with any of existing Learning Management Systems (LMS). LMSs offer support for a wide area of activities specific to e-learning, such as: communication and collaboration tools, registration and authentication tools, security features, curriculum design support (authoring tools), assessment support (on-line testing, automated grading, online gradebook), student tracking and reporting tools, student portfolio, groupwork support, instructional standards compliance [15]. All that is needed is the creation of an agent intermediary that will become an interface between the LMS and the agent-based system ([14]).

5   Resource Matching and System Adaptability

As indicated above, we plan to use an overlay model for defining resource profiles. In the case of non-human resources we will overlay the profile over the domain ontology. In the case of human resources, we will utilize not only the domain ontology, but also the organizational ontology (e.g. the fact that a given personal agent represents a human manager at a given level of system hierarchy and thus has to communicate with specified other agents). Resource profiles will provide context for directing flow of resources in the system. This will be achieved in two ways: (a) on the basis of information directly stored in the ontology of the organization, and (b) through context matchmaking, which should be broadly understood as comparing profile of a given resource to that of another resource and deciding if there is close enough match for these resources to be of interest to each other (e.g. establishing on the basis of their profiles – by comparing them and reasoning over the results – if a given learning module is appropriate to a given member of the organization).

Finally, resource profiles will be used for collaborative information processing, e.g. an agent searching for information on a topic will query other agents, profiles of which indicate that they may be interested in a given subject (and thus store pertinent resources), for useful information.

The implicit feedback method is the most desirable to adapt resource profiles, since (1) it can be applied both to human and non-human resources, and (2) it is more comfortable to the user (who doesn't need to fill questionnaires or mark resources by hand). Typically, the notion of implicit feedback is considered to be highly dependable on the environment in which it is collected. However, our innovative approach can be made environment independent, since we utilize ontology-based resource profiles. As mentioned above, in our approach any interaction between two resources constitutes an implicit feedback and a potential reason for profile adaptation. Therefore, implicit feedback resource profile conceptualization proposed here can be used in a variety of context aware information provisioning environments.

In the context of e-learning (as learning is also an important component of the system), adaptation refers to the creation of an educational experience that is dynamically changing in order to suit each learner's needs, with the purpose of maximizing the subjective learner satisfaction, the learning speed (efficiency) and the assessment results (effectiveness) [2].

There are three factors that must be taken into account when talking about adaptation in e-learning ([15]): (1) the *learner* (which is characterized by his knowledge level, technical background, learning goals, interests, motivation, cultural background, learning styles, personality traits etc.); (2) the *hardware and software platform* (PC/laptop/PDA/mobile phone etc, screen size, available input devices, connection bandwidth, processor performance, memory size, operating system, Web browser etc.); (3) the *environment* (the physical environment where interaction takes place - surrounding light, noise, geographical location and other external elements that may have a influence).

Thus matching between resources can also be based on the above criteria. For example, a non-human resource (e.g. an educational material) will be considered appropriate to a learner if there is a correspondence between the following characteristics of the non-human and human resource profiles, as described in their respective ontologies: i) the prerequisites level and the knowledge level; ii) the intended purpose and the actual learning goal; iii) the most appropriate learning style and the recorded cognitive characteristics; iv) the desirable hardware and software features of the used device and the actual platform available.

The resource matching will be managed by a dedicated infrastructure agent. The matching logic will be flexible, unlike traditional fixed-pedagogy approaches, being able to incorporate various instruction strategies (e.g. matched learning style in case of high activity level learners and mismatched learning style in case of low activity level learners [11]).

Rules for matching human resources are different and depend on the objective. In case of collaborative problem solving (group forming), a "heterogeneity rule" may be applied, since studies have shown that learners work best in mixed-ability groups [10, 17]. Slavin [17] for example recommends a group size of four: one high achiever, two average achievers, and one low achiever. Other learner characteristics can be taken into account, like personality traits (introvert or extrovert), attitude toward team work, motivation, goals, interest for the subject.

In case of offering support as peer help, a "near-peer-matching" rule could be applied [18]: when a learner needs help, she/he will be directed to a peer with a slightly higher knowledge level. This will insure a fair distribution of help demands (avoiding the situation that the highest proficiency learners will be overcome) and also provide learners with the opportunity of explaining to others what they have just understood ("learning by teaching"). The matching will be done by means of negotiations between the personal agents of learners, and will be based on their profiles.

6   Concluding Remarks

In this note we have outlined how a conceptualization of a virtual organization utilizing ontologies and software agents can be used for personalized adaptive delivery of educational content. We have discussed how the overlay model on profile instantiation allows for resource matching and thus helps establishing when a human resource is in need of training and which training module should be utilized. As future work we intend to implement the suggested approach and provide real-world validation.

References

1. Alagar, V. S., Holliday, J., Thiyagarajan, P.V., Zhou, B.: Agent Types and Their Formal Descriptions. Technical Report of SCU Computer Engineering department (2002). http:// www.cse.scu.edu/send.cgi?research/techreports/COEN-22002-09-19A.pdf
2. ALFANET - D8.2 Public final report (2005). http://rtd.softwareag.es/alfanet/.
3. Fink, J., Kobsa, A.: User Modeling for Personalized City Tours, *Artif. Intell. Rev.*, Vol. 18, No. 1, Kluwer Academic Publishers (2002) 33–74.
4. Gawinecki, M., Gordon, M., Paprzycki, M., Vetulani, Z.: Representing Users in a Travel Support System. In: KwaIJnicka, H. et. al. (eds): *Proceedings of the ISDA 2005 Conference*, IEEE Press (2005) 393–398.
5. Gawinecki, M., Gordon, M., Nguyen, N.T., Paprzycki, M., Zygmunt Vetulani, Z.: Ontologically Demarcated Resources in an Agent Based Travel Support System. In: R. K. Katarzyniak (ed.): *Ontologies and Soft Methods in Knowledge Management*, Advanced Knowledge International, Adelaide, Australia (2005) 219-240.
6. IEEE LOM. http://ltsc.ieee.org/wg1/
7. IEEE PAPI. http://edutool.com/papi/.
8. IMS MD. http://www.imsglobal.org/metadata/index.html.
9. IMS LIP. http://www.imsglobal.org/profiles/index.html.
10. Inaba, A., Supnithi, T., Ikeda, M., Mizoguchi, R., Toyoda, J.: How Can We Form Effective Collaborative Learning Groups? In: *Proc. of the Int. Conf. on Intelligent Tutoring Systems*, Springer-Verlag (2000) 282Ű-291.
11. Kelly, D., Tangney, B.: Adapting to intelligence profile in an adaptive educational system. *Interacting with computers*, No.18, Elsevier Science (2006) 385Ű-409.
12. Kobsa, A., Koenemann, J., Pohl, W.: Personalised hypermedia presentation techniques for improving online customer relationships. In: *Knowl. Eng. Rev.* 16 (2001) 111Ű-155.
13. Montaner, M., López, B., de la Rosa, J. L.: A Taxonomy of Recommender Agents on the Internet. In: *Artif. Intell. Rev.* 19 (2003) 285-Ű330.
14. Otsuka, J.L., Bernardes, V.S., Rocha, H.V.: A Multiagent System for Formative Assessment Support in Learning Management Systems. In: Anais do I Workshop Tidia, Sao Paulo, Brazil, (2004).
15. Popescu, E., Trigano, P., Bădică, C.: Evaluation of a Learning Management System for Adaptivity Purposes. In: *Proc.ICCGI'2007*, (2007).
16. Rich, E.A.: User modeling via stereotypes. In: *Cognitive Science* 3 (1979) 329Ű-354.
17. Slavin, R.E.: Developmental and Motivational Perspectives on Cooperative Learning: A Reconciliation. *Child Development*, Vol. 58, No. 5., Special Issue on Schools and Development (1987) 1161Ű-1167.
18. Sloep, P., Van Rosmalen, P., Brouns, F., Van Bruggen, J., De Croock, M., Kester, L., De Vries, F.: Agent Support for Online Learning. *BNVKI Newsletter*, August (2004) 90–92.

The Analysis and Visualization of Entries in Wiki Services

Jakub Gawryjołek and Piotr Gawrysiak

Institute of Computer Science, Warsaw University of Technology
Nowowiejska 15/19, 00-665 Warsaw, Poland
{J.Gawryjolek, P.Gawrysiak}@elka.pw.edu.pl

Summary. The use of online collaboration environments has become exceptionally widespread over the past decade. One of the most popular styles of collaboration are the "wiki" web sites. They have attracted attention because of their policy of letting anyone become an editor. This paper presents the technique for the analysis and visualization of Wikipedia - the largest wiki in existence. Specifically, it concentrates on some activity patterns of its contributors. First, a new visualization and analysis tool named JWikiVis is presented. Second, with the use of this software, some interesting user behaviors are described. Finally, text classification algorithms are applied in order to determine some patterns observed in individual wiki pages as well as in the entire service.

Keywords: information visualization, Wiki, collaboration, text categorization, text and web mining.

1 Introduction

A wiki is a type of Web site that gives every user the possibility to contribute to its content, very often without the need for registration. Such an approach makes a wiki an effective tool for collaborative authoring. Vandalism seems as a natural consequence of the open philosophy of this technology. The only outcome that we should expect is mess and vulgarism on the pages. And these things happen very often, still wikis seem to work very well. The important question is who does what and what constitutes the success of wiki technology? Furthermore, exploring multiple visual presentations, or visualizations, often helps a user make sense of a large collection [9].

Although most wikis are open to the public, an organization of roles (readers, editors, reviewers, destroyers, etc.)[1], as defined in Nupedia project, is also visible. Most of the actions like creations, mass and small deletions, corrections, and swapping of some parts of text, when visualized and analyzed, can provide useful information concerning behaviors of groups, for instance the discrepancies between anonymous and registered users. Furthermore, the statistical analysis of articles together with the graphical representation can serve as a tool for finding

K.M. Węgrzyn-Wolska and P.S. Szczepaniak (Eds.): Adv. in Intel. Web, ASC 43, pp. 118–123, 2007.
springerlink.com

some interesting trends within the service or for the prediction of the direction of a wiki evolution. Obviously the most interesting wiki for research is Wikipedia, with its 1 500 000 articles in English version only and multilingual content (see [11]).

The remainder of the paper is organized as follows. Section 2 presents related work. Section 3 explains visualization approach and algorithms used. Section 4 presents analysis of behaviors and trends observed in the system. Finally, section 5 describes possible future development directions and summarizes the paper.

2 Review of the Related Work

The philosophy of most popular wiki system - MediaWiki - is that it should facilitate correction of mistakes, rather than preventing them. MediaWiki's "diff" and "hist" features are such tools that help in restoring the article's content, viewing changes etc. They does not allow, however, to depict whole development history of the article. The research conducted by F. Viegas et al. [4] shows how such broader system - called History Flow - might be constructed, using highly visual means. Our approach, dubbed JWikiVis, has been highly inspired by History Flow, and delivers similar visualization power addressing History Flow deficiencies. Other work in the area is The ThemeRiver visualization[9] which depicts thematic variations over time within a large collection of documents. Visualizing the affective structure of a text document is the subject matter of [5] and graphical representation of interaction in an on-line collaboration environment is described in [2]. Finally, the semantic coverage of Wikipedia and its authors is dealt with in [10].

3 The Visualization System

The main corpus that we used for visualization and analysis was Polish version of Wikipedia, specifically the complete page edit history from November 2006, totaling 520882 pages and 5158509 revisions, resulting in 28GB of data. The visualization software - JWikiVis - has been written mainly in Java, with 3-D components partially implemented in C++.

3.1 JWikiVis - The Visualization Part

True text visualizations should represent textual content and meaning to analysts without them having to read it in the manner that text normally requires[6]. Main factors that determined the type of the visualization was the structure of text in Wikipedia articles. A text is a string of characters which is modified at some intervals. Such an approach limits the illustration possibilities to two dimensions with text-corresponding structures on one axis, and the flow of time on the second. In this fashion works the main part of JWikiVis visualization.

(a) 2-D visualization of the article about 'Jolanda Di Savoia'

(b) 3-D Visualization of the article about 'Jolanda Di Savoia'

Fig. 1. JWikiVis 2-D and 3-D visualization

The example in figure 1(a) shows the result for the article about Italian county, Jolanda Di Savoia.

Every single rectangle in the picture presents the part of text - paragraph or single line. In the example, rectangles correspond to the paragraphs. In figure 1(a) there are 11 paragraphs altogether - first row of the table - numbers 0-10. On the left-hand side we can see the name of the contributors and the date of their revisions. Rows represent the article at a certain point of time indicated by the revision time, and columns - parts of text. Columns are the added parts of texts in all revisions - set of all positive entries. Each part is assigned certain column in this set. When a part is placed in certain revision (row), it takes only the position (column) of a part existing in that revision keeping at the same time the correct order of the whole article's text. In last revision in fig. 1(a) the orange rectangle - first part in text - is placed in the second column - first existing part in that version. JWikiVis is not limited only to 2-D visualization. The three-dimensional representation of the above example is shown in figure 1(b). The third, z-axis, is the user axis, on which every user is placed on the separate level, what is helpful for examining certain user's contributions.

4   The Entries and Trend Analysis

Having all the necessary tools (including document multi-classifier presented in [3]) we can make some analysis of typical contributors' behaviors and trends in Wikipedia. The examination of the behaviors revealed some common patterns like anonymous versus named authorship[3], negotiation, content stability, vandalism and repair described in [4]. Additionally, evolution in subject matter of the documents and the activity of edits in some areas were observed. Here we can present only a small example of the analysis prepared in [3].

4.1   Parts Durability

There are two factors that determine the appearance of JWikiVis's charts. First, the number of parts of individual users. Second, and more important, is the time of existence of these parts. If a user added fewer parts which existed for a long time, their contribution to the chart may be more significant than those who added more, but short-living ones. The full description and visualization from the user perspective is presented in [3].

In order to better understand the durability issue we have to focus on the article as a whole, not only on the individual contributors. It would seem natural that pages would tend to stabilize over time, but pages change in size and turnover in text [4]. Figure 2(a) presents the article about 'Optical Disk'. Notice that the parsing delimiter was a single new line character. Figure 2(b), on the other hand, presents the same article but with a double new line character - a paragraph - as a delimiter.

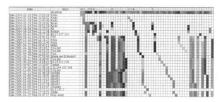
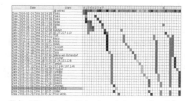

(a) 'Optical Disc' with a single new line character as a delimiter

(b) 'Optical Disc' with a double new line character as a delimiter

Fig. 2. Visualization of the same article depending on the delimiter

In fig. 2(a) noticeable is the fact that there are many long columns what indicates that some parts existed through many subsequent revisions. Moreover, initial text of a page usually exists longer and suffers fewer changes than the parts added later. Another indication of that hypothesis can be the "stairs" in the middle part of the picture. Such "stairs" inform that corresponding texts did not last for more than one or two revisions. As the consequence new parts appeared and disappeared very often. The second figure, tells another story. With the exception to three or four parts, none of the paragraphs survived more than 5 revisions. It is because a change in a sentence causes a paragraph to be marked as deleted. Hence, we can conclude that most of the changes were rather small edits within individual lines rather than large modifications of the entire article. Moreover, the graphical representation of Wikipedia articles is very often an unstructured formation. It is because people tend to delete and insert new parts rather than move already existing text. One explanation given by F.Viegeas et al. [4] is that Wikipedia editing window is small what makes it difficult to see whole article at once. Other explanation can be that users usually agree on the order of paragraphs and sentences in the documents, but they think they should be formulated in a different way. There is another issue concerning the durability of parts in the article, namely how long they have existed in time rather than in how many revisions. In other words, we want to see the revisions scaled by date[4], figure 3(a).

Some of the parts which existed in many revisions also survived significant amount of time, but there are also some that are almost invisible because of

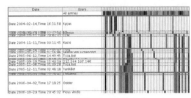

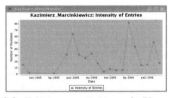

(a) 'Optical Disc' scaled by date

(b) Editing activity of Kazimierz Marcinkiewicz page

Fig. 3. Different visualization of frequency patterns

their short time of existence. Specifically, places where users seemed to disagree on the topic are now hidden. This suggests that whenever there is something controversial or users have different opinions on the matter, revisions occur more often. Figure 3(a) is also very helpful in scrutinizing frequency patterns which is the subject matter of the following subsection.

4.2   Frequency Patterns

Users become significantly active in editing certain page in the time of year which somehow corresponds to the content. This behavior is 'marked' by gray horizontal strips in the scaled version of 'Optical Disk' and peaks in figure 3(b) about Polish politician, Kazimierz Marcinkiewicz.By scrutinizing the charts we can find some important facts concerning people's lives, about whom the articles are. Secondly, the anniversaries of important history events is indicated by the increased activity in the corresponding history articles. Finally, high frequency of contributions during the whole period may indicate that the topic is highly controversial. The issue about controversial articles as well as vandalism, disagreement and negotiation patterns are examined in [3].

5   Future Work and Conclusions

One of the most important aspects for the daily use of Wikipedia is the strong interconnection of its articles through the links[7]. Two pages can be treated as neighbors if their links direct to each other or if they link to the third, different page. If two pages point to many same sites, their similarity may increase. In this fashion we could find another measure of similarity of pages and create theme maps [8]. It may be revealing, also, to detect pages having similar visualization structure. We hope that the discrete character of JWikiVis visualization, with some improvements and corrections, may turn out to be a sufficient tool for such analysis.

The evolution of a topic in Wikipedia is a complex and long process during which many patterns occur. Contributors act in different, positive and negative roles. Negotiations, disagreements, acts of vandalism are the examples of possible behaviors taking place in wiki communities. JWikiVis is a visualization software that helps to understand how collaborative documents are created and how they evolve over time. MediaWiki engine together with Wikipedia's Talk pages and forums provide tools to heal undesirable effects. However, in order to understand and possibly prevent these activities as well as to have a detailed insight into many positive patterns a visualization information is needed. We hope that JWikiVis and its future development can satisfy some of these needs.

References

1. L. Aronsson. Operation of a large scale, general purpose wiki website: Experience from susning.nu's first nine months in service. In *Verlag für Wissenschaft und Forschung*, 2002.
2. R. P. Biuk-Aghai. Visualization of interactions in an online collaboration environment. In *Collaborative Technologies and Systems, 2005. Proc. of the 2005 International Symposium*, 2005.
3. J. Gawryjołek. The analysis and visualization of entries in wiki services, 2007. Institute of Computer Science, Warsaw University of Technology, BSc. Thesis.
4. K. Dave F. Viegas, M. Wattenberg. Studying cooperation and conflict between authors with history flow visualizations. In *Proc. of SIGCHI*, 2004.
5. H. Lieberman H. Liu, T. Selker. Visualizing the affective structure of a text document. In *Conference on Human Factors in Computing Systems*, 2003.
6. K. Pennock D. Lantrip M. Pottier A. Schur V. Crow J.A. Wise, J.J. Thomas. Visualizing the non-visual: spatial analysis and interaction with information from text documents. In *Proc. on Information Visualization*, 1995.
7. M. Völkel M. Krötzsch, D.Vrandecic. Wikipedia and the semantic web - the missing links. In *Proc. of Wikimania 2005 - The First International Wikimedia Conference*. *Wikimedia Foundation*.
8. M. Brewster H. Foote N. E. Miller, P. C. Wong. Topic islands - a wavelet-based text visualization system. In *IEEE Visualization, Proc. of the Conference on Visualization*, 98.
9. P. Whitney L. Nowell S. Havre, E. Hetzler. Themeriver: visualizing thematic changes in large document collections. *Visualization and Computer Graphics, IEEE Transactions*, Jan/Mar 2002.
10. K. Börner T. Holloway, M. Božičević. Analyzing and visualizing the semantic coverage of wikipedia and its authors. In *Comlexity, Special issue on Understanding Complex Systems*.
11. C. S. Ang U. Pfeil, P. Zaphiris. Cultural differences in collaborative authoring of wikipedia. *Journal of Computer-Mediated Communication,12(1),art. 5*, 2006.

Relational Model Based Annotation of the Web Data

Fatih Gelgi[1], Srinivas Vadrevu[2], and Hasan Davulcu[3]

[1] Arizona State University, Tempe, AZ
    fagelgi@asu.edu
[2] Arizona State University, Tempe, AZ
    svadrevu@asu.edu
[3] Arizona State University, Tempe, AZ
    hdavulcu@asu.edu

Summary. In this paper, we present a fast and scalable Bayesian model for improving *weakly annotated data* – which is typically generated by a (semi) automated information extraction (IE) system from Web documents. Weakly annotated data suffers from incorrect ontological role assignments. Our experimental evaluations with the TAP and a collection of 20,000 home pages from university, shopping and sports Web sites, indicate that the model described here can improve the accuracy of role assignments from 40% to 85% for template driven sites, from 68% to 87% for non-template driven sites.

Keywords: Weakly annotated data, information extraction, classification, Bayesian models.

1 Introduction

In this paper, our focus will be on improving *weakly annotated data* which is typically generated by a (semi) automated information extraction (IE) system such as [8, 5, 4, 10] from the Web documents. In weakly annotated data, *annotations* correspond to ontological role assignments such as *Concept*, *Attribute*, *Value* or *Noise*. Weakly annotated data has two major problems; (i) might contain incorrect role assignments, and (ii) have many missing attribute labels between its various entities.

We will use the Web pages in Figure 1 to illustrate weakly annotated data that might be extracted using an IE algorithm such as [4, 10]. Each of these pages presents a single instance of the 'Digital Camera' concept. In Figure 1(a), attributes such as 'storage media', and values such as 'sd memory card' have uniform and distinct presentation. However, for an automated system it would be extremely difficult to differentiate the 'storage media type' label as an attribute and 'sd secure digital' as its value due to their uniform presentation in Figure 1(b). On the other hand, in Figure 1(c) the attribute 'storage media' does not even exist, but only its value 'sd memory card' has been reported.

For weakly annotated data, we propose a domain specific probabilistic model that utilizes contextual regularities of labels to improve its overall accuracy by

K.M. Węgrzyn-Wolska and P.S. Szczepaniak (Eds.): Adv. in Intel. Web, ASC 43, pp. 124–129, 2007.
springerlink.com

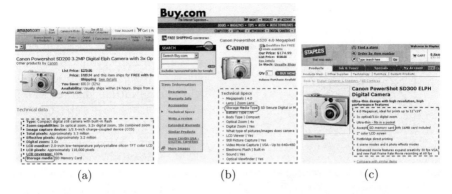

Fig. 1. The instance pages for Canon digital camera from three different Web sites. Digital camera specifications are marked by dashed boxes in each page. In page (a), attributes are explicitly given as bold whereas in (b) they are not obvious. On the other hand, the attributes are altogether missing in (c).

automatically **correcting their role assignments**. We formulate the role assignment problem as a classification problem. We use a Bayesian model due to the robustness of Bayesian models on classification tasks [9] with large number of features. Since discovery of the Bayesian network dependencies on data is a hard problem [3], we stick to the independence assumption which also ensures the scalability of the proposed model for the Web data.

The distinguishing feature of our model from the standard Bayesian models is the preservation of the hierarchical structure of the *relational graph* (see Section 2 for details) by incorporating the edge probabilities. Furthermore, the number of features (i.e. tagged labels) are not fixed, as required by some of the conventional classification methods. Due to the space limitations, the model and the experiments are briefly discussed in this paper; interested reader may refer to the technical report [7].

Probabilistic Relational Models (PRMs) [6] are powerful methods to learn the underlying structure of relational data. However, they assume strongly annotated data, and their scalability is a problem which makes it inappropriate for the Web data.

2 Probabilistic Model

From the automatically extracted data we generate a *relational graph* of the domain where nodes correspond to the labels with assigned roles, and the edges correspond to association strengths between nodes. Such a graph would capture the global occurrence statistics of the labels and their associations within a domain.

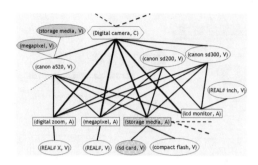

Fig. 2. A portion of the relational graph for the Shopping domain is shown. Each node is composed of a ⟨*label, role*⟩ pair. The thickness of an edge is proportional to the *association strength* between its nodes.

We will briefly explain how the system operates on the example given in Figure 1. A fragment of the corresponding relational graph is depicted in Figure 2. Consider the label 'storage media' marked in Figure 1. A collection of Web pages such as those in Figure 1(a) will yield a strong association between 'canon sd200' as an object and 'storage media' as an attribute. Whereas the incorrect annotation extracted from Figure 1(b) will yield a weak association between 'canon a520' as an object and the 'storage media' as a value. Hence, the Bayesian classifier presented here would be able to re-assign the attribute role to the 'storage media' label by using the statistics in the relational graph within its context.

The notation used for formalization is given as follows:

- The set of all labels in the domain is denoted as $\mathcal{L}$.
- The **ontological roles** $\mathcal{R}$ is the set of *Concept, Attribute, Values* or *Noise*. Formally, $\mathcal{R} = \{C, A, V, N\}$.
- A **term** is a pair $\langle l, r \rangle$ composed of a label $l \in \mathcal{L}$ and the role $r \in \mathcal{R}$.
- We denote a **Web page** as the set of its terms. Formally, assuming m labels in the Web page $\mathcal{W}$; $\mathcal{W} = \{\langle l_1, r_1 \rangle, \langle l_2, r_2 \rangle, \ldots, \langle l_m, r_m \rangle\}$.
- The **relational graph** $\mathcal{G}$ is a weighted undirected graph where the nodes are the terms in the domain, and the weights on the edges represent the *association strength* between the terms.

The node weights of $\mathcal{G}$ are initialized to be the counts of the corresponding terms in the domain corpus, and the edge weights are initialized with the counts of the corresponding edges between its terms.

2.1 Label Role Inference

The problem of role assignment for each label can now be formally defined as follows;

Definition 1. *Given a Web page $\mathcal{W}$, the probability of a term $\langle l, r \rangle$ where $l \in \mathcal{L}$ and $r \in \mathcal{R}$ is $P(\langle l, r \rangle | \mathcal{W})$.*

And, the role with the maximum probability will be the role assignment for the particular label l that is, $\arg\max_r P(\langle l, r \rangle | \mathcal{W})$. For simplicity we use the *naive assumption*, which states that,

Assumption 1. *All the terms in $\mathcal{G}$ are independent from each other but the given term $\langle l, r \rangle$.*

Furthermore, we only utilize the first order relationships of a term in its context, i.e, neighbors of the term in $\mathcal{G}$. During the role assignment probability calculation of a term, since we would like to utilize only the label's context we also assume,

Assumption 2. *The prior probabilities of all the roles of a label l are uniform.*

Note that, the priors of the roles of the labels other than l in the Web page are their support values as determined by their frequencies. To motivate the idea, consider the label 'Instructor'. 'Instructor' rarely occurs as a concept in the *Courses* domain. Some of its attributes can be listed as 'Phone', 'Fax', 'E-mail'. However, since 'Instructor' usually appears as an attribute of a course in other documents. Thus, the prior probability, i.e., $P(\langle Instructor, A \rangle) >> P(\langle Instructor, C \rangle)$, might strongly bias the the role assignment towards its more common role. This would yield an incorrect tagging for the label as an attribute.

Now, with the above assumptions, we can state the following theorem.

Theorem 1. *Let* $\mathcal{W} = \{t_1, t_2, \ldots, t_m\}$. *Then,*

$$\arg\max_r P(\langle l, r \rangle | \mathcal{W}) = \arg\max_r \prod_{i=1}^{m} P(\langle l, r \rangle | t_i). \tag{1}$$

Proof. Omitted due to the space limitation. For details, see [7].

As shown in Figure 2, a conditional probability such as $P(t|t_i)$ depends on the association strength between the terms t and t_i in the relational graph $\mathcal{G}$. That is, $P(t|t_i) = \frac{P(t,t_i)}{P(t_i)} = \frac{w_{tt_i}}{w_{t_i}}$ by Bayes's rule where w_{tt_i} is the weight of the edge (t, t_i) and w_{t_i} is the weight of the node t_i. Our probability model is based on the methodology of association rules [1]. Hence, the initialization for the above conditional probabilities is defined analogous to $P(t|t_i) \equiv Confidence(t_i \to t)$ [2]. This formulation is consistent with Assumption 2 since it is independent from the prior, $P(t)$.

3   Experiments

The two data sets used in the experiments are *TAP* and *Homepage* data sets.

3.1   TAP Dataset

Stanford *TAP Knowledge Base 2* [8] is a well-known template driven data set. We selected categories including *AirportCodes, CIA, FasMilitary, GreatBuildings, IMDB, MissileThreat, RCDB, TowerRecords* and *WHO*. These categories alone comprise 9, 068 individual Web pages.

To test our probabilistic method, first we converted the RDF files in TAP into triples. Then we applied {5%, 20%, 40%, 60%} distortions to obtain four test cases of weakly annotated data. For each tagged label in the data set, one of the two distortion types, *deletion* or *role change*, was chosen randomly during the distortion process.

For each Web site, a Bayesian classifier was trained (since domains are different). The evaluations of TAP data set are conducted in an automated way assuming that the original TAP data set has the correct annotations. As a baseline method, we used a

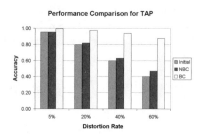

Fig. 3. Performances of the Bayesian (BC) and the naive Bayes (NBC) classifiers for label accuracies

naive Bayes classifier. Figure 3 depicts the average final accuracies of Web sites for different test cases.

Overall results show that there is a large gap between the Bayesian and the naive Bayes classifiers. Bayesian model can improve the accuracy from 40% (60% distortion) to 85% whereas naive Bayes only slightly improved the initial accuracies (less than 10%). The individual accuracies of different Web sites don't vary significantly. We observed that the performance is usually better with the Web sites containing large number of Web pages due to the high consistency and regularity among the Web pages.

3.2   Homepage Dataset

We prepared a data set which is composed of *faculty, course* home pages, *shopping* and *sports* Web pages consisting of 225 Web sites and more than 20,000 individual pages.

In this experiment, we used the semantic partitioner [10] to obtain initial annotations of the labels from Web pages. The semantic partitioner system transforms a given Web page into an XML-like structure by separating and extracting its meta-data and associated data. For the evaluations, we created a smaller data set containing randomly se-

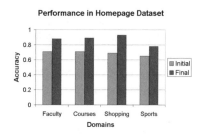

Fig. 4. Performance of the Bayesian classifier in homepage data set

lected 160 Web pages from each domain, and they are manually evaluated by 16 non-project member computer science graduate students.

The average accuracies of initial bootstrapping by the semantic partitioner and of the corrected data by the Bayesian classifier is presented in Figure 4. The recovery rate is highest (24%) in shopping domain due to the more common usage of the meta-data. Conversely, the overall accuracy of the sports domain is the lowest among the four since the variation of the jargon of the domain is much more than the others.

4 Conclusion

In this paper, we proposed a fast and scalable probabilistic model to improve the Web data annotations that are generated through (semi) automated IE systems. Our method can be distinguished by its capability of reasoning with contextual information. Although the initial data contains many incorrect annotations, the Bayesian model presented here was shown to substantially improve the Web data annotations for both template driven and non-template driven Web site collections.

References

1. R. Agrawal, T. Imielinski, and A. N. Swami. Mining association rules between sets of items in large databases. In *ACM SIGMOD*, pages 207–216, Washington, D.C., 1993.
2. E. Alpaydin. *Introduction to Machine Learning*, chapter 3, pages 39–59. MIT Press, 2004.
3. D. M. Chickering. Learning bayesian networks is NP-complete. *Learning from Data: Artificial Intelligence and Statistics V*, 1996.
4. V. Crescenzi and G. Mecca. Automatic information extraction from large web sites. *Journal of ACM*, 51(5):731–779, 2004.
5. S. Dill, N. Eiron, D. Gibson, D. Gruhl, R. Guha, A. Jhingran, T. Kanungo, K. S. McCurley, S. Rajagopalan, A. Tomkins, J. A. Tomlin, and J. Y. Zien. A case for automated large-scale semantic annotation. *Journal of Web Semantics*, 1(1):115–132, 2003.
6. N. Friedman, L. Getoor, D. Koller, and A. Pfeffer. Learning probabilistic relational models. In *IJCAI*, pages 1300–1309, 1999.
7. F. Gelgi, S. Vadrevu, and H. Davulcu. Automatic extraction of relational models from the web data. Technical Report ASU-CSE-TR-06-009, Arizona State University, April 2006.
8. R. Guha and R. McCool. TAP: A semantic web toolkit. *Semantic Web Journal*, 2003.
9. K. Murphy. A brief intro. to graphical models and bayesian networks, 1998.
10. S. Vadrevu, F. Gelgi, and H. Davulcu. Semantic partitioning of web pages. In *WISE*, pages 107–118, New York, NY, USA, 2005.

Automated Ontology Learning and Validation Using Hypothesis Testing

Michael Granitzer[1], Arno Scharl[1,3], Albert Weichselbraun[2],
Thomas Neidhart[3], Andreas Juffinger[3], and Gerhard Wohlgenannt[2]

[1] Know-Center Graz, Austria
    mgrani@know-center.at
[2] Vienna University of Economics and Business Administration, Austria
    albert.weichselbraun@wu-wien.ac.at, wohlg@ai.wu-wien.ac.at
[3] Graz University of Technology, Austria
    {scharl,tneidhart,ajuffinger}@tugraz.at

Summary. Semantic Web technologies in general and ontology-based approaches in particular are considered the foundation for the next generation of information services. While ontologies enable software agents to exchange knowledge and information in a standardized, intelligent manner, describing todayŠs vast amount of information in terms of ontological knowledge remains a challenge.

In this paper we describe the research project AVALON - Acquisition and VALidation of ONtologies, which aims at reducing the knowledge acquisition bottleneck by using methods from ontology learning in the context of a cybernetic control system. We will present techniques allowing us to automatically extract knowledge from textual data and formulating hypothesis based upon the extracted knowledge. Based on real world indicators, like for example business numbers, hypotheses are validated and the result is fed back into the system, thereby closing the cybernetic control system's feedback loop. While AVALON is currently under development, we will present intermediate results and the basic idea behind the system.

1 Introduction

Real-world applications that provide shared meaning understandable for machines and humans alike require semantic technologies, whose increasing importance is reflected by the Gartner Group's prediction that lightweight ontologies will be part of 75% of application integration projects. Tim Berners-Lee's vision of the Semantic Web [1] goes beyond lightweight ontologies and requires a network of trust, in which technology is capable of distinguishing reliable knowledge from collections of trivial data. Implementing this vision, however, involves the transformation of massive amounts of existing information, as well as the validation of the extracted semantics reliability and validity with regards to the real world they describe. Semantic services not only encompass the World Wide Web, but also smaller corporate intranets and their integration into a global scheme. To ensure trust in the extracted knowledge, semantic technologies also need to cope with the highly dynamic contextual change of information.

K.M. Węgrzyn-Wolska and P.S. Szczepaniak (Eds.): Adv. in Intel. Web, ASC 43, pp. 130–135, 2007.
springerlink.com   © Springer-Verlag Berlin Heidelberg 2007

Addressing these challenges, the research project AVALON develops a new generation of adaptive knowledge acquisition and management services that test semantic hypotheses by (i) automatically extracting knowledge from heterogeneous, unstructured information sources, (ii) discovering semantic associations within the automatically extracted knowledge base, and (iii) validating the extracted knowledge on observable real-world indicators. The ontology-supported testing of semantic hypotheses will increase the credibility and usefulness of the continuously evolving knowledge base. AVALON's evaluation and quality assurance processes utilise (semi-)automatic feedback loops to align extracted knowledge with external indicators and the expertise of individuals. Such feedback loops are particularly useful in highly dynamic environments like the World Wide Web. Therefore AVALON's methodology is based on a cybernetic control system [6]: A system with an internal knowledge base monitors real-world indicators and uses the knowledge base to recommend a particular action. If the decision-maker accepts the systemŠs recommendation, his or her action affects the real world. By measuring changes in the real-world indicators resulting from taken actions, the knowledge base can be updated and refined automatically.

AVALON implements such an adaptive process for Web-based resources, automatically extracting semantics from unstructured and structured information sources, and validating the extracted knowledge on real-world indicators.

2   Conceptual Approach

AVALON starts by extracting and populating ontologies[1] from textual resources using state of the art techniques (see Section 3). Afterwards, hypothesis can be formulated upon the extracted ontology and tested statistically against the observable real-world indicators that describe out-comes of the decision making process. Based on this testing the knowledge base and subsequently the ontology can be validated and refined (see also Figure 1 for an overview.).

The formulation of ontology-based hypotheses deserves particular attention. The granularity of the hypothesis directly relates to the granularity of the ontology. In general, we distinguish two types of hypotheses:

- Testing relationships stored in the ontology verifies whether the internal representation is accurate; the results strengthen or weaken dependencies among ontological concepts. This type of test can be performed automatically by choosing various hypotheses and comparing them to measured indicators. Rather than following a random or brute force approach, AVALON is based on heuristics for choosing the most relevant hypotheses for advancing the knowledge base.

---

[1] We distinguish here between a knowledge base and an ontology as stated in [9]. An ontology defines the domain schema, while the knowledge base contains instances of schema concepts.

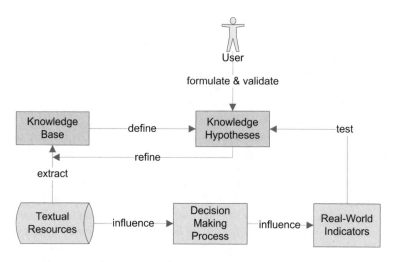

Fig. 1. Conceptual Model of AVALON

- Testing of new relationships reveals new knowledge and refines the internal representation. If users analyzing the knowledge base find irregularities or patterns of interest, hypotheses can be formulated interactively to test those patterns. Through this process, AVALON does not only support knowledge discovery, but also the immediate validation of this knowledge based on a dynamic data set.

As an example, consider AVALON is crawling RSS feeds and blog entries of recent developments and trends in IT industries. After pre-processing the textual resources, relevant concepts (e.g. Persons, Topics etc.) and relations are extracted, together with instances of companies populated in the knowledge base. Taking into account stock quotes as real world indicators, AVALON is capable of combining concepts associated with companies and their loss/gain in stock quotes. This allows validating hypotheses of the form "*Innovative products represent success factors of Web 2.0 companies*". Extending the extraction and population methods to recognize persons and job titles (which can be easily done by means of information extraction) allows us, for example, to include Chief Executive Officers (CEO's) of companies, and to validate hypotheses that postulate that CEOŠs significantly influence corporate success.

More formally, we have an ontology O in the form of a graph, a set of instances I assigned to parts of the ontology and real world indicators R as properties of parts of the instances. A hypothesis H is formulated in terms of concepts and relations of the ontology O and therefore can be seen as sub graph of the ontology $H \subset O$. To support user-defined hypothesis (see above) additional relations may be defined by the user of the system, to extend the hypothesis. Evaluating a hypothesis is done by selecting a hypothesis H and by splitting the instances into a set of instances $I_H \subset I$ satisfying the hypothesis and a set of instances hypothesis $I_{\overline{H}} \subset I$ not satisfying the hypothesis. Applying statistical significance tests on the indicators R estimate the correctness of a hypothesis.

In the above scenario for example, companies, properties describing a company and employees are defining the ontology. Thus, the knowledge base consists of instances of people who work for a company with specific properties. The occurrence function now counts, how often particular relationships or further on patterns exist. Statistically testing the assigned indicators of companies w.r.t to properties of the company and/or people working in the company, leads to a validation of the pattern and therefore of the hypothesis.

Iterating these steps allows creating an accurate and trustworthy knowledge base with a minimal amount of user interaction. Users will have access to knowledge facts and analysts will have the opportunity to formulate hypotheses and test them automatically. Thus, AVALON is capable of dynamically mapping external processes into an internal representation.

3   Technical Approach and Algorithms

From an algorithmic point of view, AVALON rests upon three pillars:

- *Determining the domain structure via ontology learning from text*: Unstructured information sources such as Web sites, Wikis, RSS Feeds and Blogs have to be analysed and converted into an ontology. Similar systems like for example the weblyzard [8] KIM platform [10] or Text2Onto [2] demonstrate the feasability to enrich unstructured information. In AVALON, the user defines relevant domain concepts via a seed ontology. AVALON extends this ontology in multiple iterations (c.f. [8]).
- *Populating the Knowledge Base via Information Extraction*: AVALON uses methods from information extraction for populating the knowledge base and attaching indicators to knowledge base instances. Our approach here is similar to [3, 7], where gazetteers are defined via instances in the knowledge base and rules consider ontological relationships (e.g. a CEO IS-A Person). Thus, we can bootstrap information extraction easily and consider domain relations during the extraction process. Finally, results are fed back implicitly to improve the ontology's accuracy and enlarge the knowledge base.
- *Selecting hypotheses via graph mining*: Hypothesis selection is a critical task in the context of the systems accuracy and performance. Selecting all possible hypothesis is computational intractable, therefore requiring intelligent selection mechanisms. Finding groups of instances sharing similar structures in the ontology can be seen as clustering of graph structures [5] and will allow us to reduce the number of necessary hypothesis tests. For example, assume that we have populated our ontology with companies and attached indicators to them. By clustering those companies based on their relationship to other concepts (e.g. topics, persons etc.), groups of similar companies can be defined. Those groups may serve as starting point for testing hypothesises based on for example comparing stock quote indicators of one group to all other groups of similar companies.

From a technical point of view, computational power and storage capabilities are critical for the success of AVALON. Large scale computing is necessary in order to enable ontology learning on very large document sets, and to allow successive hypothesis testing. Fortunately, current developments in the Open Source community allow us to use distributed approaches originally pioneered by Google. The Hadoop project[2] provides a Open Source computing platform based on Google's Map&Reduce algorithm[4]; a functional programming model allowing large scale distributed computing on several hundreds of machines. Especially pre-processing and natural language processing of documents can be greatly enhanced using this platform.

In the current stage of our project, we are focusing on ontology learning and population using distributed computing based on Map&Reduce. Machine learning algorithms for learning the type of relations are in development and provide promising first results. In the next phase, AVALON will mine relevant sub-graphs to assist in formulating meaningful hypotheses. Finally, integrating real world indicators and user testing will reveal the capabilities of our conceptual model.

4   Conclusion and Outlook

Acquiring and validating knowledge is essential in any knowledge management system. In this paper we have presented the AVALON approach to acquiring and validating knowledge in a dynamic, iterative fashion. Combining textual resources with real-world indicators will help to continually improve the acquired knowledge. As defined in the theory for cybernetic control systems, combining very heterogeneous sources of information which have a relationship in the real world can lead to a accurate internal representation of the world. Based on this feedback process, changes in the real world can be reflected effectively in the internal representation. We consider the conceptual model of AVALON as a step towards flexible and dynamic knowledge-based systems that avoid the cumbersome process of manually scanning large document sets. This will enhance the user's access to existing archives, and facilitate the deduction of new knowledge.

Acknowledgement

The project results have been developed in the AVALON (Acquisition and Validation of Ontologies) project. AVALON is financed by the Austrian Research Promotion Agency (http://www.ffg.at) within the strategic objective FIT-IT under the project contract number 810803 (http://kmi.tugraz.at/avalon). The Know-Center is funded by the Austrian Competence Center program Kplus under the auspices of the Austrian Ministry of Transport, Innovation and Technology (www.ffg.at), and by the State of Styria.

---

[2] http://lucene.apache.org/hadoop/

References

1. T. Berners-Lee, J. Hendler, and O. Lassila. The semantic web. *Scientific American*, 284(5):34–43, 2001.
2. P. Cimiano and J. Völker. Text2onto - a framework for ontology learning and data-driven change discovery. In A. Montoyo, R. Munoz, and E. Metais, editors, *Proceedings of the 10th International Conference on Applications of Natural Language to Information Systems (NLDB)*, volume 3513 of *Lecture Notes in Computer Science*, pages 227–238, Alicante, Spain, JUN 2005. Springer.
3. F. Ciravegna, A. Dingli, Y. Wilks, and D. Petrelli. Adaptive information extraction for document annotation in amilcare. In *SIGIR '02: Proceedings of the 25th annual international ACM SIGIR conference on Research and development in information retrieval*, pages 451–451, New York, NY, USA, 2002. ACM Press.
4. J. Dean and S. Ghemawat. Mapreduce: Simplified data processing on large clusters. November 2004.
5. L. Getoor and C. P. Diehl. Link mining: a survey. *SIGKDD Explor. Newsl.*, 7(2):3–12, 2005.
6. F. Heylighen and C. Joslyn. Cybernetics and second-order cybernetics. In R. A. Meyers, editor, *Encyclopedia of Physical Science & Technology*. Academic Press, 3 edition, 2001.
7. J. Iria, F. Ciravegna, P. Cimiano, A. Lavelli, E. Motta, L. Gilardoni, and E. Mönch. Integrating information extraction, ontology learning and semantic browsing into organizational knowledge processes. In *Proceedings of the EKAW Workshop on the Application of Language and Semantic Technologies to support Knowledge Management Processes, at the 14th International Conference on Knowledge Engineering and Knowledge Management*, OCT 2004.
8. W. Liu, A. Weichselbraun, A. Scharl, and E. Chang. Semi-automatic ontology extension using spreading activation. *Journal of Universal Knowledge Management*, 0(1):50–58, 2005.
9. A. Maedche and S. Staab. Ontology learning for the semantic web. *IEEE Intelligent Systems*, 16(2):72–79, 2001.
10. B. Popov, A. Kiryakov, D. Ognyanoff, D. Manov, and A. Kirilov. Kim a semantic platform for information extraction and retrieval. *Nat. Lang. Eng.*, 10(3-4): 375–392, 2004.

Attacking the Web Cancer with the Automatic Understanding Approach

R.S. Grewal, R. Janicki, T. Kakiashvili, K. Kielan, W.W. Koczkodaj, K. Passi, and R. Tadeusiewicz

wkoczkodaj@cs.laurentian.ca

Summary. The new method, based on automatic understanding is proposed for fighting spam in web information exchange (especially email correspondence). The web cancer term is used to reflect the variety and sophistication of web contaminations. The notable oncology achievements in medicine could inspire more research towards finding solutions to what can easily turn into an analogous civilization crisis. Automatic understanding is appropriate for the semantic-level content analysis and is expected to substantially reduce the wasted user time for semi-automatic analysis needed for the massive processing as most filters are too tight or too loose.

Keywords: spam, automatic understanding, Internet, security, privacy, knowledge base.

1 Introduction

There is very little doubt that we are fast approaching or even most likely passed a point in time where our personal computers are busier attacking web cancer (malware, malignant software, is related only to software) than conducting useful computations. We may even wonder, if computers are so good in defending themselves why even bother to do anything about it. Unfortunately, it is not only our computer time which is wasted but our own as well. Each of us needs to spend probably close to one hour per day verifying whether or not our defense actually works. Our wasted time is ever increasing since illegal attacks become more and more sophisticated. Unless something drastic is done soon, we may face a real disaster since massive computer contamination can become a dangerous weapon in the hands of terrorists.

It is generally agreeable in medicine to characterize cancer as some kind of malignancy of potentially unlimited growth that expands by invasion. Individual computers connected to the internet resemble cells of a living organism. Cancer is still considered to be incurable in most cases; however, medicine has already developed treatments and coping mechanisms which we may need to look at to learn from past experience.

As with everything else, the beginning is not easy but if we are able to cut spam by, say by 90%, with a close to 100% certainty, we will save millions of hours of time otherwise wasted. This in turn will allow the computer community to develop more sophisticated techniques. We are fully aware that perpetrators,

K.M. Węgrzyn-Wolska and P.S. Szczepaniak (Eds.): Adv. in Intel. Web, ASC 43, pp. 136–141, 2007.
springerlink.com

like all criminals, can adjust their actions after reading this and subsequent contributions but when we are ahead of them even one inch, we will win eventually.

2   Automatic Understanding Approach

Spam is inexpensive to generate. Millions of email messages can be sent in minutes at a fraction of the cost of regular mail or a phone call. Unlike a phone or regular mail, potential buyers are just a click away from a purchase and it works this way more often than for conventional spam methods. The California legislature found that spam cost United States organizations alone more than $10 billion in 2004 and the amount of spam email messages still grows exponentially. The costs include lost productivity, the additional equipment, software, and manpower needed to process the extra load. Big companies usually have an army of volunteers often at the level of 100,000 allowing them to develop a knowledge base (KB). However, smaller internet service providers (ISPs) face a serious crises. It is not uncommon for them to get 30 to 40MB of random spam traffic per day. It is not a problem for a big company as presented in [2]. However, we fear that it may be easier to fight spam with an army of 100,000 volunteers than for smaller ISPs. We are also afraid that smaller ISPs may share the fate of dinosaurs unless the developed anti-spam knowledge base, described in [2], is released to the public domain. After all, the authors in [2] confirm that this knowledge base was created with the help of volunteers (hence, the public) and their effort would be more rewarded if it were used *pro publico bono*.

The most recent shift of SPAM from text to graphics shouts for such actions. An automatic understanding (AU, coincidentally also stands for gold in the periodic table which we take for a good sign) of images may be needed in this case. The most successful application of AU is for medical images.

Before going any further, let us examine whether spam is really a case of malware. Software is considered malware based on the perceived intent of the creator rather than any particular features. It is the legal issue of the intent to harm rather than any particular characteristic of the tool used for any harmful action. Malware includes all the computer threats such as viruses, worms, trojan horses, spyware, adware, and other malicious and unwanted software. Although spam, in particular its most harmful form, email spam, is not harming a computer, it is originated by a code intended to inflict a harm and as such is a case of malware. Malware can send email spam. It can infect email address books. Once this occurs, spam will be sent to every single person in the address book. By doing it, spam will usually advertise some webpage and attempt to trap a person who browses for an action which she/he may not be willing to do without seeing this web page. The malware disseminator can also profit from it by stealing important information. For example, it can steal contact information, social insurance number, bank account data, credit card numbers, or software licensing data. Malware can also take control of the computer modem for making expensive long distance telephone calls. Once infected, a malware computer is hard to clean and it may require a complete system deletion to restore the system to its normal state of operation.

3   "The Law and Order" Approach

In law, malware is sometimes referred to as a computer contaminant. The origin of the computer contaminant term can be traced to the legal codes of California, West Virginia, and several other U.S. states, and the most important of all, the CAN-SPAM Act of 2003. As a federal law, it supercedes all state laws in the USA and many other countries may use it to model their computer contamination laws. CAN-SPAM Act includes:

Section 2. CONGRESSIONAL FINDINGS AND POLICY, (a) FINDINGS- The Congress finds the following:(1) Electronic mail has become an extremely important and popular means of communication, relied on by millions of Americans on a daily basis for personal and commercial purposes. Its low cost and global reach make it extremely convenient and efficient, and offer unique opportunities for the development and growth of frictionless commerce.

(2) The convenience and efficiency of electronic mail are threatened by the extremely rapid growth in the volume of unsolicited commercial electronic mail. Unsolicited commercial electronic mail is currently estimated to account for over half of all electronic mail traffic, upfrom an estimated 7 percent in 2001, and the volume continues to rise.Most of these messages are fraudulent or deceptive in one or more respects.

The anti spam laws are unenforceable in practice because of the amount of crime involved. Ironically, it would not help much even if we placed a policeman in front of each computer. A single user would not be able to trace all unwanted email messages to their sources.

We propose the use of AU as a long-term solution. In fact, the advancement of AU may contribute even to saving lives since it can and should be used to the automatic medical image understanding, from which it has originated completing the circle. SMTP (Simple Mail Transfer Protocol) has not been designed to verify the validity of the sender's address. Most recently, it has been remedied by a promising technology known as the sender address verification (SAV). SAV software sends an email to senders for verification. Once verified, the sender's address is added to the *"whitelist"* and will not need to be verified any more until the email address is removed from the whitelist. Mail is delivered to everyone on the whitelist without confirmation. In other words, SAV eliminates the guesswork of the possibly contaminated contents by merely asking first-time senders to follow a well established custom of "knocking before entering".

The basic SAV assumption is that spam is not a content problem but the lack of a sender address verification problem. However, SAV is not entirely an ideal solution. It contributes to the increase of email traffic and in some rare case may result in not delivering perfectly acceptable messages. What is more important, in some case urgent and correct messages may not be delivered on time although their notice, due to confirmation, is delivered on time. Although it is highly unlikely that web offenders will respond to the verification requests, assuming that the return email address they provide is actually real, such possibility cannot be totally ruled out in the light of total disinterred or even inability to fight back by the above mentioned law enforcement.

4 Semantic Analysis Is the Way to Go

Current methods, such as filters, are either "too tight" or "too loose" and the use of AU is evident to avoid occurrence of "false positives". to have our legitimate messages delivered at possibly the first attempt. There are promising technologies on the horizon, such as Sender Policy Framework (SPF) and Domain Keys. However, at this time the efficacy of either approach cannot be accurately measured due to lack of widespread industry adoption.

It also means that we should not give up the idea of accepting the messages by analyzing their contents. It needs to be done, however in a somewhat more sophisticated manner. We propose the use of automatic understanding. Automatic understanding is not easy to define although it is intuitively understandable. In fact, the "natural" understanding has no good definition and some philosophers claim that it cannot be defined because "everything depends on something" and it can be even true. Understanding of any kind brings us immediately to psychological processes, cognition, and philosophy. Unfortunately, addressing concepts of knowledge, epistemology, and meaning goes beyond this presentation. However, doing nothing with computer contaminants, until philosophers become glad with their findings, is not an option for us so we will use the intuitive mining of the automatic understanding. An intuitive meaning of the understanding will be assumed as a psychological process related to an abstract or physical object, such as, a computer contaminant, circumstance of criminal intent of unwanted email messages, whereby we are able to analyze and use concepts to deal adequately with that object or situation. Similarly, the automatic understanding is assumed, as the intuition dictates, to be done by a computer program rather than a human computer operator. It is important to note that AU differs from the automatic image recognition (to which it is close in nature and spirit) in at least one important aspect. We do not need to "see", or even imagine what is in the image to take proper actions. In our case, we need to be sure; apparently close to 100% that something is attempting to harm us to initiate protective or even destructive actions against the threat.

AU can help us to determine the meaning of the analyzed data of any format including images. It has roots in cognitive methods related to psychological and neurophysiological processes of understanding the analyzed data. AU complements both Unified Modeling Language (UML) and the Rational Unified Process (RUP).

The automatic understanding of web cancer is based on the linguistic description of the potential harm. For spam, such linguistic descriptions need to be developed for the known images and phases by the image content description language (*ICDL*). Based on a specialized graph-grammar, the *ICDL* structure is designed for the computation of a demerit index. The detection precision depends on the available web cancer knowledge (e.g, CRCs, popular images or text phases). The structure of the designed graph-grammar is enhanced as more experience is gained. It is a difficult issue and can be perceived as a bottleneck of the proposed method. However, it has been demonstrated as working in practice by numerous applications developed for the medical image understanding. Having an accurate knowledge-based graph-grammar, every particular web cancer case can be converted to a demerit index based on its linguistic description. This, in turn, allows us to perform an automatic parsing. The parsing process used in our method of automatic understanding of web cancer is very similar to the process of automatic translation of a computer program source code into the binary code for a direct execution by the CPU. During the parsing process, two streams of information are combined and in some sense collide with each other. The first stream begins with the suspected email contents. It brings all needed details of the potential harm (e.g., unwanted advertisements or obscene images). The second stream of information begins with the web cancer knowledge represented in a form of the specified graph-grammar and the developed parser. This second stream of information brings some demands as to the format which can be observed in the contents when the demerit index of the contents reaches a certain level. The confrontation, between demands taken from the knowledge base and real parameters and features extracted from the input image, is very similar to the interaction between two waves (for example, during the light interference). Some of the facts derived from the knowledge match some features disclosed on the input image. Other features of the image can disclose conflict with some expectations based on the knowledge. It may lead to changing the working hypothesis about the understanding of the contents. This iterative parsing process of the linguistic content description is called a cognitive resonance. The cognitive resonance is based on the knowledge incorporated into the parser structure.

5   Conclusions

It is time to realize the extent of disaster since it costs us our time and plenty of it. It is a known fact that some, we admit less reputable companies producing anti-virus software, inflict or at least contribute to the infliction of contaminants to increase their market by offering free versions of their software which installs malware to detect it. We need to be better organized by creation of repositories of algorithms, methods and even CRC of contaminating images. We should also influence professional associations, such as IEEE, ACM, or The World Wide Web Consortium (W3C) to develop new methodologies (specifications, guidelines, software, good practices, and tools) to fight the web cancer. In fact, W3C has declared itself as a forum for information, commerce, communication, and collective understanding and as such should be encouraged to create task forces and committees to deal with the ever growing web cancer problem.

The challenges in research are:

- further improvement of the reliability of AU detection algorithms in consideration of ever evolving contamination methods,
- better methods for the automatic reasoning and decision making,
- embedding the software into existing web software (email and browser),
- design of a repository for algorithms, heuristics, CRC of the offending graphics, methodologies (such as SAV),
- creation of a knowledge base.

The proposed method (without any modifications) can also be used for fighting one of the biggest threats of our civilization, terrorism. Only a different knowledge base needs to be developed.

A successful obtaining of the crucial semantic content of the unwanted web contaminant imbedded into email message may contribute considerably to the initiation of a protective or destructive action. Like as in a science fiction novel, the draft of this paper about spam ended up in spam folder of one of the co-authors. As of now, a constructive skepticism is better than a hurray optimism as there are no clear signs that we, regular computer "bread eaters" are winning a struggle with web criminals.

Acknowledgement. This work has been supported by the Ministry of Science and Higher Education, Republic of Poland, under project number N 519 007 32/0978; decision no. 0978/T02/2007/32.

References

1. Tadeusiewicz, R. (2004) Automatic Understanding of Signals. Intelligent Information Systems: 577-590
2. Goodman, J., Cormack, G. V., Heckerman, D. (2007) Spam and the Ongoing Battle for the Inbox, Commun. ACM 50(2): 24-33

A Trustworthy Resource Selection Approach in Grid with Fuzzy Reputation Aggregation

Chunmei Gui, Quanyuan Wu, and Huaimin Wang

School of Computer Science Institute National University of Defense Technology
410073, Changsha, China
plantsperfume@yahoo.com.cn

Summary. An important challenge regarding resource selection in Grid is how to cope with the wide range of selection and the high degree of strangeness. In this paper, reputation evaluation mechanism, as a fundamental building block technique, is introduced to leverage the guarantee of trustworthiness and reliability. Guided by the reputation evaluation ideas and fuzzy logic, the proposed approach can better handle uncertainty, fuzziness, and incomplete information, so the security of information service is highly strengthened. By applying the approach using eBay transaction statistical data, the paper demonstrate the final integrative decision order in various conditions. Compared with other methods, this approach has better overall consideration, accord with human selection psychology naturally.

1 Introduction

Grid has emerged as one of the strategically fundamental establishment in high-performance computing and information service, which enable the achievement of researches in dominant science problems through the connection and integration of various resources wide distributed and pervasively heterogeneous. Resource sharing and accessing in grid have broken the boundary of administrative domain, spanning from closed, acquaintance-oriented and relatively static intro-domain computing environment to the open, decentralized and highly dynamic inter-domain computing environment. The wide scale of resources and the high strangeness among entities complicate the decision of resource selection. It is challenging to make a reliable and trustworthy selection in such wide distributed and pervasively heterogeneous computing environment.

Reputation mechanism provide a way for building trust through social control by utilizing community based feedback about past experiences of entities to help making recommendation and judgment on quality and reliability of the transactions [1]. For the similarity of reputation relations between real society and virtual computing environment, reputation is promising to perform well in Grid. Meanwhile, reputation is multi-faceted concept [2], which means the reputation status of resource often has multiple aspects, such as capability, honesty, recommending, history value, fresh behavior evaluation and so on.

Confronted with so multi-facets reputation conditions and multi-objectives selection request, how to scientifically evaluate them, reasonably integrate

K.M. Węgrzyn-Wolska and P.S. Szczepaniak (Eds.): Adv. in Intel. Web, ASC 43, pp. 142–147, 2007.
springerlink.com © Springer-Verlag Berlin Heidelberg 2007

information, and further make the final reliable selection? It is just what we focus on solving in this paper.

The rest of this paper is structured as follows: in Section 2, fuzzy optimal solution models of evaluation are introduced. In Section 3, analyze metrics of evaluation and explain trustworthy resource selection approach by means of a case study. In Section 4, related work is briefed and compared. Finally in Section 5, we summarize future work and conclude the whole paper.

2 Fuzzy Optimal Solution Models of Evaluation and Multi-objective Decision Making

Generally speaking, evaluation means the behavior of specifying the objectives, measuring entity's attributes, and turning them into subjective effect (which will satisfy what the evaluator demands to a certain degree). The theory of fuzzy set is an efficient tool to solve those complex decision problems which contain fuzzy uncertainty factors. Guided by fuzzy set theory [3], according to the given evaluation metrics and measure value, synthetically considering these objectives which might conflict one another, we can evaluate entities and finally provide the most satisfied scheme to decision maker.

According to the entities' different features and forms provided in evaluation, the typical evaluation model might be described as: $\max_{x \in X} \{f(x)\}$. Where X stands for decision making space or feasible field, x is a evaluation variable, and $f(x) = (f_1(x), f_2(x), \cdots, f_m(x))^T$ is vector function of objective (the total objective number is m, and m is positive integer). As different objective might conflict each other, decision maker's fuzzy partial information is necessary to be considered when selecting the satisfied solution.

Definition 1. Deem that $\tilde{A}_i$ is a fuzzy subset in $[m_i, M_i]$ $(i = 1, 2, \cdots, m)$, where m_i and M_i are the lowest and the highest boundary of $f_i(x)$ in decision space X respectively, the membership degree of $\tilde{A}_i$ on y is $\mu_{\tilde{A}_i}(y)(y \in [m_i, M_i])$. If $\mu_{\tilde{A}_i}(y)$ is a strict monotone increasing function of $y(y \in [m_i, M_i])$, and $\mu_{\tilde{A}_i}(M_i) = 1$, then $\tilde{A}_i$ is a fuzzy optimal set of $f_i(x)$. Correspondingly, $\mu_{\tilde{A}_i}(y)$ is the optimal membership degree of $y(y \in \tilde{A}_i)$.

Decision maker's partial information can be embodied through selecting membership degree $\mu_{\tilde{A}_i}(y)$. Fuzzy multi-objectives decision model can be converted to: $\max_{x \in X} \{\mu(x)\}$, where $\mu(x) = (\mu_1(x), \mu_2(x), \cdots, \mu_m(x))^T \in [0, 1]^m \subseteq R^m$. Each membership degree of objective is a value in $[0, 1]$.

Definition 2. Deem that $\tilde{F}$ is a fuzzy set on domain $\bar{X} = \bigcap_{i=1}^{m} X_i$, whose membership degree on x is $\mu_{\tilde{F}}(x)(x \in \bar{X})$. If there is a fuzzy optimal points set $\tilde{f}_i(i = 1, 2, \cdots, m)$, which satisfies $\mu_{\tilde{F}}(x) = h(\mu(x))$, where function $h(t)$ is strict monotone increasing in $t \in [0, 1]^m$ and there is $h(t, t, \cdots, t) = t$ for any $t \in [0, 1]$, then $\tilde{F}$ is the fuzzy optimal points set for evaluation model, Accordingly, $\mu_{\tilde{F}}(x)$

is the optimal membership degree of fuzzy optimal points $x \in \bar{X}$. The optimal membership degree of x_j about objective f_i is μ_{ij}.

Definition 3. Deem that $\tilde{F}$ is the fuzzy multi-objective optimal points set for evaluation model. If $x^* \in \bar{X}$ satisfies $\mu_{\tilde{F}}(x^*) = \max_{x \in \bar{X}} \{\mu_{\tilde{F}}(x)\}$, then x^* is the fuzzy optimal solution of evaluation model about $\tilde{F}$. Accordingly, $\mu_{\tilde{F}}(x^*)$ is the optimal membership degree of x^*.

The optimal solution of $\max_{x \in \bar{X}} \{\mu_{\tilde{F}}(x)\}$ is the solution of evaluation and decision making, and the membership degree according with which stands for the extent of decision maker's satisfactory degree. The objective style, real problem characteristics, and decision maker's request are basis factors when determining relative optimal membership. Usually, objectives includes: benefit style objective, cost style objective, fixation style objective (the nearer to some certain value, the better), and interval style objective (within a certain interval is good).

3 Trustworthy Resource Selection Based on Relative Optimal Membership Degree

To guarantee the trustworthiness and reliability of resource selection, entity's reputation is a key factor to determine our selection, no matter who is provider or consumer. Built on top of idea of SOA, based on fuzzy logic methods, our approach is efficient to deal with uncertainty, fuzziness, and incompleteness of information in systems, and finally builds instructive decision. In this section, we first sum up the typical, influential, and reputation-correlative evaluation objectives, which embody both the entities' multi-facets reputation conditions and the decision maker's multi-objectives in application background, and then build up optimal membership degree based trustworthy resource selection method by means of a case study. In the case, different methods for different decision making psychology are sufficiently demonstrated.

3.1   Evaluation Metrics

When evaluating entities' reputation, we should take the metrics below into unified consideration:

- Selection overhead: overhead of selecting optimum entity for providing service to consumer. Adopting the selection made by reputation evaluation and decision making system, consumer takes on overhead as little as possible in usual situation.
- Service overhead: the necessary cost an entity must pay to provide knight service, such as bandwidth, capacity, man-hour, material and energy etc. The value should satisfy the consumer and accord with real industry condition. Too higher value might mean service provider costs too much, and the costs might be converted as unnecessary burden to consumer. Conversely, too lower value might mean QoS can't reach consumer's anticipation.

- Service performance: A popular metric, including quality of service, service efficiency, maintenance after sale etc.
- Fresh density of reputation: Perfect service amount of an entity during late unit time, which is one of the most representative reputation metric to embody entity's latest reputation conditions, and mostly is attached more importance by enterprising consumers.
- Perfect rate of reputation in history statistic: It can provide relative comprehensive data to embody entity's entire reputation condition, which is interested in by those traditional consumers.
- Operating ability of resisting disaster: It embodies the ability that an entity could recover to its former nice reputation condition when reputation value collapses or shakes acutely (for example: feed back of market aroused by entity's subjective or objective nonstandard actions, entity suffered from malicious attack etc.), which is an important representation of entity's immanent consciousness and capability.

3.2   A Case Study

Deem that $X = \{x_1, x_2, x_3, x_4, x_5\}$ stands for 5 computing resource providers, consider 6 aspects for evaluation: selection overhead (f_1), service overhead (f_2), service performance (f_3), fresh density of reputation (f_4), perfect rate of reputation in history statistic (f_5), operating ability of resisting disaster (f_6). The 5 providers' aggregated reputation ratings in 6 aspects are given in Table 1, which is the to-be-evaluated information system:

Table 1. Resource providers' aggregated reputation information in 6 aspects

F	x_1	x_2	x_3	x_4	x_5
f_1	1250	750	1370	1250	2200
f_2	250	984	766	1861	2161
f_3	0.34	0.23	0.39	0.36	0.29
f_4	83	110	130	234	176
f_5	14	25	10	26	14
f_6	middle	good	poor	good	poor

Using the linear mode of $\mu_{ij} = (f_{ij}/f_{i\max})^{p_i}$ for benefit style objectives f_3, f_4, and f_5, using linear mode of $\mu_{ij} = (f_{i\min}/f_{ij})^{p_i}(f_{i\min} \neq 0)$ for cost style objective f_1, using $\mu_{ij} = f_i^*/(f_i^* + |f_{ij} - f_i^*|)(i = 2)$ for fixation objective f_2 and considering the optimal value $f_2^* = 1340$ requested in special background, choosing optimal membership degree 1.0, 0.75, 0.50 for fuzzy judgments of good, middle, and poor, we convert table 1 into the relative optimal membership degree matrix μ (equation (1)).

In order to embody personal partialness or expecting of decision maker and field expert, we select the weighted vector $\omega = (0.24, 0.18, 0.18, 0.12, 0.12, 0.16)^T$

for $f_i (i = 1, 2, \cdots, 6)$. According to matrix μ and weighted form of $\mu_{ij}^{\omega_i}$, we get weighted relative optimal membership degree matrix $\bar{\mu}$ (equation (2)).

$$\mu = \begin{pmatrix} 0.60 & 1.0 & 0.55 & 0.60 & 0.34 \\ 0.55 & 0.79 & 0.70 & 0.72 & 0.62 \\ 0.87 & 0.59 & 1.0 & 0.92 & 0.74 \\ 0.35 & 0.47 & 0.56 & 1.0 & 0.75 \\ 0.54 & 0.95 & 0.38 & 1.0 & 0.54 \\ 0.75 & 1.0 & 0.5 & 1.0 & 0.5 \end{pmatrix}. \tag{1}$$

$$\bar{\mu} = (\bar{\mu} ij)_{6 \times 5} = \begin{pmatrix} 0.885 & 1.0 & 0.866 & 0.885 & 0.772 \\ 0.898 & 0.958 & 0.938 & 0.943 & 0.918 \\ 0.975 & 0.909 & 1.0 & 0.985 & 0.947 \\ 0.882 & 0.913 & 0.933 & 1.0 & 0.966 \\ 0.929 & 0.995 & 0.890 & 1.0 & 0.929 \\ 0.955 & 1.0 & 0.895 & 1.0 & 0.895 \end{pmatrix}. \tag{2}$$

We get the total order: 1) with Maximin Method, $x_4 \succ x_2 \succ x_3 \succ x_1 \succ x_5$(considering w) and $x_2 \succ x_4 \succ x_1 \succ x_3 \succ x_5$(not considering w).2) with Maximax Method, $x_4 \approx x_2 \approx x_3 \succ x_1 \succ x_5$(considering w) and $x_4 \approx x_2 \approx x_3 \succ x_1 \succ x_5$(not considering w).3) with Minimum Membership Degree Deviation Method, $x_4 \succ x_2 \succ x_3 \succ x_1 \succ x_5(q = 1, q$ is distance parameter), $x_2 \succ x_4 \succ x_1 \succ x_3 \succ x_5(q = 2)$, and $x_2 \succ x_4 \approx x_1 \succ x_3 \succ x_5(q \to \infty)$.4) With Maximum Membership Degree Deviation Method, $x_4 \succ x_2 \succ x_3 \succ x_1 \succ x_5(q = 1)$, and $x_2 \succ x_4 \succ x_3 \succ x_1 \succ x_5(q = 2$ or $q \to \infty)$.5) with Relative Ratio Method, $x_4 \succ x_2 \succ x_3 \succ x_1 \succ x_5(q = 1)$, and $x_2 \succ x_4 \succ x_1 \succ x_3 \succ x_5(q = 2$ or $q \to \infty)$. Different method embodies different request of decision maker who has special characteristics, and final selection can be done according to the order.

4   Related Work

Undoubtedly, reputation is not only of great helpful to subjective selection in humanities, but also important as a formalizing computational concept in scientific computing field. In [4], for the first time, formalizing trust as a computational concept is proposed, which provides a clarification of trust and present a mathematics model for precise trust discussion. In [5], the conception of trust management is used to explain the fact that security decision needs accessorial security information. In [6], "personalized similarity" is adopted to evaluate an entity's credibility. In [7], "the propagation of distrust", an interesting idea, which allows the proactive dissemination of some malicious entity's bad reputation and maintains positive trust values for peers at the meanwhile. Model and framework are also the main field in some papers, yet resource selection method is scarce.

However, currently available resource selection methods, including some commerce system, neither adequately respect multi-facets of resource reputation nor address the problem of decider's multi-objectives in grid environment.

5   Conclusions and Future Work

With the blend of Grid and SOA, grid application is increasingly abundant and extensive. The guarantee of high trustworthiness holds the balance for secure sharing and efficient collaboration among entities in wide distributed and dynamic domain. In this paper, resource selection and reputation mechanism are unified considered. As reputation is uncertain and the selection is multi-objective, we build relative optimal membership to model resource providers' inferior and superior relationship, and by means of information integration we provide the final order, which is used to guide the final resource selection.

For the future, we suggest that many-person decision should be taken into consideration, as it is universal to evaluate and make decision by various decision makers. Converting the approach provided in this paper to a production would be also great helpful for whatever Grid or society life.

References

1. P. Resnick, R. Zeckhauser, E. Friedman, and K. Kuwabara Reputation Systems. Communications of the ACM, 43(12), December 2000: 45–48.
2. Yao Wang and Julita Vassileva. Trust and Reputation Model in Peer-to-Peer Networks. In Proceedings of the 3rd IEEE International Conference on Peer-to-Peer Computing. Linkoping: IEEE Computer Society, 2003. 150-158.
3. LI Dengfeng. Fuzzy Multiobjective Many-person Decision Makings and Games[M].National Defense Industry Press, Beijing, 2003.
4. Marsh Stephen: Formalising trust as a computational concept. PhD Thesis. Scotland, University of Stirling, 1994.
5. M. Blaze, J. Feigenbaum, J. Lacy. Decentralized trust management. In: J. Dale, G. Dinolt,eds., in Proceedings of the 17th Symposium on Security and Privacy. Oakland, CA: IEEE Computer Society Press, 164-173, 1996.
6. Li Xiong and Lin Liu, "PeerTrust: Supporting Reputation-Based Trust in Peer-to-Peer Communities", IEEE Transactions on Knowledge and Data Engineering, Special Issue on Peer-to-Peer Based Data Management, 16(7), July, 2004.
7. R. Guha et al,"Propagation of Trust and Distrust,"*Proc. ACM World Wide Web Conference* (WWW2004),ACM Press, 2004, pp. 403–412.

Inferring Trust from Recommendations in Web-Based Knowledge Sharing Systems

Weisen Guo[1,2] and Steven Kraines[2]

[1] Institute of Systems Engineering of Dalian University of Technology, Dalian 116024, China
guows@dlut.edu.cn
[2] Division of Project Coordination of the University of Tokyo, Tokyo 277-8568, Japan
{gws,sk}@cb.k.u-tokyo.ac.jp

Summary. Conventional web-based systems for knowledge sharing cannot help users determine the reliability of an unknown knowledge source on the web. This paper introduces an approach for using the concepts of trust and recommendation from social networks in web-based knowledge sharing systems. A simulation study of three algorithms for calculating trust inferred from recommendations on a social network is presented. The results show that our proposed algorithms can calculate inferred trust values for over 99% of the entities in the network with a high degree of accuracy. Finally, a prototype MAS system that uses trust and recommendation in knowledge sharing is described[1].

1 Introduction

The Web is an important source of knowledge for day to day life, such as shopping, study, and travel. However, much of this knowledge obtained from Web search engines such as Google is from sources that are not directly known to the user, e.g. information on services provided by an unknown hotel or car dealer. This paper introduces trust and recommendation into the process of knowledge sharing to help people determine the reliability of a knowledge source on the web.

Consider a social network of Web users Ann, Brad, Chad, Diana, Eve, Fred, Jane, Kay, Lucy, and Sam (Figure 1). Each user is characterized as having high reliability (H), moderate reliability (M), low reliability (L), or as being dishonest (D). In this example, Sam is a dishonest car dealer who often sells unreliable cars. However, his advertisement on the Web contains false information that may lead a naive person to think that he offers a better service.

The trust relationships between the users connecting Ann to Sam are shown in Figure 1 as arrows pointing from a trusting user to a trusted user. The arrows

---

[1] Steven Kraines is the corresponding author. This work is supported by the Japan Science and Technology Agency through the Shippai Chishiki Project, the National Natural Science Foundation of China (70431001) and the Liaoning Province Science and Technology Agency (20061063).

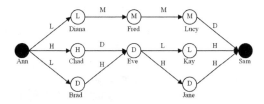

Fig. 1. the trust chains from Ann to Sam in the scenario given in the text

are labeled with the trust or reliability level believed to be correct for the trusted user by the trusting user. For example, Ann knows and trusts Brad, Chad, and Diana, but she trusts Chad most. Diana, in addition to knowing Ann, knows Fred, whom she trusts moderately. Lucy, who has bought a car from Sam, knows that Sam is a dishonest car dealer. Kay also knows Sam, but because she is new to the Web and has not bought a car from Sam, she believes Sam when he tells her that he is a trustworthy car dealer. Finally, Jane is a dishonest user who is in cohorts with Sam, so she tells everyone she knows that Sam is a highly trustworthy car dealer.

There are 12 relationships between the 10 users, forming five chains of trust links that connect Ann to Sam. How can Ann best use the information available to her in order to make the most accurate estimate of Sam's trust value? This question forms the basis of the work presented in this paper.

The rest of the paper is organized as follows. Section 2 discusses some related work. Section 3 presents the algorithms and a simulation study. Section 4 introduces a prototype MAS system to implement the trust network. Finally, section 5 concludes the paper with suggestions for future work.

2   Related Work

Trust and recommendation are social concepts that have received much attention in the computing field recently, particularly in the context of the goals that have been proposed for the Semantic Web [2, 3, 5]. Initially, most research on the topic of trust focused on security issues. However, recently more social notions of trust, such as reputation-based trust, are gaining attention [1, 2, 3, 4, 5]. Proposed methods for providing information to users on which websites to trust are dominated by formal methods for rating the reputation of a site or user. For example, in the eBay system, after a user conducts a transaction with another user, she can publicize a trust value for the transaction partner together with a comment, which other users can use to learn about the trustworthiness of the transaction partner. Other systems, such as Amazon and Epinions, adopt similar methods.

Massa and Bhattacharjee [1] studied the use of trust in recommender systems. They pointed out several weaknesses of recommender systems and presented a Trust-aware system to alleviate the weaknesses. They analyzed data from the Epinions website and showed that any two users usually have few rated items in common. By propagating trust in their system and inferring additional information for the trustworthiness of other users, they showed that useful recommendations can be computed against a larger number of users. Our approach is also based on propagating trust and inferring trust values for other users. However, while their system does not evaluate the accuracy of the inferred trust values, we calculate a measure of the accuracy of the inferred trust values for other users.

Golbeck and Hendler [2] addressed the accuracy of metrics for inferred trust and reputation in semantic web-based social networks. They described an algorithm for generating locally-calculated reputation ratings from a semantic web social network and showed mathematically and experimentally that the algorithm can accurately infer the trust value of an entity on the network. However, while Golbeck and Hendler's algorithm ignores trust chains that contain entities that have been given negative ratings, we propose an approach that uses negative rating information to improve the accuracy of the inferred trust value. For example, in Figure 1 Ann trusts Chad highly and Brad less. Brad trusts Eve, but Chad distrusts Eve. Eve is dishonest, and so she recommends Jane as being trustworthy even though she knows that Jane is dishonest. If the negative rating of Eve by Chad is ignored, Ann will use the information provided by Eve to infer a high trust value for Jane, which is incorrect.

3 The Trust-Recommendation Network Analysis

We describe and assess through a simulation study the ability of three algorithms to evaluate the reliability of a large number of unknown entities using trust information from a social network such as that shown in Figure 1. We rate trust and reliability on a discrete scale with five levels, as described in the introduction: high trust/reliability (H), moderate trust/reliability (M), low trust/reliability (L), unknown (U), and distrust/dishonesty (D). In the simulation we assume that entities with high, moderate, and low reliabilities will give correct recommendations 90%, 80%, and 70% of the time, respectively, and a dishonest entity will give opposite recommendations 90% of the time.

When a user wants to estimate the reliability of an inferred trust value for another agent, the user can translate the recommended trust values of the entities forming a trust chain into values between 0 and 1 and then multiply the values to get the reliability of the final recommendation. For example, consider the chain from Ann to Sam through Diana, Fred, and Lucy in Figure 1. Ann can estimate the reliability of Lucy's recommended trust value "distrust" for Sam as $0.7 \times 0.8 \times 0.8 = 45\%$.

There are five chains of trust recommendations from Ann, the source agent, to Sam, the sink agent, in Figure 1:

1. Ann -(L)->   Diana -(M)->   Fred -(M)->   Lucy -(D)->   Sam
2. Ann -(H)->   Chad -(D)->   Eve -(L)->   Kay -(H)->   Sam
3. Ann -(H)->   Chad -(D)->   Eve -(H)->   Jane -(H)->   Sam
4. Ann -(L)->   Brad -(H)->   Eve -(L)->   Kay -(H)->   Sam
5. Ann -(L)->   Brad -(H)->   Eve -(H)->   Jane -(H)->   Sam

For the first trust inference algorithm, the agent chooses the trust recommendation from the chain having the highest reliability. The method described above is used to calculate the reliability of the trust recommendation from each chain. If there is any agent in a chain that distrusts the next agent, the reliability of that chain is zero. For the scenario in Figure 1 the highest reliability is given by the 5th trust chain (57%), so using this algorithm, Ann will believe that Sam is highly trustworthy, which we know to be incorrect.

The second algorithm is based on the observation that if there is more than one trust chain leading to the same inferred trust value, the total reliability of that inferred trust value should higher than the reliability of each chain. For example, the 4th and 5th chains both recommend that Sam is highly trustworthy. Combining the two chains, Ann can estimate the reliability of the trust value "high trust" for Sam as $1-(1-0.44)(1-0.57)=76\%$.

In the third algorithm, we include the information from negative ratings of agents. While the previous algorithms consider each chain independently, the third algorithm accounts for the possibility that an agent may be encountered in more than one chain from source agent to sink agent. If an agent appears in more than one chain, the source agent should give it the same trust value in each chain, and the inner agent should recommend a single trust value for the sink agent. For example, Eve is in all chains except the first one. So, Ann should estimate the trust value of Eve first. Chad recommends that Eve is dishonest. Brad recommends that Eve is highly reliable. Ann trusts Chad more highly than Brad, so she has reason to believe that Eve is dishonest. Therefore, she will discard the chains through Eve and get the correct inferred trust value of "distrust" for Sam from Lucy through the 1st trust chain. This algorithm differs from Golbeck and Hendler's algorithm where, because the information that Chad distrusts Eve is ignored, Ann will get the inferred trust value of "high trust" for Sam from Jane through the 5th trust chain.

In the third algorithm, first all chains from the source agent to the sink agent are found up to a given maximum chain length (6 in the study here). However, it is possible that the same agent appears in different chains with different trust values from its preceding agents. In order to accurately estimate the trust value of the sink agent, we establish a unique trust value for each agent in the trust chains, starting with the agents closest to the sink agent. The process continues recursively until the source agent has a single trust value for all the agents in the chains.

In all three algorithms, it is possible that two recommendations for different trust values will have the same highest reliability. We use the following tie-breaking rule: if one recommends "distrust" and another recommends "high trust", "moderate trust", or "low trust", then the inferred trust value is unknown (U). Otherwise, the lowest recommended trust value is used as the inferred trust value.

We create a network populated by two kinds of agents: those connecting with six agents, and those connecting with twenty-four agents, to assess our algorithm. 80% of the agents are of first kind, and 20% are of the second kind. There are 100,000 agents in the network, of which 10% are dishonest, 40% have low reliability, 40% have moderate reliability, and 10% have high reliability. We then randomly select one agent from the network and use each of the three algorithms to try to find all of the dishonest agents. We set the maximum length of the trust chains to be six, based on the concept of six degrees of separation. Table 1 shows the results.

Table 1. The simulation results

	Total	First Algorithm			Second Algorithm			Third Algorithm		
		D	H+M+L	U	D	H+M+L	U	D	H+M+L	U
LR	9931	7388	2303	240	7912	1983	36	8228	1689	14
MR	9955	7453	2315	187	7721	2208	26	8231	1720	4
HR	9845	7380	2264	201	7718	2075	52	8136	1691	18
Average	9910	7407	2294	209	7784	2089	38	8198	1700	12
Accuracy		75%	23%	2%	79%	21%	0%	83%	17%	0%

"LR", "MR", and "HR" means that the source agent has low reliability, moderate reliability, and high reliability respectively. "Total" is the total number of dishonest agents found out of the 10,000 dishonest agents in the network. "D", "H", "M", "L", "U" shows the number of dishonest agents found with the inferred trust value "distrust", "high trust", "moderate trust", "low trust", and "unknown" respectively. "Average" is the average of "LR", "MR", and "HR", and "Accuracy" is the average in the column divided by the average total number of dishonest agents found.

From the results in the table, we obtain the following conclusions. First, using the trust-recommendation network, one agent can get inferred trust values for more than 99% of the other agents, in agreement with Massa and Bhattacharjee's work. Second, the third algorithm yields a significant improvement in accuracy over the first and second, where an agent can identify a dishonest agent 83% of the time. Because the third algorithm uses the "distrust" information, the reliability of inner agents can be inferred more accurately. We have assessed the algorithms on several other networks created under different conditions and obtained the same conclusions.

4 Prototype System

We have developed a prototype MAS system on the Java Agent Development Framework (JADE). In the network, each node represents an agent that can provide or search for some knowledge source. Each link represents the level at which one agent trusts another. When one agent wants to know how trustworthy a particular unknown agent is, it sends out a request for information to the agents that it trusts. Those agents give recommendations about the unknown agent if they have such information. Otherwise they ask the agents that they trust. The initiating agent then uses the recommendations to decide the trust level for the unknown agent.

We use a two level architecture for the MAS to protect the transmission of trust information. The trust for security is managed at the JADE platform level. We use https to create a secure and authenticated communication channel between platforms. Because all of the open messaging on the Internet takes place between the platforms, all communication between agents within a single platform can be protected from attackers outside of the platform, so it is sufficient to ensure that the communication between platforms is secure. Trust for specific knowledge services of individual agents is managed at the service level.

The prototype system supports dynamic trust in the following way. When an agent gets new trust information about another agent, it checks local data to find the agents to which it has recommended the trust information. The agent then sends the new trust information to all of those agents. This process results in a push style of trust information transfer. In this way, a recommendation from a particular agent can be updated quickly among highly trusted peers when the recommender gets a new trust value.

5 Future Work

We are developing models to create trust networks that exhibit behaviors closer to real social networks, such as small world behavior and clustering. We will study the effectiveness of our algorithm by conducting our simulation study on these networks and networks based on real data from the web. Finally, we are investigating how to integrate the trust network with a knowledge searching and matching application that we have developed in other work [6].

References

1. Massa, P., Bhattacharjee, B., Using Trust in Recommender Systems: an Experimental Analysis. LNCS, Vol.2995/2004, pp. 221-235.
2. Golbeck, J., Hendler, J., Accuracy of Metrics for Inferring Trust and Reputation in Semantic Web-based Social Networks. LNCS, Vol.3257/2004, pp. 116-131.
3. Golbeck, J., Parsia, B., Hendler, J., Trust Network on the Semantic Web. Proc. of Cooperative Intelligent Agents 2003, August 27-29, Helsinki, Finland.
4. O'Donovan, J., and Smyth, B., Trust in Recommender Systems, In IUI '05: Proc. 10th Intl Conf. on Intelligent user interfaces, pp.167-174.
5. Richardson, M., Agrawal, R., Domingos, P., Trust Management for the Semantic Web, Proc. of ISWC 03, Sanibel Island, Florida.
6. Kraines, S., Guo, W., Kemper, B., and Nakamura, Y., EKOSS: A Knowledge-User Centered Approach to Knowledge Sharing, Discovery, and Integration on the Semantic Web. Proc. of ISWC 06, LNCS 4273/2006, pp. 833-846.

Improving Text Classification by Web Corpora[*]

Rafael Guzmán[1,2], Manuel Montes[3], Paolo Rosso[2], and Luis Villaseñor[3]

[1] FIMEE, Universidad de Guanajuato, Mexico
[2] DSIC, Universidad Politécnica de Valencia, Spain
[3] LTL, Instituto Nacional de Astrofísica Óptica y Electrónica, Mexico

Summary. A major difficulty of supervised approaches for text classification is that they require a great number of training instances in order to construct an accurate classifier. This paper proposes a semi-supervised method that is specially suited to work with very few training examples. It considers the automatic extraction of unlabeled examples from the Web as well as an iterative integration of unlabeled examples into the training process. Preliminary results indicate that our proposal can significantly improve the classification accuracy in scenarios where there are less than ten training examples available per class.

1 Introduction

Nowadays there is a lot of digital information available from the Web. This situation has produced a growing need for tools that help people to find, filter and analyze all these resources. In particular, text classification [4], the assignment of free text documents to one or more predefined categories based on their content, has emerged as a very important component in many information management tasks.

The state-of-the-art approach for automatic text classification considers the application of a number of statistical and machine learning techniques, including regression models, Bayesian classifiers, support vector machines (SVM), nearest neighbor classifiers (k-NN) and neuronal networks [4]. A major difficulty with this kind of supervised techniques is that they commonly require a great number of labeled examples (training instances) to construct an accurate classifier. Unfortunately, because a human expert must manually label these examples, the training sets are extremely small for many application domains. In order to overcome this problem, recently many researchers have been working on semi-supervised learning algorithms (for an overview see [5]). It has been showed that augmenting the training set with additional information it is possible to improve the classification accuracy using different learning algorithms such as naïve Bayes [3], SVM [1], and k-NN [7].

In this paper we propose a new method for semi-supervised text classification. This method differs from previous approaches in three main concerns. First, it is

[*] This work was done under partial support of CONACYT-Mexico (43990), MCyT-Spain (TIN2006-15265-C06-04) and PROMEP (UGTO-121).

K.M. Węgrzyn-Wolska and P.S. Szczepaniak (Eds.): Adv. in Intel. Web, ASC 43, pp. 154–159, 2007.
springerlink.com

specially suited to work with very few training examples. Whereas previous methods consider groups of ten and even hundreds of training examples, our method allows working with less than ten labeled examples per class. Second, it does not require a predefined set of unlabeled examples. It considers the automatic extraction of related untagged data from the Web. Finally, given that it deals with very few training examples, it does not aim including a lot of additional information in the training phase; on the contrary, it only incorporates a small group of examples that considerably augment the dissimilarities among classes.

It is important to point out that the Web has been lately used as a corpus in many natural language tasks [2]. In particular, Zelikovitz and Kogan [8] proposed a method for mining the Web to improve text classification by creating a background text set. Our method is similar to this approach in that it also mines the Web for additional information (extra-unlabeled examples). Nevertheless, our method applies finer procedures to construct the set of queries related to each class and to combine the downloaded information.

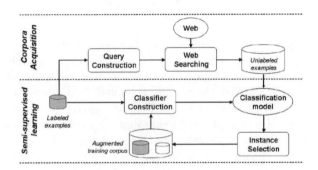

Fig. 1. General overview of the method

2   Proposed Method

Figure 1 shows the general scheme of the proposed method. It consists of two main processes. The first one deals with the corpora acquisition from the Web, while the second one focuses on the semi-supervised learning problem. The following sections describe in detail these two processes.

2.1   Corpora Acquisition

This process considers the automatic extraction of unlabeled examples from the Web. It first constructs a number of queries by combining the most significant words for each class; then, using these queries it looks at the Web for some additional training examples related to the given classes.

Query Construction. In order to form queries for searching the Web, it is necessary to previously determine the set of relevant words for each class in the training corpus. The criterion used for this purpose is based on a combination of the frequency of occurrence and the information gain of words. We consider that a word w_i is relevant for class C if:

1. The frequency of occurrence of w_i in C is greater than the average occurrence of all words (happening more than once) in that class. That is:

$$f_{w_i}^C > \frac{1}{|C'|} \sum_{\forall w \in C'} f_w^C, \text{ where } C' = \{w \in C | f_w^C > 1\}$$

2. The information gain of w_i with respect to C is positive ($IG_{w_i}^C > 0$).

Once obtained the set of relevant words per class, it is possible to construct the corresponding set of queries. Founded on the method by Zelikovitz and Kogan [8], we decide to construct queries of three words. This way, we create as many queries per class as all three-word combinations of its relevant words. We measure the significance of a query $q = w_1, w_2, w_3$ to the class C as:

$$\Gamma_C(q) = \sum_{i=1}^{3} f_{w_i}^C \times IG_{w_i}^C$$

Web Searching. The next action is using the defined queries to extract from the Web a set of additional unlabeled text examples. Based on the observation that most significant queries tend to retrieve the most relevant web pages, our method for searching the Web determines the number of downloaded examples per query in a direct proportion to its Γ-value. Therefore, given a set of M queries $q_1, , q_M$ for class C, and considering that we want to download a total of N additional examples per class, the number of examples to be extracted by a query q_i is determined as follows:

$$\Psi_C(q_i) = \frac{N}{\sum_{k=1}^{M} \Gamma_C(q_k)} \times \Gamma_C(q_i)$$

2.2 Semi-supervised Learning

As we previously mentioned, the purpose of this process is to increase the classification accuracy by gradually augmenting the originally small training set with the examples downloaded from the Web. Our algorithm for semi-supervised learning is an adaptation of a method proposed elsewhere [6]. It mainly considers the following steps:

1. Build a weak classifier (C_l) using a specified learning method (l) and the training set available (T).
2. Classify the downloaded examples (E) using the constructed classifier (C_l). In order words, estimate the class for all downloaded examples.
3. Select the best m examples ($E_m \subseteq E$) based on the following two conditions:
   a) The estimate class of the example corresponds to the class of the query used to download it. In some way, this filter works as an ensemble of two classifiers: C_l and the Web (expressed by the set of queries).
   b) The example has one of the m-highest confidence predictions.

4. Combine the selected examples with the original training set ($T \leftarrow T \cup E_m$) in order to form a new training set. At the same time, eliminate these examples from the set of downloaded instances ($E \leftarrow E - E_m$).
5. Iterate σ times over steps 1 to 4 or repeat until $E_m = \emptyset$. In this case σ is a user specified threshold.
6. Construct the final classifier using the enriched training set.

3   Experimental Evaluation

3.1   Experimental Setup

Corpus. It is a set of Spanish newspaper articles about natural disasters. It consists of 210 documents grouped in four different categories: forest fires (C1), hurricanes (C2), inundations (C3), and earthquakes (C4). For experimental evaluation we organized the corpus as follows: four different training sets (formed by 1, 2, 5 and 10 examples per class respectively) and a fixed test set of 200 examples (50 per class).

Searching the Web. We used Google as search engine. We downloaded 1,000 additional examples (snippets for these experiments) per class.

Learning methods. We selected two state-of-the-art methods for text classification, namely, support vector machines (SVM) and naïve Bayes (NB) [4].

Evaluation measure. The effectiveness of the method is measured by the classification accuracy, which indicates the percentage of documents that have been correctly classified from the entire document set.

Baseline. Baseline results correspond to the application of the selected classifiers on the test data. Table 2 shows these results for the four different training conditions. They evidence that traditional classification approaches achieve poor performance levels when dealing with *very* few training examples.

3.2   Experimental Results

This section presents some results related to the main processes of the proposed method, namely, the corpora acquisition from the Web and the semi-supervised learning approach.

   The central task for corpora acquisition is the automatic construction of a set of queries that express the relevant content of each class. Table 1 shows some numbers on this task. It is noticeable that, because the selection of relevant words relies on a criterion based on their frequency of occurrence and their information gain, there is not the same number of queries per class even thought there were used the same number of training examples. In addition, it is also visible that an increment on the number on examples not necessarily represents a growth on the number of built queries.

   Nevertheless, it is important to clarify that using more examples allows to construct more general and consequently more relevant queries. For instance,

Table 1. Some numbers about query construction

Number of training examples	Relevant words per class C1 C2 C3 C4				Queries per class C1 C2 C3 C4			
1	5	5	7	3	10	10	35	1
2	4	5	6	2	4	10	20	1
5	5	5	6	5	10	10	20	10
10	4	5	5	5	4	10	5	10

Table 2. Experiment result using $m = 1$ and $m = |T|$

Training examples	Baseline SVM	NB	m-value	Our method					
				1st iteration SVM	NB	2nd iteration SVM	NB	3rd iteration SVM	NB
1	50.0	51.7		49.1	**78.3**	51.0	77.3	55.3	76.0
2	58.3	56.7		62.3	70.0	68.1	**86.0**	67.0	**86.1**
5	77.1	80.4	$m = 1$	76.4	82.2	80.1	85.1	87.0	**92.1**
10	80.4	77.1		82.1	83.1	85.2	87.2	90.1	**91.3**
1	50.01	51.72		49.0	**78.2**	51.5	77.5	55.2	76.5
2	58.33	56.71		68.2	86.5	74.0	**87.6**	74.5	86.5
5	77.14	80.41	$m = |T|$	93.5	**97.0**	92.5	96.5	96.0	95.6
10	80.42	77.14		96.5	97.2	96.1	**97.5**	95.1	96.5

using only two examples about hurricanes we constructed queries such as $<Baja + California + hurricane>$, whereas using ten examples we could obtain queries such as $<hurricane + kilometers + storm>$.

Using these queries we collected from the Web a set of 1,000 snippets per class, obtaining a total of 4,000 additional unlabeled examples. Then, we added some of these examples to the original training set. Mainly, we performed three different experiments by varying the parameter m of the algorithm of Section 2.2.

1. At each iteration we added to the training set one additional example per class (i.e., we set $m = 1$).
2. At each iteration we added to training set as many unlabeled examples as the number of instances in the original set (i.e., we set $m = |T|$).
3. In one single step we added to the training set all unlabeled examples satisfying the condition (a) of the algorithm.

Table 2 shows the results of the first two experiments. They indicate that our method outperformed all base configurations especially when using the naïve Bayes classifier. In particular, setting $m = |T|$ lead to accuracy improvements on the range of 30%. On the other hand, the results of the third experiment do not favor the proposed method. They showed a fall in accuracy around 5 to 25%. In some way these results confirms our intuition that in scenarios having very few training instances it is better to include a small group of unlabeled examples that considerably augments the dissimilarities among classes than including a lot of doubtable-quality information.

4 Conclusions and Future Work

In this paper we proposed a method for semi-supervised text classification that is specially suited to work with very few training examples. This method differs from previous approaches in that: (i) it automatically collects from the Web the set of unlabeled examples and, (ii) it only incorporates into the training phase a small group of unlabeled examples.

The experimental results on a set of newspaper articles about natural disasters demonstrate the viability of the method. In some way, they confirm our hypothesis that when dealing with very few training instances it is better to add a selected set of unlabeled examples (those that considerably augments the dissimilarities among classes) than incorporate a lot of doubtable-quality information. In particular, our method obtained the best results when we added to the training set as many unlabeled examples as the number of original labeled instances. It was also noticeable that our method achieved the best results only after two or three iterations.

References

1. Joachims T., Transductive inference for text classification using support vector machines, Sixteenth International Conference on Machine Learning, 1999.
2. Kilgarriff A., and Greffenstette G., Introduction to the Special Issue on Web as Corpus, Computational Linguistics, 29(3), 2003.
3. Nigam K., Mccallum A. K., Thrun S., and Mitchell T., Text classification from labeled and unlabeled documents using EM, Machine Learning, 39(2/3), 2000.
4. Sebastiani F., Machine learning in automated text categorization, ACM Computing Surveys, 34(1), 2002.
5. Seeger M., Learning with labeled and unlabeled data. Technical report, Institute for Adaptive and Neural Computation, University of Edinburgh, UK, 2001.
6. Solorio T., Using unlabeled data to improve classifier accuracy, Master Degree Thesis, Computer Science Department, INAOE, Mexico, 2002.
7. Zelikovitz S., and Hirsh H., Integrating background knowledge into nearest-Neighbor text classification, Advances in Case-Based Reasoning (ECCBR), 2002.
8. Zelikovitz S., and Kogan M., Using Web Searches on Important Words to Create Background Sets for LSI Classification, 19th FLAIRS conference, Florida, USA, 2006.

On Clustering Visitors of a Web Site by Behavior and Interests

Natascha Hoebel[1] and Roberto V. Zicari[2]

[1] Database and Information Systems, Computer Science Institute, J.W. Goethe
University Frankfurt, Germany
hoebel@dbis.cs.uni-frankfurt.de
[2] Database and Information Systems, Computer Science Institute, J.W. Goethe
University Frankfurt, Germany
zicari@cs.uni-frankfurt.de

Summary. This paper addresses the issue of how to define clusters of web visitors
with respect to their behavior and supposed interests. We will use the non-obvious
user profiles (NOPs) approach defined in [10], and present a new clustering algorithm
which is a combination of hierarchical clustering together with a centroid based method
with priority, which allows to cluster web users by similar interest in several topics.

1 Introduction

Companies today operate in an increasingly competitive environment. Therefore,
finding and retaining customers is a major success factor for most businesses,
off-line and online. One of the keys to building effective E-customer relationships
is an understanding of consumer behavior online [12]. However, analyzing the
behavior of customers online is not necessarily an indicator of their declared
interest. In order to measure the supposed interest of a web user, we have introduced in [10] the concept of *non-obvious profiles* (NOPs).

This paper presents a new algorithm to define clusters of web visitors with
respect to their behavior and supposed interests using the NOP approach defined
in [10] and implemented by the Gugubarra Engine [8], [9]. The rest of the paper
is structured as follows: In Sect. 2 we present our cluster algorithm. An example
is given in Sect. 3. In Sect. 4 we briefly describe the current implementation and
evaluation results. Related work is presented in Sect. 5 and conclusion in Sect. 6.

2 Clustering Algorithm

2.1 Initial Considerations

In this subsection we will look at how to define clusters, how to give an interpretation of a cluster, and how to partition users into clusters.

K.M. Węgrzyn-Wolska and P.S. Szczepaniak (Eds.): Adv. in Intel. Web, ASC 43, pp. 160–167, 2007.
springerlink.com   © Springer-Verlag Berlin Heidelberg 2007

We will start with the following pre-considerations: To recognize trends in interests, we will not distinguish between interest weight values such as for example 0.3 and 0.35, but rather we will use a nominal scale: such as {*no interest, little interest, strong interest, total interest*} or {*small interest, medium interest, high interest*} etc.

For that we introduce the parameter g (granularity), which indicates how fine the owner would like to differentiate in the interest scale. For example, $g = 4$ means an interest scale composed of {no interest, little interest, strong interest, total interest}. We then split the range from Zero to One into g intervals and every interval has a centroid which tells us how to interpret the resulting cluster. The centroids are calculated with the function $G(x) = x/(g-1)$ with $0 \leq x \leq g$. Note that these are not centroids in traditional sense, because they are fixed and are never recalculated as balance point of the users. They are centroids in the sense of representative of each cluster.

Supposed we want to define clusters based on only one topic Tp_i, using $g = 4$, we will get the four clusters with fixed centroids shown in Table 1.

Table 1. Concern topic Tp_i; Intervals for $g = 4$ and scale of interest

Name	x Centroid G(x)		Interval	Code	Interest
Cluster 1	0	0	[0.00; 0.16]	00	no
Cluster 2	1	0.33	[0.17; 0.49]	01	little
Cluster 3	2	0.66	[0.50; 0.83]	10	strong
Cluster 4	3	1	[0.84; 1.00]	11	total

For clustering over several topics $|Tp|$, we need an amount of maximum $g^{|Tp|}$ clusters with $|Tp|$ is the amount of topics. For example if we have $|Tp| = 2$ and $g = 4$ it results in 16 centroids. Note that this is the maximum number, that will normally be not reached by our algorithm. In our approach, besides defining the number of clusters, we also associate a *meaning* to each cluster in correlation to the scale of interest (see the cluster interpretation in Fig. 1).

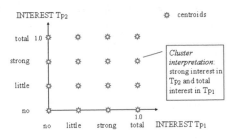

Fig. 1. 16 centroids for $g = 4$ and 2 topics Tp_1 and Tp_2

2.2   The Takahe Algorithm

In this section we present a new *cluster-by-priority algorithm*, that we call *Takahe*. Takahe is a divisive clustering algorithm, which creates a tree structure top down visualized in a dendrogram. A dendrogram is a tree diagram to visualize the arrangement of the clusters produced by a cluster algorithm. Every vertex of the dendrogram represents a cluster.

The algorithm sorts in the initialization the set of n topics, $Tp_1, \ldots, Tp_n$, according to their priority. A higher priority is assigned to a topic if the interest of all current registered users u for this topic is on the average higher. The algorithm produces a sorted vector of n elements, each of which is the average value of the interest of all users for that topic. We call such vector, *the Community NOP (CNOP(t))*. The running time for the initialization is $O((log\ u + |Tp|) * u)$.

The algorithm then determines the number of clusters which should be created, depending on the number and interest of the users and a predefined threshold-value. A threshold-value is used to reduce the number of topics considered by the algorithm. In the following we call the border topic *the threshold topic* Tp_T. In this way, we do not cluster users for topics that are not considered relevant. The running time for the termination condition is $O(1)$.

The algorithm then divides the set of users by sequential inspection of the sorted topics in the $CNOP(t)$. At the beginning, all users are placed into one cluster (top vertex in the dendrogram). Then we calculate the nearest centroid for the topic with highest priority in the $CNOP(t)$ for every user. This is a combination of hierarchical clustering with the k-means algorithm [6], as we search the nearest centroid instead of the nearest user. For the new centroid, we then add a child to the current cluster and a code for the centroid as an edge label.

Until the termination condition is fulfilled, we continue to partition the clusters successively (level for level to construct the dendrogram). We therefore have a single path clustering algorithm in relation to one topic because we do only one iteration. As we continue the algorithm under inclusion of a further topic until Tp_T, we keep the computation time of $O((log\ u + |Tp|) * u)$.

The algorithm terminates when all topics in $CNOP(t)$ above the given threshold are considered and the users are distributed in the clusters created by the dendrogram. The pseudo code for the Takahe algorithm is as follows:

```
initialize(){
    calculate vector CNop;
    calculate MaxCentroids; }

determineTerminationCondition(){
    if(thresholdTopic not reached) terminate=false;
    else if(centroidsCount < maxCentroids) terminate=false;
    else terminate=true; return terminate; }
```

```
ClusterUsersByPriority(clusters, t_x){
    for each cluster in clusters do
        for each user in cluster do
            centroidCode=getNearestCode(w_x of user);
            if(child centroidCode not exists in cluster)
                add new child centroidCode to cluster;
                add new child centroidCode to children;
                add user to child centroidCode;
            if(!determineTerminationCondition())
                ClusterUserByPriority(children, t_{x+1}); }
```

When the algorithm is finished, it visualizes the extraction of topic Tp_T (see for example Fig. 3). Then, if necessary the owner can display the clusters upwards in the tree (summary) or continue the division of the displayed clusters by inclusion of the next topic Tp_{T+1}.

The Takahe algorithm works well with high number of topics, because the threshold allows us to discard topics at the beginning that are considered not relevant for the clustering. A further way to reduce the topics is to build a hierarchy of related topics as e.g. used by [11].

3 An Example

We will now show with an example how the algorithm works. With $g = 4$ we define the interest scale (total, strong, little, no) interest. Thus we get the four clusters shown in Table 1 (see Sec. 2.1). Since $g = 4$, $x \in \{0, 1, 2, 3\}$, we can calculate the centroids, using the formula $G(x) = x/(g-1)$. We obtain four centroids, with values 1, 0.66, 0.33 and 0. From the centroids we calculate the corresponding four intervals. We associate a code to each interval, that will be used to build the dendrogram. As result, each interval has a coded meaning in the interest scale e.g. *total interest*, defined by interval 0.84-1.00.

Let's assume we want to cluster 10 users (including Tom and Joe) visiting the web site www.frankfurt.de. For this web site we consider 5 topics: "Frankfurt at a Glance" (Tp_1), "Culture" (Tp_2), "Sports" (Tp_3), "Fair" (Tp_4) and "Zoo" (Tp_5). The NOPs for these users are shown in Table 2, where w_i corresponds to the value of the NOP for topic Tp_i. A value 0 means no interest; a value 1 means total interest.

In the initialization we calculate $CNOP(t)$, shown in Table 3. In the example, we have set a threshold value of 0.1. Therefore topic Tp_1 is the threshold topic Tp_T. If we look at Table 3, we see that topic Tp_5 will not be considered by the clustering algorithm. This is the mechanism we use to reduce the number of topics taken into account by the algorithm. The threshold is set by the owner of the web site and depends on the application requirements.

The owner might consider a subset of registered users for the calculation of $CNOP(t)$. Indeed the algorithm always works on a set of users, as the amount of users is a dynamic value and changes with time.

Table 2. NOPs at time t with 5 topics for 10 users

USERID	w'_1	w'_2	w'_3	w'_4	w'_5
1	0.3	0.9	0.1	0	0
2	0.27	0.84	0	0	0.1
3 (Joe)	0.3	0	0.65	0.24	0.3
4	0.9	0.5	0.5	0.4	0
5	0.2	0.68	0.12	0	0
6 (Tom)	0.29	0.15	0.45	0.16	0
7	0.3	0.7	0	0.4	0.24
8	0.05	0.05	0.6	0.9	0.1
9	0.3	0.6	0.1	0.2	0.1
10	0	0	0.55	1	0

Table 3. CNOP(t), sorted

	w'_2	w'_4	w'_3	w'_1	w'_5
$CNOP(t)$	0.44	0.33	0.31	0.29	0.08

We can now proceed to construct the dendrogram (see Fig. 2). We start by creating a root node (one cluster) in which we place all 10 users. We then consider the topic which has the highest value in $CNOP(t)$, that is Tp_2, with $w'_2 = 0.44$. For all 10 users we look at their NOP value for Tp_2 defined in Table 2, and we check in which interval of Table 1, this value fits in.

In the dendrogram, we create an edge from the root with the appropriate code if we find at least one user who has an interest value which fits into the corresponding interval. For topic Tp_2, we create three edges with code 11, 10 and 00 ending into the nodes A, B and C, in depth one. The nodes A, B and C are clusters of users with "similar" interest in topic Tp_2. The algorithm proceeds by taking into account the next topic, which is Tp_4 until the threshold topic Tp_1. Afterwards the algorithm ends. Fig. 2 shows the resulting dendrogram of the algorithm.

The algorithm visualizes the resulting clusters like shown in Fig. 3. A, B, C denote clusters for level 1 of the dendrogram. D, E, F, G, H denote clusters for level 2 and I, J for level 3.

In particular, when considering Joe and Tom: Joe is placed in cluster B, which means no interest in Tp_2 ("Culture"). Joe is also in the inner cluster E, which means little interest in Tp_4 ("Fair") and Tp_1 ("Frankfurt"), strong interest in Tp_3 ("Sports"). Tom is also placed in cluster B, so Tom and Joe share this property. However, Tom is placed in another inner cluster F, which means no interest in ("Fair") and little interest in "Sports" and "Frankfurt".

Once we have clustered web users by NOPs, we now have a breakdown of users that can be targeted offering them personalized information and services,

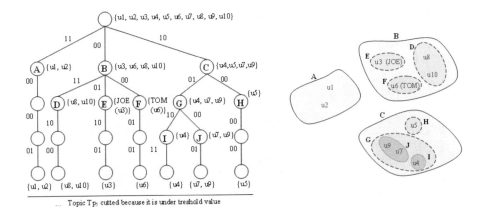

Fig. 2. Dendrogram of the clusters produced by the Takahe Algorithm with $g = 4$

Fig. 3. Visual Extraction for Tp_1 (level 4) with $g = 4$

if and only if they request so. For example, as a result of this clustering we could avoid offering Tom information on "Culture" and/or "Fair" in "Frankfurt at a Glance", but rather we could target him for a personalized information related to a major "Sports" event in "Frankfurt at a Glance": the "World Soccer Cup 2006".

Choosing the number of clusters and the scale of interest is a key for the clustering. Our approach is flexible and gives the owner of a web site the possibility to tune the clusters. The $G(x)$ does not necessarily have to be linear. The definition of function $G(x)$ is application dependent, it reflects the requirements of the application, and of course it influences how users at the end are clustered.

4   Implementation and Evaluation

We have implemented a prototype to evaluate the Takahe algorithm [5]. The module produces clusters of users based on the criteria and rules defined from the web site owner. It is a Java rich client application based on Eclipse 3.1.2 [3]. We used data sets with 1,000 (Set A), 2,500 (Set B), 5,000 (Set C), 50,000 (Set D) and 100,000 (Set E) users and 5 topics. We have benchmarked the Takahe algorithm ($g = 5$) against the k-means algorithm ($k = 20$) and against agglomerative hierarchical clustering using manhattan distance and complete linkage (HC) [2], [5]. We used these algorithms because they are widely known. In Table 4 the corresponding real execution times are shown using the 5 data sets. The test results confirmed the theoretical computation time of Takahe. The positive result is that Takahe depends $u \, log \, u$ on the number of users. On the contrary, when using k-means on larger data sets, the algorithm produces an out-of-memory exception (OoM). The same happened with HC on Set C.

We have also compared the computation time of the three algorithms based on the number of topic. The results are presented in Fig. 4. Takahe has a computation time $O((log\ u + |Tp|) * u))$, therefore it works well as long as the number of topics used is not too large.

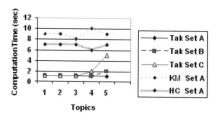

Fig. 4. Performance related to $|Tp|$

Table 4. Performance Results (sec)

	Takahe	K-Means	HC
SetA	1	9	7
SetB	2	OoM	86
SetC	8	OoM	OoM
SetD	22	-	-
SetE	37	-	-

As a result the Takahe algorithm can scale well for applications such as Web portals with a large set of registered visitors (e.g. 1,000,000) and hundreds of topics.

5   Related Work

Our clustering algorithm differs from most related work as it takes into account the time spent by the user on a page. Furthermore it looks at the topics associated to a set of pages, and clusters users based on their supposed interest in these topics. It gives an interpretation to the clusters as a value of a predefined scale of interest. Many related work cluster users on the base of sequences by visited pages, often also without inclusion of the time spent on it [4], [7]. Banerjee and Ghosh [1] include the time in a similarity function.

In some related work [4], [7], [13], [14], users are clustered over sequences of pages without looking at the content of the pages. We do not use sequences of pages; instead we look at each page's content to find the interests of the visitor. Instead of proceeding from a semantic structuring of a web site in folders [1], [4], [13], [14], we have automatically a reduction of the dimension by the allocation of the pages to topics.

6   Conclusion

We have presented a new algorithm to cluster web users based on their profiles. A word of caution is needed here, the authors are aware that this research area is sensitive in several issues, from the ethic perspective to the privacy and data protection issues. The technology presented in this paper needs to be carefully used and not misused. Users need to be aware of the existence of the profile, and most importantly they need to trust and know the usage of such profiles. This relationship of trust between the web site owner and his community is a of importance. Improper use of profiles will potentially harm rather than help. It is the responsibility of all of us in the research community to raise the awareness of such issues.

Acknowledgments

The authors would like to thank the members of the Gugubarra team: B. Brandt, D. Guettinger, S. Kaufmann, N. Mustaq and K. Tolle, who gave valuable feedback, helped with the implementation of the algorithm and reviewed earlier drafts of this paper.

References

1. Banerjee A, Ghosh J (2001) Clickstream Clustering using Weighted Longest Common Subsequences. In: Proceedings Web Mining Workshop, Chicago
2. Chakrabarti S (2003) Mining the Web. M. Kaufmann, San Francisco
3. Eclipse Foundation. web resource www.eclipse.org
4. Fu Y, Sandhu K, Shih M (1999) Fast Clustering of Web Users Based on Navigation Patterns. In: Proceedings of SCI/ISAS'99, Orlando
5. Guettinger D (2006) Rich-Client-Entwicklung am Beispiel von Clusterverfahren und Benutzerprofilen. MA thesis, DBIS, JW Goethe University, Frankfurt
6. Hartigan J, Wong M (1979) K-Means Clustering Algorithm. Applied Statistics
7. Hay B, Wets G, Vanhoof K (2001) Clustering navigation patterns on a website using a Sequence Alignment Method. Limburg University Centre
8. Hoebel N, Kaufmann S, Tolle K, Zicari R (2006) The Design of Gugubarra 2.0: A Tool for Building and Managing Profiles of Web Users. In: Proceedings of Web Intelligence Conference, 18-22 December, Hong Kong
9. Hoebel N, Kaufmann S, Tolle K, Zicari R (2006) The Gugubarra Project: Building and Evaluating User Profiles for Visitors of Web Sites. IEEE Workshop on Hot Topics in Web Systems and Technologies, 13-14 November, Boston
10. Mushtaq N, Tolle K, Werner P and Zicari R (2004) Building and Evaluating Non-Obvious User Profiles for Visitors of Web Sites. CEC 04, 6-9 July, San Diego
11. OpenDirectory Project. web resource www.dmoz.org
12. Turban E et al. (2004) Electronic Commerce 2004. Pearson Prentice Haal
13. Wang Q, Makaroff D, Keith Edwards H (2004) Characterizing Customer Groups for an E-Commerce Website. In: Proceedings of the 5th ACM Conference on Electronic Commerce, ACM Press, New York
14. Wang W, Zaïane O (2002) Clustering Web Sessions by Sequence Alignment. In: Proc. of the 13th Int. Workshop on DB and Expert Systems Applications

Providing Private Recommendations Using Naïve Bayesian Classifier

Cihan Kaleli and Huseyin Polat

Department of Computer Engineering, Anadolu University, 26470, Turkey
{ckaleli,polath}@anadolu.edu.tr

Summary. Today's CF systems fail to protect users' privacy. Without privacy protection, it becomes a challenge to collect sufficient and high quality data for CF. With privacy protection, users feel comfortable to provide more truthful and dependable data. In this paper, we propose to employ randomized response techniques (RRT) to protect users' privacy while producing accurate referrals using naïve Bayesian classifier (NBC), which is one of the most successful learning algorithms. We perform various experiments using real data sets to evaluate our privacy-preserving schemes.

1 Introduction

Online vendors employ collaborative filtering (CF) techniques for recommendation purposes. CF systems predict the preferences of active user (a), based on the preferences of others. Naïve Bayesian classifier (NBC) [6] is widely employed for CF. Providing accurate referrals are advantageous to online vendors because customers prefer returning to stores with better referrals and they search for more products to buy. CF has many important applications in e-commerce, direct recommendations, and search engines [1].

CF systems have a number of disadvantages [1]. They are a serious threat to individual privacy. They pose various, severe privacy risks like unsolicited marketing, price discrimination, being subject to government surveillance, and so on [2]. Due to privacy concerns, a significant number of people are not willing to divulge their information and users dislike data transfer [2]. Moreover, customer data is a valuable asset and it has been sold when some e-companies suffered bankruptcy [1]. If privacy is protected, people feel comfortable to give private data and contribute more truthful data.

We investigate how to achieve private recommendations efficiently based on the NBC using the randomized response techniques (RRT). We want to answer the following questions: *How can users contribute their personal information for CF purposes without compromising their privacy? How can the server provide referrals efficiently with decent accuracy without deeply jeopardizing users' privacy?* Our goal is to prevent the server from learning the true values of users' ratings and which items are rated by users including a.

K.M. Węgrzyn-Wolska and P.S. Szczepaniak (Eds.): Adv. in Intel. Web, ASC 43, pp. 168–173, 2007.
springerlink.com

2   Related Work

Canny proposes privacy-preserving collaborative filtering (PPCF) schemes [1] in which users control all of their own private data, where homomorphic encryption is used. In [8], achieving private referrals on item-item similarities is discussed. Users' privacy is achieved using the RRT. Although we employ the RRT for data perturbation, we use the NBC for referrals rather than using item-based CF algorithms. Partitioned data-based PPCF is discussed by [7]. They propose schemes to produce private referrals on integrated data without violating privacy. They discuss privacy-preserving protocols for providing predictions on vertically or horizontally partitioned data. Although such schemes are for partitioned data-based referrals, in our scheme here, we assume an existing central database, which consists of perturbed ratings from users.

Miyahara and Pazzani [6] employ the NBC for producing recommendations. The "naïve" assumption states that features are independent given the class label. Therefore, given its n feature values, the probability of an item belonging to $class_j$, where $j \in \{like, dislike\}$, is:

$$p(class_j | f_1, f_2, \ldots f_n) \propto p(class_j) \prod_{i}^{n} p(f_i | class_j), \tag{1}$$

where both $p(class_j)$ and $p(f_i | class_j)$ can be estimated from training data and f_i corresponds the feature value of target item for user i. To assign a target item (q) to a class, the probability of each class is computed, and the example is assigned to the class with the highest probability.

Warner [10] introduced the RRT as a technique to estimate the percentage of people in a population that has attribute A. The interviewer asks each respondent two related questions, the answers to which are opposite to each other. Using a randomizing device, respondents choose the first question with probability θ and the second question with probability $1-\theta$, to answer. The interviewer learns responses but does not know which question was answered.

3   Providing Private Recommendations Using the NBC

We use the RRT to disguise private data. Since the NBC-based CF is based on aggregate values of a data set, we hypothesize that *by combining the RRT with the NBC-based CF algorithms, we can achieve a decent degree of accuracy for PPCF*. We implement the RRT for an NBC-based algorithm [6]. To perturb the ratings vector for user u, V_u, u generates a random number (r_u) using uniform distribution over the range $[0, 1]$. If $r_u \leq \theta$, then u sends the true data, V_u. Otherwise, he/she sends the false data (exact opposite of the ratings vector). With probability θ, true data is sent while false data is sent with probability $1-\theta$. With privacy concerns, our goal is to achieve private referrals efficiently with decent accuracy. Since accuracy, privacy, and efficiency conflict, we want to have a good balance between them. Thus, we propose to use both one-group and multi-group schemes. Since CF systems perform two tasks (prediction for a single item and top-N recommendations), our proposed privacy-preserving schemes should achieve such tasks using the NBC.

One-Group Scheme. In this scheme [3], all ratings are put into the same group and all of them are either reversed together or left unaltered. The conditional probabilities estimated from masked data are the same as the ones computed from original data because all ratings are either reversed together or left the same. Thus, we achieve the same accuracy on masked data as with the original one. However, the privacy level is very low. If the server learns the true rating for only one item, it can obtain true votes for all items. Users can partition the items into M groups, where the RRT is used to perturb each group *independently* and $1 < M < m$. The decision is the same for all items in the same group, but the decisions for different groups are independent.

Multi-Group Schemes. Users group the items in the same way and disguise their ratings in each group independently. Although privacy improves, accuracy decreases due to increasing randomness. Since the server can calculate $p(class_j)$ values from a's data, the problem is how to compute $p(f_i|class_j)$ values from masked data. The server knows that the users send true or false data with probabilities θ and $1-\theta$, respectively. If we call the perturbed data Y_k and the true data X_k, and $\overline{X}_k$ represents the exact opposite of X_k (or false data), where $k = 1, 2, \ldots, M$, and k shows the group name, the server needs to find $p(X_k|Y_k = X_k)$ and $p(\overline{X}_k|Y_k = X_k)$ for each group, where $p(X_k|Y_k = X_k) + p(\overline{X}_k|Y_k = X_k) = 1$. $p(X_k|Y_k = X_k)$ can be calculated using the Bayes' rule, as follows:

$$p(X_k|Y_k = X_k) = \left[p(Y_k = X_k|X_k)p(X_k)\right]/p(Y_k = X_k), \qquad (2)$$

where $p(Y_k = X_k|X_k)$ is θ. The value of $p(Y_k = X_k)$ can be calculated from disguised data, while the value of $p(X_k)$ can be computed, as follows, using the facts that $p(Y_k = X_k|X_k) = \theta$ and $p(Y_k = \overline{X}_k|X_k) = 1 - \theta$:

$$p(Y_k = X_k) = \theta p(X_k) + (1 - \theta)p(\overline{X}_k). \qquad (3)$$

Eq. (3) can be solved for $p(X_k)$, using the fact that $p(X_k) + p(\overline{X}_k) = 1$:

$$p(X_k) = \left[p(Y_k = X_k) + \theta - 1\right]/(2\theta - 1). \qquad (4)$$

We get the following after replacing $p(X_k)$ with its equivalent in Eq. (2):

$$p(X_k|Y_k = X_k) = \left[\theta^2 + \theta p(Y_k = X_k) - \theta\right]/\left[2\theta p(Y_k = X_k) - p(Y_k = X_k)\right].$$

Since X_k and Y_k are ratings vectors, to find $p(Y_k = X_k)$, the server finds posterior probabilities for all items in each group k, selects the best one, and uses it as $p(Y_k = X_k)$. After finding $P = p(X_k|Y_k = X_k)$ values for each group, the server can now use them for providing predictions. The server needs to consider all possibilities to find the conditional probabilities. Since the disguised data can be true or false in each group, the ratings vector that the server received from a user can be one of the 2^M possible vectors of that user. Therefore, the server can estimate $C = p(f_i|class_j)$ values, as follows:

$$C = C_{(Y_1=T \wedge ... \wedge Y_M=T)}P^M + C_{(Y_1=T \wedge ... \wedge Y_{M-1}=T \wedge Y_M=F)}P^{M-1}(1-P) + ...$$
$$+C_{(Y_1=T \wedge Y_2=F \wedge ... \wedge Y_M=F)}P(1-P)^{M-1} + C_{(Y_1=F \wedge ... \wedge Y_M=F)}(1-P)^M,$$

where $Y_k = T$ and $Y_k = F$ mean the server considers the data in group k is true and false, respectively. We describe our results up to five-group because undesirable performance for schemes beyond five-group makes them not very useful. Our scheme can be extended to provide top-N recommendations. To prevent the server from learning rated items, users randomly select some unrated items' cells to be filled with fake ratings. Each user u finds the number of unrated items (m_{ut}) and uniformly randomly creates an integer, m_{ur}, over the range $(1, \gamma)$. They then choose f number of cells, and fill them, where $f = m_{ur} \times m_{ut}/100$. Each user u fills $\lfloor (m_{ur} \times m_{ut})/200 \rfloor$ randomly selected items' cells with 1 and the remaining cells with 0. To protect a's data, we use the 1-out-of-n Oblivious Transfer protocol [4]. a sends Y-1 randomly generated vectors and his/her true ratings vector to the server. After finding referrals, the server uses the 1-out-of-n Oblivious Transfer protocol to send them.

The server does not know the rated items due to fake ratings. Privacy can be measured with respect to the reconstruction probability (p) with which the server can obtain the true ratings vector of a user given disguised data. Thus, we can define the privacy level (PL) in terms of p, as follows [9]: PL $= (1-p) \times 100$, where p can be written in terms of $p(X_k|Y_k = X_k)$ and M:

$$p = \left[p(X_k|Y_k = X_k) \right]^M = \left[(\theta^2 + \theta Y - \theta)/(2\theta Y - Y) \right]^M, \tag{5}$$

where $Y = p(Y_k = X_k)$. With increasing p, PL decreases. To decrease p, the randomness should be increased, which makes accuracy worse. With increasing M, p decreases, while PL increases. The value of p depends on θ, M, and the value of Y or X, where $X = p(X_k)$.

4 Experiments

We used Jester [5] and MovieLens Million (MLM), which was collected at the University of Minnesota (www.cs.umn.edu/research/Grouplens), in our experiments. Using classification accuracy (CA) and F-measure (FM), we measured accuracy. Using the similar methodology conducted by [6], we first transformed numerical ratings into two labels (*like*, *dislike*). We then randomly selected 500 test and 1,000 training users who have rated at least 80 movies from MLM. We also randomly selected 500 test and 1,000 training users who have rated at least 60 jokes from Jester. Finally, we randomly selected 60 rated items for MLM and 40 for Jester as a training set, and 20 items for MLM and Jester as a test set. For each a from the test set, we found referrals randomly selected 5 rated items. We ran data disguising 10 times.

To show how number of features or users (n) affects our result, we performed testings while changing n from 100 to 1,000 for both data sets. We fixed the θ at 0.70 and employed three-group scheme. Since CA and FM values are similar, we only showed CAs in Table 1. As expected, the results, based on masked data, become better with increasing n values. They also converge to the results on original data with increasing n because aggregate data can be estimated with decent accuracy if enough data is available.

Accuracy varies for different θ values because randomness differs. We performed trials, where we used 200 training users from Jester and MLM. We set M at 3, where we varied θ from 0.51 to 1.00 because complementary θ values give the same results. We showed CAs and FMs in Table 2. When θ is 1, we achieve the same accuracy with original data because users send true data. However, when θ is 0.51, we add the largest randomness; and with decreasing θ values towards 0.51, accuracy worsens. Accuracy is more likely to improve when more features are used because we only employed 200 features.

To show how data partition affects our results, we performed testings with varying M. We used 200 training users from Jester and MLM, where $\theta = 0.70$. We performed experiments for up to five-group scheme. Since our results show similar trends for both data sets, we only showed results for MLM in Table 3. As seen from the table, our results become better with decreasing M values because we add less randomness to original data. Up to five-group scheme, it is still possible to provide accurate private referrals.

Table 1. CA With Varying n Values

	Jester				MLM			
n	100	200	500	1,000	100	200	500	1,000
Original Data	68.28	68.56	69.45	69.68	74.24	77.30	79.80	80.28
Masked Data	58.45	61.23	63.92	65.56	72.40	75.34	78.40	79.58

Table 2. Accuracy With Varying θ Values

	Jester				MLM			
θ	0.51	0.70	0.85	1.00	0.51	0.70	0.85	1.00
Classification Accuracy	55.52	61.23	63.23	68.56	75.00	75.34	76.96	77.30
F-Measure	57.98	62.45	62.89	73.68	85.27	86.94	89.78	90.89

Table 3. Accuracy With Varying M Values

	Classification Accuracy				F-Measure			
M	1	2	3	5	1	2	3	5
	77.30	77.12	75.34	65.45	90.89	89.54	86.94	76.34

5   Conclusion and Future Work

We presented solutions to achieving private referrals on the NBC using the RRT. Our solutions make it possible for servers to collect private data without greatly compromising users' privacy. Experiment results show that our schemes allow providing referrals with decent accuracy. To obtain a balance between accuracy, privacy, and efficiency, the parameters of our schemes can be adjusted. We will study whether we can still provide accurate, if users perturb their data using different θ values and group schemes.

References

1. J. Canny. Collaborative filtering with privacy via factor analysis. In *Proceedings of the 25th ACM SIGIR'02*, pages 238–245, Tampere, Finland, August 2002.
2. L. F. Cranor. 'I didn't buy it for myself' privacy and E-commerce personalization. In *Proceedings of the 2003 ACM Workshop on Privacy in the Electronic Society*, pages 111–117, Washington, DC, USA, October 2003.
3. W. Du and Z. Zhan. Using randomized response techniques for privacy-preserving data mining. In *Proceedings of the 9th ACM SIGKDD'03*, Washington, DC, USA, August 2003.
4. S. Even, O. Goldreich, and A. Lempel. A randomized protocol for signing contracts. *Communications of the ACM*, 28:637–647, 1985.
5. D. Gupta, M. Digiovanni, H. Narita, and K. Goldberg. Jester 2.0: A new linear-time collaborative filtering algorithm applied to jokes. In *Proceedings of the Workshop on Recommender Systems, ACM SIGIR'99*, Berkeley, CA, USA, August 1999.
6. K. Miyahara and M. J. Pazzani. Improvement of collaborative filtering with the simple Bayesian classifier. *IPSJ Journal*, 43(11), November 2002.
7. H. Polat and W. Du. Privacy-preserving top-N recommendation on horizontally partitioned data. In *Proceedings of the 2005 IEEE/WIC/ACM International Conference on Web Intelligence*, Paris, France, September 2005.
8. H. Polat and W. Du. Achieving private recommendations using randomized response techniques. In *Proceedings of the 10th Pacific-Asia Conference on Knowledge Discovery and Data Mining*, Singapore, April 2006.
9. S. J. Rizvi and J. R. Haritsa. Maintaining data privacy in association rule mining. In *Proceedings of the 28th VLDB Conference*, Hong Kong, China, 2002.
10. S. L. Warner. Randomized response: A survey technique for eliminating evasive answer bias. *Journal of the American Statistical Association*, 60(309):63–69, 1965.

A Technique for Representing Course Knowledge Using Ontologies and Assessing Test Problems

Javed Khan and Manas Hardas

Kent State University, Kent, Ohio 44240, USA
{javed,mhardas}@kent.edu

Summary. In this paper we present a novel method for qualitative assessment of educational resources, specifically 'test problems'. For test problems, the basic elements of design are in the form of concepts arranged in a hierarchy. Course concept knowledge can be represented in the form of prerequisite relation based ontology using which, assessment and information extraction from test problems is possible. Using a schema based on Web Ontology Language (OWL), course ontologies can be represented in a standard and sharable way. Some synthetic parameters for the assessment of a test problem in its concept space are introduced. The assessment system can be further extended to analyze any type educational resource.

Keywords: ontology, assessment, complexity, knowledge, test problems.

1 Introduction

The web has greatly facilitated online sharing of course material. There have been many organized attempts to create large digital courseware libraries to promote sharing like NIST's Materials Digital Library Pathway, NSDL Digital Libraries, OhioLink, ACM Professional Development Center etc. MITSs Open Course Ware (OCW) project has more than 1000 course materials freely available, Universia maintains translated versions of OCW courses in 11 languages, China Open Resources for Education (CORE) has a goal to include Chinese versions of the OCW. The amount of digital courseware content available online is huge. Surprisingly, the real sharing of the materials among the educators is still very low. In OCW it has been noted that only 16% of the users are educators out of which not more than 26% use it for planning their course or teach a class [1]. Most courseware today, on the web or otherwise is not accompanied with a conceptual design. There is no well formed encoding principle for capturing and sharing the schema associated with course materials. To make this digital content reusable, the associated meta data should be consistently represented. Traditionally, concepts maps are used to represent the concept space for the course knowledge like for biology courses [2] and many other areas however they are too expressive and consequently contain more semantic relationships than necessary for effective computation. Ontologies provide a means to effectively map this knowledge into concept hierarchies. Standardization of semantic

K.M. Węgrzyn-Wolska and P.S. Szczepaniak (Eds.): Adv. in Intel. Web, ASC 43, pp. 174–179, 2007.
springerlink.com © Springer-Verlag Berlin Heidelberg 2007

representation standards like RDF and OWL offers great technical platform to represent the concept knowledge space symbolized by ontologies and greatly improve its machine usability. In this paper we present an approach to course knowledge representation using ontology in an expressible and computable format using has-prerequisite relationships where concepts involved in teaching a course are arranged in an hierarchical order of learning. Another original approach for specifically pointing out areas in ontologies of maximum relevance called as *CSG extraction* is given. We investigate the properties of test problems by following a purely knowledge based approach for assessment using course ontologies. The system has the potential to make the already available test-ware resources on the web reusable.

2 Course Ontology

Most ontologies today are so extensive in the breadth of knowledge that processing of these ontologies becomes a gargantuan computation task. There needs to be a way to efficiently process the relevant information to give results in minimum time and complexity of computation. Ontologies are made up of individuals, classes, attributes and relationships. The has-prerequisite relationship in a course ontology refers to the prerequisite understanding of the child node needed to understand the parent node. On the whole the course ontology is constructed in such a hierarchical fashion that the children of node represent the knowledge required to understand the parent node. A node is characterized by two values namely, self-weight and prerequisite weight, Figure 1. The self-weight of a concept node is the amount of knowledge required to grasp the concept. To understand the concept entirely, knowledge of the prerequisite concepts is also required, which is given by the prerequisite weight of the node. Another value which characterizes the course ontology representation is the link weight. The link weight again is the value of the semantic importance of child concept to the parent concept. Child concepts imperative in the understanding of parent concepts will have a greater link weights than the others.

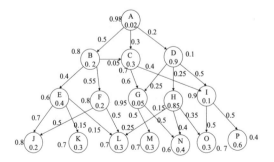

Fig. 1. Example Concept space graph, T(A)

2.1   Representing Course Ontologies

The course ontology is mathematically defined in the form of a concept space graph (CSG). A CSG is a view of the concepts space distribution in the domain of a particular course.

A concept space graph $T(C, L)$ is a projection of a semantic net with vertices C and links L where each vertex represents a concept and each link with weight $l(i, j)$ represents the semantics that concept c_j is a prerequisite for learning c_i, where $(c_i, c_j) \in C$ and the relative importance of learning c_j for learning c_i is given by the weight. Each vertex i in T is further labeled with self-weight value $W_s(i)$ and cumulative prerequisite set weight $W_p(i)$.

A CSG with root A is represented as T (A) in Figure 1. For any node in the CSG, the sum of self-weight and prerequisite weights and the sum of the link weights for all children is 1.

2.2   Prerequisite Effect of a Node

The notion of node path weight is introduced to compute the effect a prerequisite node has on a root node through a specific path. A single node can have different prerequisite effect on a root through different paths.

When two concepts x_0 and x_t are connected through a path consisting of nodes given by the set $[x_0, x_1, ..., x_t]$ then the node path weight between these two nodes is given by:

$$\eta(x_0, x_t) = W_s(x_t) \prod_{m=t}^{1} l(x_{m-1}, x_m) * W_p(x_{m-1}) \tag{1}$$

In the Figure 2, concept L is connected to B through E and F. Therefore the prerequisite effect it has on B is dependent on the prerequisite effect both E and F have on B respectively. From the node path weight calculations we can see that L has a stronger prerequisite effect on B through F rather than E. This is because, L is more important to F (0.5) than E (0.15), prerequisite importance of L is more to F (0.8) than E (0.6) and subsequently F (0.55) is more important to B than E (0.4). Thus node path weights takes into consideration not only the singular effect a node has on its immediate parent but also the combined prerequisite effect a node would have to a root, B in this case, along a certain path.

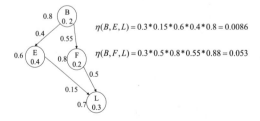

Fig. 2. Calculating prerequisite effect of a node along a path; Node path weight

3 Assessment Approach and Projection Graph

The assessment process is essentially a two step approach. The first main step is the extraction of the relevant concepts from the CSG and is called as *ŞCSG extraction Ť*. A generalized CSG can be vast and therefore it is irrational to process the whole ontology. We define a pruned sub-graph called as *projection graph* which cuts the computation based on a limit on propagated semantic significance. The pruning is achieved by introducing a variable called as the projection threshold coefficient (λ). By varying the threshold coefficient the size of the computable projection graph can be varied and thus the semantic significance. The threshold coefficient can be thought of as a parameter which can inversely set the depth to which the topic has been taught.

3.1 Projection Graph

Given a CSG, $T(C, L)$, with local root concept x_0, and projection threshold coefficient λ, a projection graph $P(x_0, \lambda)$ is defined as a sub graph of T with root x_0 and all nodes x_t where there is at least one path from x_0 to x_t in T such that node path weights $\eta(x_0, x_t)$ satisfies the condition: $\eta(x_0, x_t) \geq \lambda$.

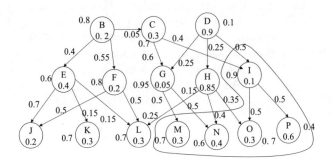

Fig. 3. Projection of concepts B, D and overlapping region

The projection set for x_0 is $[x_0, x_1, ..., x_t]$ represented as $P(x_0, \lambda) = [x_0^{x_0}, x_1^{x_0}, ..., x_t^{x_0}]$ where x_i^j represents the ith element of the projection set of node j. The projections for B and D and the overlapping region of their projections is shown in Figure 3. All nodes that satisfy the condition of node path weights greater than threshold coefficient (through any path) are included in the projection.

4 Assessment Parameters

The second step in the assessment approach is to apply the parameters algorithms to the extracted projection graph to obtain the assessment values.

4.1   Coverage

The coverage of a question gives a cumulative prerequisite effect of the projection graph on the knowledge required to answer a particular question. Coverage of a concept is a direct indicator to the scope of the question in context of the concept space of the course. Formally, *ςcoverage of a node x_0 with respect to the root node r is defined as, the product of the sum of the node path weights of all nodes in the projection set $P(x_0, \lambda)$ for the concept x_0 , and the incident path weight $\gamma(r, x_0)$ from the root r$\check{T}$. If the projection set for concept node x_0, $P(x_0, \lambda)$ is given by then the coverage for node x_0 about the ontology root r is defined as,*

$$\alpha(x_0) = \gamma(r, x_0) * \sum_{m=0}^{n} \eta(x_0, x_m) \tag{2}$$

where $\gamma(r, x_0)$ is called as the *Incident Path Weight* and

$$\gamma(r, x_0) = \frac{\eta(x_0, x_n)}{W_s(x_n)} = \frac{\eta(x_0, x_n)}{\eta(x_n, x_n)} \tag{3}$$

Total coverage of multiple concepts in a problem given by set $[C_0, C_1, ..., C_n]$ is, $\alpha(T) = \alpha(C_0) + \alpha(C_1) + ... + \alpha(C_n)$.

The node path weight defines the prerequisite effect of a node to its designated root. Therefore the summation of the node path weights of all the nodes in the projection set gives the cumulative prerequisite effect of the nodes in the projection graph on their respective mapped concept roots. The concepts in the projection graph in turn are the concepts which are required to understand a particular concept, controlled by the threshold coefficient. The *coverage* is thus, the amount of knowledge required to answer or rather understand a particular concept.

4.2   Diversity

Diversity is calculated by measuring the effect of common and uncommon prerequisite concepts from the projections of the mapped concepts. Diversity is formally defined as *$\varsigma$the ratio of summation of node path weights of all nodes in the non-overlapping set to their respective roots, and the sum of the summation of node path weights of all nodes in the overlap set and summation of node path weights of all nodes in the non-overlap set.$\check{T}$* Consider a problem maps to set of concepts $C = [C_0, C_1, ..., C_n]$ with projections $P(C_0, \lambda), P(C_1, \lambda), ..., P(C_n, \lambda)$ and the non-overlapping and overlapping sets given by $N = [N_0, N_1, ..., N_p]^i$ and $O = [O_0, O_1, ..., O_q]^j$ where i and j are the local root parents of any element from N and O and $\forall i, j \in C$. Diversity is given by,

$$\Delta = \frac{\sum_{m=1}^{p} \eta(i, N_m^i)}{\sum_{m=1}^{p} \eta(i, N_m^i) + \sum_{m=1}^{q} \eta(j, O_m^j)} \tag{4}$$

4.3   Conceptual Distance

Conceptual distance is a measure of distance between two concepts with respect to the ontology root. Alternatively conceptual distance measures the similarity between two concepts by quantifying the distance of the concepts from the ontology root. Formally it is defined as *Şthe log of inverse of the minimum value of incident path weight (maximum value of threshold coefficient) which is required to encompass all the mapped concepts from the root conceptŤ*. The conceptual distance parameter is designed in such a way that it should be sensitive to the depth of the concepts. Hence it is a function of maximum threshold coefficient required to cover all the nodes from the ontology root. Incident path weight (γ) of a concept to the root is equivalent to the threshold coefficient (λ) required to encompass the node. If question asks concept set $C = [C_0, C_1, ..., C_n]$ then the conceptual distance from the root concept r is,

$$\delta(C_0, C_1, ..., C_n) = \log_2\left(\frac{1}{min[\gamma(r, C_0), \gamma(r, C_1), ..., \gamma(r, C_n)]}\right) \qquad (5)$$

References

1. 2004 MIT OCW Program Evaluation Findings Report (June 2006). $http : //ocw.mit.edu/NR/rdonlyres/FA49E066 - B838 - 4985 - B548 - F85C40B538B8/0/05_{Prog_Eval_Report_Final}.pdf$
2. Edmondson, K., Concept mapping for Development of Medical Curricula. Presented at the annual meeting of the American Educational Research Association (Atlanta, GA, April 12-16, 1993). 37p.
3. Li, T and S E Sambasivam.ăQuestion Difficulty Assessment in Intelligent Tutor Systems for Computer Architecture.ăIn The Proceedings of ISECON 2003, vă20 (San Diego): ğ4112. ISSN:ă1542-7382. (Also appears in Information Systems Education Journal 1:ă(51). ISSN:ă1545-679X.)
4. Rita Kuo, Wei-Peng Lien, Maiga Chang, Jia-Sheng Heh, Difficulty Analysis for Learners in Problem Solving Process Based on the Knowledge Map. International Conference on Advanced Learning Technologies, 2003, 386-387.
5. Silva, L., Oliveira, J.P., (2004). Adaptive Web Based Courseware Development using Metadata Standards and Ontologies. AH 2004, Eindhoven.

Adaptive Decision Support System Using Web-Users Profile Data

W. Kosiński and D. Kowalczyk

[1] Polish–Japanese Institute of Information Technology
   ul. Koszykowa 86 , PL-02-008 Warszawa, Poland
   wkos@pjwstk.edu.pl
[2] Department of Mathematics, Physics and Technology
   Kazimierz Wielki University in Bydgoszcz
   ul. Chodkiewicza 30, PL-85-064 Bydgoszcz

1 Introduction

For most of us the term *Web user profile* should contain user's URL address, sometimes user's name and e-mail address, and some other components such as employeeŠs region and home organization if the user belongs to some company. However, when we think on a Web-based patient education that allows for the delivery of educational content to the patient, or on e-learning communities and on a system that serves each individual of the community, this information may not be sufficient. In the latter case the user profile should contain personal preferences of the individual, goals and needs, as well as the history of activity in the community, while in the former - patient parameters describing user's history, results of recent examinations, type of disease and so on. If mobile communicators are used (i.e. WAP) the CC/PP profile appears which is a more complex collection of capabilities and preferences associated with the user such as the user's position (which can change with time), and the characteristic of the user's hardware (e.g.phone).

In the examples above one of main issues is to construct appropriate algorithms and an adaptive (so-called intelligent)decision support information system that supply the Web user with the required information or educational materials. Most of construction methods are based on some heuristic approaches. On the other hand, one can construct such system copying the construction process of an *approximator*[7, 8] in which unknown function relationship is looked for when a set of training data TRE relevant for the wanted relationship is given. For our purpose we assume that an expert has supplied us with a number, say P, of examples which form TRE, each element of which is an ordered pair: a given user profile and appropriate decision. Each user profile has been encoded in an n dimensional vector from $\mathbf{R}^n$ while the decision - in a number. So, our user profile-decision pairs $(\boldsymbol{x}, y)$ form the database TRE, called at other occasions [7] the *training set*. It is a subset of $\mathcal{X} \times \mathcal{Y}$, n-dimensional input space, say $\mathcal{X}$, and 1D output one, say $\mathcal{Y}$.

K.M. Węgrzyn-Wolska and P.S. Szczepaniak (Eds.): Adv. in Intel. Web, ASC 43, pp. 180–185, 2007.
springerlink.com

Different users are characterize by different profiles, however, some of them are similar and one can think on clustering them. In the construction of the system an appropriate cluster analysis should proceed the training, the adaptation process, while the use of two different tools: an adaptive fuzzy inference system and feed-forward neural networks, may be more promising [7, 11].

In the paper the construction procedure begins with a kind of data mining of profiles in which two families of clusters are constructed on TRE. Then a module of two-conditional fuzzy rules consequent parts of which are outputs of artificial feed-forward neural networks, is designed. Neural networks from consequent parts of the rules should be trained on corresponding pairs of clusters as on their training sets, and in this way users profiles supply a knowledge in the final designing stage.

2 Knowledge Extraction from Users Profiles

In the standard approach dealing with approximation problems the knowledge about the function to be approximated is contained in the set TRE, which represents a discrete number of points from the graph of unknown (i.e. to be looked for) function relationship between values of profiles and corresponding decisions since in each pair $p^q = (x^q, y^q) \in$ TRE the value y^q is the so-called desired value corresponding to the user profile represented by (or better to say - encoded in) given input vector x^q.

We describe two methods of extracting knowledge from the set TRE: a seed growing approach for clustering problem of large numerical multidimensional data set [2] and the evolutionary approach [3, 9] for inference systems.

Clustering via seed growing algorithm. The approach used in [2] is based on image segmentation known in the medical, digital picture segmentation to extract the significant information from images and to improve the interpretation process realized by the end-user, e.g. a physician. However, in medical images the number of possible classes is given by the physician explicitly together with the seed pixel for each class. Moreover, in the set TRE it is a lack of a natural neighboring topology of the images that is based on the concept of the 4- or 8-connectedness.

The seed growing algorithm is composed of two parts:

Rough clustering according to y in which a fuzzy histogram of the variability of y is used, and is divided into the following steps:

1. create a histogram of the variability of y;
2. fuzzyfy it by calculating its convolution with a Gauss-like function;
3. use the minima of the fuzzy histogram as boundaries of subintervals in y;
4. divide all the elements (pairs (x, y)) of TRE into clusters; pairs with y's belonging to the same subinterval form the same cluster.

Exact clustering according to x and y in which the seed growing approach is developed after the rough clustering [2].

At the end of this clustering analysis we get a covering $\{\mathcal{K}_{1h}, h = 1, 2, ..., M_1\}$, i.e. a union of clusters that covers the whole set TRE.

Clustering via evolutionary algorithm. In [9] an evolutionary algorithm for extracting the knowledge from the database by splitting it into clusters was proposed and implemented in [3]. Here we sketch only its main features.

We are distinguishing two types of evolutions: external, at the level of clusters, and internal, at the level of training pairs.

We are introducing: three types of *selection operators* for the population of clusters: roulette, tournament selection, proximity selection, as a combination of two others, and four types of genetic operators (cf.[9]) for the population of clusters: unification operator that acts on a pair of clusters and produces a new cluster as a union of both parents, crossover operator that exchanges parts of two clusters, separation operator that produces two other clusters by splitting a cluster into two, and global mutation operator that acts on an individual covering, regarded as a family of clusters producing a new covering.

In the first stage m independent evolutionary processes of m coverings by creating m initial coverings are performed. It is done with the help of the histogram as in the previous approach. During the evolutionary process for each generation one of the genetic operators that acts on clusters (or pairs of clusters) are applied. Then one of the selection operators to the whole population is used. In this way new covering is constructed that forms the population for the next generation. After a fixed number of generations a global fitness (evaluation) function to each covering is applied and then a population of all coverings is formed.

3 Adaptive Fuzzy Inference System

At the end two coverings of TRE by clusters are ready, i.e. two families of clusters $\{\mathcal{K}_{\alpha 1}, \mathcal{K}_{\alpha 2}, ..., \mathcal{K}_{\alpha M_\alpha}\}$, $\alpha = 1, 2$, such that

$$\mathsf{TRE} = \bigcup_{h=1}^{M_1} \mathcal{K}_{1h} \text{ and } \mathsf{TRE} = \bigcup_{k=1}^{M_2} \mathcal{K}_{2k} , \qquad (1)$$

where each $\mathcal{K}_{1h}$, $\mathcal{K}_{2k} \subset \mathbf{R}^{n+1}$. The number of elements in each covering may be different, i.e. $M_1 \neq M_2$, and each cluster may contain different number of training pairs from TRE, i.e. by $N_{1h} \neq N_{2k}$, where N_{1h} is the number of points of $\mathcal{K}_{1h}$ and N_{2k} of $\mathcal{K}_{2k}$, where $h = 1, 2, ..., M_1, k = 1, 2, ..., M_2$.

Now the projection of each $\mathcal{K}_{ih}, \mathcal{K}_{2k} \subset \mathbf{R}^{n+1}$ on the input space $\mathcal{X} \subset \mathbf{R}^n$ forms two families $\{\mathcal{X}_1, \mathcal{X}_2, ..., \mathcal{X}_{M_\alpha}\}, \alpha = 1, 2$, of subdomains (input clusters) that form two coverings of the input data $\boldsymbol{x}'s$ from $\mathcal{X}$. To each cluster we can relate its centroid $\boldsymbol{a}^{1h}$ by

$$\boldsymbol{a}^{1h} = (N_{1h})^{-1} \sum_{j=1}^{N_{1h}} \boldsymbol{x}_j^{1h} , \qquad (2)$$

where points $x_j^{1h}, j = 1, 2, ..., N_{1h}$, belong to $\mathcal{X}_{1h}$; in similar way we define centroid to each cluster $\mathcal{X}_{2k}$.

In the input domain to each cluster we will relate a fuzzy set, forming in this way two fuzzy coverings [7] of the input domain $\mathcal{X}$. Corresponding membership function of each set will take into account the structure of those clusters, their spread, thanks to two families of scatter (variance-covariance) matrices S_{1h} and S_{2k} corresponding to each input cluster $\mathcal{X}_{1h}$ and $\mathcal{X}_{2k}$, where $h = 1, 2, ..., M_1, k = 1, 2, ..., M_2$, by the formula [1]:

$$S_{1h} = \frac{1}{N_{1h}} \sum_{j=1}^{N_{1h}} (x_j^{1h} - a^{1h}) \otimes (x_j^{1h} - a^{1h}), S_{2k} = \frac{1}{N_{2k}} \sum_{j=1}^{N_{2k}} (x_j^{2k} - a^{2k}) \otimes (x_j^{2k} - a^{2k})$$
(3)

where $\otimes$ denotes the tensor product of two vectors. Matrices S_{1h}, S_{2k}, are symmetric and positive semi-definite;

For simplicity we assume the matrices $S_h, S_k, h = 1, 2, ..., M_1, k = 1, 2, ..., M_2$ are nonsingular.

One can go beyond the case used in [8] and admit more general extraction method following the case described by the second author somewhere else.

Then we construct one family $\{\mathcal{R}_m : m = 1, 2, ..., Q\}$, with $Q = M_1 \cdot M_2$, of two-conditional generalized fuzzy rules of Takagi–Sugeno–Kang type

$$\textbf{if } x \textbf{ is } A_h \textbf{ and } x \textbf{ is } B_k \textbf{ then } y \textbf{ is } C_m,$$
(4)

with consequent parts C_m as an output of a single mapping neural network (SIMNN)[7]. This leads to the **adaptive fuzzy-neural inference system (AFNIS)**.

Premise part of each $\mathcal{R}_m$ contains a pairs of fuzzy sets A_h, B_k, for each pair $(1h, 2k)$, defined through the matrices S_{1h}, S_{2k} and centroids (a^{1h}, a^{2k}) (2)-(3) by their membership functions assumed as generalized Gaussian ones

$$\mu_{A_h}(x) = d^{1h} \exp(-0.5((x - a^{1h}) \cdot S_{1h}^{-1}(x - a^{1h}))^{b^{1h}}),$$
(5)

and similar form for B_k. Here S_{1h}^{-1} denotes the inverse of S_{1h}. In the final stage of the training process on the whole set TRE the adaptation will undergo the parameters d^{1h}, d^{2k} and b^{1h}, b^{2k} where $h = 1, 2, ..., M_1, k = 1, 2, ..., M_2$. Adaptable parameters d's and b's make the system more flexible [7].

The membership functions of A_h and B_k with a composition operation $\star$ defined, e.g. by a t-norm or by an algebraic operation in the case of ordered fuzzy sets (cf. [4, 5]), lead to the definition of the so-called *normalized level of activity* $v_m(x)$ of the m-th fuzzy rule $\mathcal{R}_m$ at x, used in the aggregation,

$$v_m(x) = \mu_{A_h}(x) \star \mu_{B_k}(x) \{ \sum_{h'=1}^{M_1} \sum_{k'=1}^{M_2} \mu_{A_{h'}}(x) \star \mu_{B_{k'}}(x) \}^{-1}.$$
(6)

However, if the fuzzy sets and their membership functions involved will be regarded as ordered fuzzy sets proposed by the first author in [4, 5], then all algebraic operations are for our disposal and t-norm are no more necessary[1].

Each consequent part C_m will be a function relation of the type

$$z_m = f_m(\boldsymbol{x}, \Omega_m) = \sum_{j=0}^{l} \omega_{mj}^{II} \sigma_m (\sum_{i=0}^{n} \omega_{mji}^{I} x_i) , \qquad (7)$$

where $\Omega_m = (\boldsymbol{\omega}_{mj}^{II}, \boldsymbol{\omega}_{mji}^{I})$ are weight vectors of the m-th **SIMNN**; the zero component x_0 of $\boldsymbol{x}$ is equal to 1 and incorporates the bias ω_{mj0}^{I}. The networks have been designed for the training pairs $\boldsymbol{p}$ from the union of clusters $\mathcal{X}_{1h} \cup \mathcal{X}_{2k}$. Notice that in this paper we make the next step beyond the known Takagi-Sugeno-Kang fuzzy rules that was proposed in previous papers [8, 11].

Each **SIMNN** contains one hidden layer of l nodes and two–parameter family of generalized sigmoidal activation functions (cf. [7, 8]): $\sigma_m(z) = (r_m)^{-1}(1 + \exp(-\delta_m z))$. Parameters r_m and δ_m give flexibility in the adaptation process. Output layer nodes have the same linear activation function.

The overall output $z(\boldsymbol{x})$ of the constructed system of two-conditional rules will be the convex (due to (6)) combination of all Q, given by

$$z = f(\boldsymbol{x}, \Theta, \Omega) = \sum_{m=1}^{Q} v_m(\boldsymbol{x}) f_m(\boldsymbol{x}, \Omega_m) . \qquad (8)$$

Here Ω is a collection of all vectors $\Omega_m, m = 1, 2, .., Q$, and Θ containing all d's and b's parameters, forms a vector of parameters.

4 Final Stage of Training and Conclusions

Each individual output $z_m = f_m, m = 1, 2, ..., Q$ from the single neural network 3 trained on the corresponding clusters $\mathcal{K}_{1h}$ and $\mathcal{K}_{2k}$ gives rise to the overall output $z = f(\boldsymbol{x}, \Theta, \Omega)$. Constructed in Sec. 3 **AFNIS** presented in (8) needs the last stage of adaptation of the components of Θ representing free parameters of membership functions in (5). To this end the new error function

$$E(\Theta, \boldsymbol{x}, y) := \frac{1}{Q}|f(\boldsymbol{x}, \Theta, \Omega) - y|^2 = \frac{1}{Q}|\sum_{m=1}^{Q} v_m(\boldsymbol{x}) f_m(\boldsymbol{x}, \Omega_m) - y|^2, \qquad (9)$$

will be minimize over all points $(\boldsymbol{x}, y)$ taken from **TRE**. The gradient descent method with some generalization or an evolutionary (genetic) algorithm with non-vanishing mutation for its convergence [6] can be implemented for this purpose, since the error function is non-quadratic in the variables Θ.

---

[1] In his Ph.D. Thesis [10] Prokopowicz has invented different aggregation procedures and methods of determination of the level of activation of two–conditional fuzzy rules with ordered fuzzy numbers, which can be adapted for the case of ordered fuzzy sets, as well.

Acknowledgement. The work on the paper was partially done by W.K. in the framework of Ministry of Science and Higher Education Project No. 3 T11 C007 28.

References

1. Anderberg MR(1973) Cluster Analysis for Applications. Probability and Mathematical Statistics, Academic Press, New York
2. Kieś P, Kosiński W , Weigl M (1997) Seed growing approach in clustering analysis. In Intelligent Information Systems. Proceedings of the VI-th International Workshop on Intelligent Information Systems, Zakopane, 9-13 June, 1997, Instytut Podstaw Informatyki, PAN, Warszawa, pp. 7–15
3. Koleśnik R , Koleśnik L , Kosiński W(1999) Genetic operators for clustering analysis, In: Intelligent Information Systems VIII, Proc. of the Workshop held in Ustroń, Poland, 14-18 June, 1999, Instytut Podstaw Informatyki, PAN, Warszawa, pp. 203–208
4. Kosiński W (2006) On fuzzy number calculus. Int J Appl Math Comput Sci 16: (1), 51–57
5. Kosiński W, Prokopowicz P, Ślęzak D (2003) Ordered fuzzy numbers. Bulletin of the Polish Academy of Sciences, Sér Sci Math 51: (3), 327–338
6. Kosiński W , Kotowski S , Socała J (2006) On asymptotic behaviour of a simple genetic algorithm, In: Kłopotek M., Wierzchoń S., Trojanowski K., (eds) Intelligent Information Processing and Web Mining, Proceedings of the International IIS: IIPWM"06 Conference held in Ustroń, Poland, June 19-22, Springer, Advances in Soft Computing, Heidelberg, pp. 55–64
7. Kosiński W , Weigl M (1998) General mapping approximation problems solving by neural networks and fuzzy inference systems. Systems Analysis Modelling Simulation 30: (1), 11–28
8. Kosiński W , Weigl M (2000) Adaptive information systems for data approximation problems. In : Fuzzy Control Theory and Practice, Hampel R., Wagenknecht M., Chaker N.(eds); Physica-Verlag, Springer, Heidelberg New York, pp. 109–120
9. Kosiński W , Weigl M , Michalewicz Z (1998) Evolutionary domain covering of an inference system for function approximation. In: Evolutionary Programming VII Port V. W., Saravanam N., Waagen D. and Eiben A. E. (eds), Proceedings of the 7th International Conference, EP'98, San Diego, California, USA, March 25-27, 1998, NCS, vol. 1447, New York (1998), pp. 167–180
10. Prokopowicz P (2005) Algorithmization of operations on fuzzy numbers and its applications (Algorytmizacja działań na liczbach rozmytych i jej zastosowania), Ph. D. Thesis (in Polish), IPPT PAN, Warszawa, kwiecień
11. Weigl M , Kosiński W (1996) Approximation of multivariate functions by generalized adaptive fuzzy inference network. In: Proceedings of the 9-th International Symposium on Methodologies for Intelligent Systems ISMIS'96, Zakopane, June, 1996, IPI PAN, Warszawa, pp. 120–133

Partial Similarity Based Retrieval of Images in Distributed Databases

Juliusz L. Kulikowski and Malgorzata Przytulska

Institute of Biocybernetics and Biomedical Engineering Polish Academy of Sciences
jlkulik@ibib.waw.pl, gosia@ibib.waw.pl,
IBIB@ibib.waw.pl

Summary. The proper balance between a symmetric and asymmetric cell division is crucial for the neural stem cell maintenance both *in vitro* and *in vivo*. These conditions are provided by specific regions of the brain called neural stem cell niches and *in vitro* occur in neurospheres or adherent clones. A method and a tool for cell culture growth monitoring applied in the investigation of the clonally growth of HUCB-NSC (Human Umbilical Cord Blood derived Neural Stem Cells) line, as an *in vitro* model of the neural stem cell niche, is proposed.

1 Introduction

Development of computer networks in the last decades opened new perspectives in wide access to the resources of experimental data stored in scientific, medical and/or technological databases distributed over the world. A great deal of experimental data has the form of images of various natural objects: from traces of nuclear particles through biological specimens, human organs, aerial photos of earth surface, up to astronomical objects. They have been stored for many years in numerous laboratories, clinics, commercial data banks etc. and they contain a big amount of information that in principle can be used in scientific investigations or for other professional purposes. However, when the number of available files or records reaches many thousands or millions units it arises a problem of effective retrieval and selection of information valuable from the users' point of view. In such case computer-based information retrieval machines may help to solve the problem [1]. In certain application areas the image retrieval process is facilitated due to standardization of the corresponding documents. As an example the DICOM standard widely used in medical picture archiving and communication systems (PACS) can be mentioned [2]. In DICOM the forms of normalized and of composed information entities (IE) containing textual, image and/or graphical fields are defined. Textual fields contain data identifying the patient, the date, site and type of medical examination, characterizing the modality(-ies) of medical imaging(-s) and the corresponding technical parameters etc., as well as short description of the content of attached images and based on them medical conclusions. In most cases the above-mentioned data are sufficient for effective image retrieval. However, in certain cases users' requirements are going deeper, for example, when

K.M. Węgrzyn-Wolska and P.S. Szczepaniak (Eds.): Adv. in Intel. Web, ASC 43, pp. 186–191, 2007.

- it is given a fragmentary image of an object and it is necessary to find out in the database all images containing somewhere in them appearing similar fragments;
- in a given collection of images it is necessary to find out subsets if images containing any identical or similar in a given sense fragmentary images of a given class [3].

Therefore, one should distinguish between the first and second type of image retrieval problems. In both cases the problem consists in selection of subsets of images satisfying a condition of partial similarity. In the first case partial similarity is based on a given fragmental reference image, while in the second case the fragmental reference images being a priori unknown, we deal with a much more difficult problem. Solution of both tasks needs looking many times over the set of available images in order to find out subsets of images satisfying the partial similarity criteria. However, such action may contain a great deal of redundant operations making the retrieval process ineffective. In this paper a solution of the problem based on a multi-step retrieval process using morphological spectra as a tool of partial similarity of images assessment is proposed. The paper is organized as follows. In Sec. 2 basic properties of morphological spectra are shortly described. In Sec. 3 the concept of partial image similarity based on morphological spectra is presented. In Sec. 4 the method of using morphological spectra to image retrieval is illustrated by an example. Sec. 5 contains short conclusions.

2   Basic Properties of Morphological Spectra

Morphological spectra (MS) of a given 2D discrete monochromatic image are defined as a hierarchical system of matrices representing the image as a collection of standard patterns formed by pixel values within fixed-size square windows, the size of which being strongly connected with the MS-level. It is assumed that the original image U is given in the form of a rectangular bit-map of $2^k \times 2^l$ size, k and l being some integers >1; for the sake of simplicity square images $(k = l)$ below will be considered. The bitmap U is assumed to be also the 0th order morphological spectrum of the original image, $S^{(0)} \equiv U$.

Higher-order morphological spectra are defined on the basis of the following four operators acting on lower-level spectral matrices:

$$\Sigma - \text{sum}, \qquad\qquad V - \text{vertical difference},$$
$$H - -\text{horizontal difference}, \quad X - \text{cross difference}.$$

If $S_m^{(\kappa)}$ is a mth component of κth order morphological spectrum then it generates four components of a higher-order morphological spectrum: $S_{\Sigma m}^{(\kappa+1)}$, $S_{Vm}^{(\kappa+1)}$, $S_{Hm}^{(\kappa+1)}$, $S_{Xm}^{(\kappa+1)}$. Before explaining the principles of generation let us remark that $S^{(\kappa)}$, $\kappa=0,1,2,\ldots,k$, consists of 4^κ matrices of $2^{(k-\kappa)} \times 2^{(k-\kappa)}$ size. For calculation of a component of $S^{(\kappa+1)}$ on the basis of $S_m^{(\kappa)}$:

1/ one should divide $S_m^{(\kappa)}$ into $2^{2(k-\kappa)}$ sub-matrices of 2×2 size;
2/ if the elements of a typical sub-matrix of this size are:

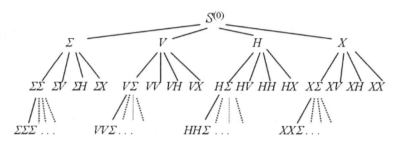

Fig. 1. Hierarchical structure of morphological spectra

$$\begin{bmatrix} s_p & s_q \\ s_r & s_t \end{bmatrix}$$

then the corresponding elements of $S^{(\kappa+1)}$ are:

$$s_p + s_q + s_r + s_t \quad \text{for } S^{(\kappa+1)}_{\Sigma m},$$
$$-s_p + s_q - s_r + s_t \quad \text{for } S^{(\kappa+1)}_{V m},$$
$$-s_p - s_q + s_r + s_t \quad \text{for } S^{(\kappa+1)}_{H m},$$
$$-s_p + s_q + s_r - s_t \quad \text{for } S^{(\kappa+1)}_{X m}.$$

The components of morphological spectra can be represented by a rooted tree whose nodes on a κth level represent spectral components of the given level, as shown in Fig. 1.

Each element of a spectral component $S^{(\kappa)}_m$ is a weighted sum (linear polynomial) of pixel values taken from the corresponding square window of $2^\kappa \times 2^\kappa$ size. The weights (polynomial coefficients) taking only the values +1 or −1 makes a simplification of calculation of the components possible. For this purpose special masks indicating the weights that to the corresponding pixel values should be assigned can be used. For calculation of all morphological spectra components the following numbers of masks are needed: 4 masks for $S^{(1)}$, 16 masks for $S^{(2)}$, 64 masks for $S^{(3)}$, 256 masks for $S^{(4)}$, etc. The masks for morphological spectra up to the 4th level have been calculated and presented in [4,5].

The method of spectral component calculation by using a mask can be illustrated by the following example. Let us assume that it is necessary to calculate a component $S^{(2)}_{HX}$ of a 2nd level morphological spectrum. The corresponding masks cover the 4×4 square windows. In Fig. 2 a mask for the HX spectral component is shown beside a corresponding sub-bitmap indicating enumeration of its pixels. The elements white-square in the mask correspond to the weight +1 while black-square to the weight −1.

3   Assessment of Partial Similarity of Images

Let us assume that there is given a set C of N images of $m \times n$ size and a fragmental reference image V of $i \times j$ size, where $i \leq m$, $j \leq n$. All tested images

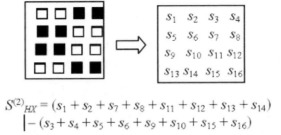

$$S^{(2)}_{HX} = (s_1 + s_2 + s_7 + s_8 + s_{11} + s_{12} + s_{13} + s_{14})$$
$$\left| - (s_3 + s_4 + s_5 + s_6 + s_9 + s_{10} + s_{15} + s_{16}) \right.$$

Fig. 2. Principle of of a spectral component

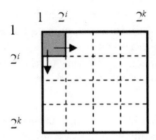

Fig. 3. Consecutive positioning of a reference fragmental image within an image tested for its partial similarity detection

containing the given reference image constitute a partial similarity class $C_V \subseteq C$. In a formal sense image retrieval tasks thus consist in selection in given sets of tested images partial similarity classes defined by (given a priori or assumed) fragmental reference images. For the given $2^k \times 2^k$ size of tested images and $2^i \times 2^i$ of a reference image proving belonging of a given tested image to C_V needs:

1/ consecutive fixing the position of the reference image within the image in all $(2^k - 2^i + 1)^2$ possible ways (when taking into account parallel translations only), as shown in Fig. 3:

2/ Testing of a local conformity of the tested image and the reference image needs calculation of distances between $(2^k - 2^i + 1)^2$ pairs of 2^i-component vectors. Calculation of each distance needs, in the case of using an absolute distance measure, calculation of 2^j differences of vector components' and $2^j - 1$ additions. The cost of calculations increases proportionally to the number N of tested images.

3/ Partial similarity is detected if in at least one position of the reference image within the tested image the distance between the corresponding vectors is below a fixed positive threshold value ε.

A calculation cost of a second-type partial similarity detection is approximately $1/2 N (2^k - 2^i + 1)^2$ times larger then this mentioned in 2/ because each $2^i \times 2^i$-size partial window of a tested image is used as a fragmental reference image.

In order to reduce the calculation costs of image retrieval based on partial similarity detection the following multi-step retrieval procedure is proposed:

1. For the given i (reference window's size) take the spectrum $S^{(i)}$ into consideration; select a component $S_m^{(i)}$ and calculate its value for the given reference image and all images to be tested;
2. fix a positive threshold value ε, calculate the absolute values $|\Delta|$ of the differences Δ between $S_m^{(i)}$ of the given reference image and of the consecutive sub-images of tested images;
3. reject all sub-images for which it is $|\Delta| > \varepsilon$, take into account the remaining sub-images in further calculations; if no such sub-images exist, conclude that no partial similarity has been detected;
4. select another component of $S^{(i)}$, say $S_n^{(i)}$, and repeat points 2 and 3 for the sub-images remaining in 3, until all components of $S^{(i)}$ have been proved;
5. tested images containing fragmental sub-images for which full conformity of their morphological spectra with this of the reference image has been detected are the ones satisfying the criterion of partial similarity to the reference image.

In the consecutive steps of this procedure the calculation costs are reduced due to the reduction of the number of sub-images subjected to the testing procedure. The conformity of morphological spectra implies a conformity of the corresponding fragmental images, because morphological spectra of any level contain complete information about the original image under examination.

4 Detection of Partial Similarity of Real Images

The above-described concept of a method of partial similarity of images detection has been proven on several images. An illustrations is given below.

In Fig. 4 a reference image a) and three different tested images b), c), and d) are shown. The 3rd order spectral components $S_{\Sigma\Sigma\Sigma}^{(3)}$, $S_{\Sigma\Sigma V}^{(3)}$, $S_{\Sigma\Sigma H}^{(3)}$, and $S_{\Sigma\Sigma X}^{(3)}$ have been calculated for the images as shown in Table 1.

Table 1. Morphological spectral components $S_m^{(3)}$ of images shown in Fig. 4

Spectral comp.	a)	b)	c)	d)
$S_{\Sigma\Sigma\Sigma}^{(3)}$	147	202, 212, 173, 84	175, 182, **137**, **140**	97, 160, 49, **147**
$S_{V\Sigma\Sigma}^{(3)}$	-375	18, 8 , 16, 541	40, 23, 22, 128	315, 142, -262, **-375**
$S_{H\Sigma\Sigma}^{(3)}$	-725	-40, -50, -34, 249	26, -35, -32, 116	-539, -284, 136, **-725**
$S_{X\Sigma\Sigma}^{(3)}$	-71	4, 10, 2, -69	-80, 9, -12, -36	29, 76, -40, **-71**

At the first step a threshold level $\varepsilon + 10$ and the component $S_{\Sigma\Sigma\Sigma}^{(3)}$ are used. Its value for a) is 147; sub-images with similar component values: 137 and 140 in image c) and 147 in image d) are detected. At the next step a spectral component

Fig. 4. Partial similarity of images: a) reference image, b),c),d) three different tested images

$S_{V\Sigma\Sigma}^{(3)}$ is used for testing three sub-images. Its value for a) is -375, for c) there are 22 and 128 and for d) it is -375. This shows that image c) should be rejected and we calculate the last two spectral components, $S_{H\Sigma\Sigma}^{(3)}$ and $S_{X\Sigma\Sigma}^{(3)}$ for image d). Their values, -725 and -71, are identical to those of the reference image a). This shows that image d) satisfies the criterion of partial similarity to a).

5   Conclusions

Morphological spectra of images are a flexible tool that can be used to a stepwise selection of images in databases by detection of their partial similarity to given reference images. Calculation of spectral components is simplified due to component masks that formerly have been prepared.

References

1. Baeza-Yates R.: Modern Information Retrieval. ACM Press, New York, 1999.
2. Pietka E.: Implementation of a Hospital Information System. Proc. CARS/ EuroPACS, London 2003.
3. Kulikowski J. L.: Computer-Aided Analysis of Serial Images. Machine Graphics and Vision, vol. 7, No 1/ 2, 1998, pp. 135-149.
4. Kulikowski J.L., Przytulska M., Wierzbicka D.: Recognition of Textures Based on Analysis of Multilevel Morphological Spectra. World Congress on Medical Physics and Biomedical Engineering, Seoul, 2006. IFMB-Proceedings, Springer, vol. 14, p. 2164.
5. Computer-Aided Analysis of Textures. XI Conference "Medical Informatics & Technologies" MIT 2006, Wisla. Conf. Materials, pp. 3-8.

Semantic Community in a Peer-to-Peer Network

Jianlin Li, Xueli Yu, Xiaobo Wu, Rui Wang, and Jingyu Sun

College of Computer and Software, Taiyuan Univ. of Tech.
lijianlin_2005@hotmail.com

Summary. The problem of knowledge sharing is eminent in the P2P area. In this paper, we propose a general framework, called semantic community, which is conceived for supporting dynamic ontology-based knowledge sharing and evolution in P2P systems. The knowledge sharing and evolution processes in semantic community are based on peer ontologies which are describing the knowledge of each peer, and on interactions among peers. This paper exploits an ontology matching algorithm to identify semantic affinity of two concepts .According to the semantic affinity threshold, the founder has to evaluate which peers are admitted in the community.

Keywords: Semantic community, P2P, Ontology.

1 Introduction

The self-formation and management of semantic communities and the availability of advanced techniques for query propagation on a semantic basis if a challenging issue in the current stage of development of open networked system architectures and schema-based P2P networks, to enforce sharing of distributed resources and semantic collaboration in an effective way. To this end, ontologies are generally employed for describing the knowledge to be shared, and appropriate techniques are required to deal with the different concept meanings in the ontologies provided by different peers for community formation [1]. The peer-to-peer (P2P) approach, which has become popular in the context of file-sharing systems such as Gnutella or KaZaA, allows handling huge amounts of data in a distributed and self-organized way. In such a system, all peers are equal and all of the functionality is shared among all peers so that there is no single point of failure and the load is evenly balanced across a large number of peers. These characteristics offer enormous potential benefits for search capabilities powerful in terms of scalability, efficiency, and resilience to failures and dynamics. Additionally, such a search engine can potentially benefit from the intellectual input, e.g., bookmarks and query logs, of a large user community. One of the key difficulties, however, is to efficiently select promising peers for a particular information need, given the total number of relevant peers in a network is not known a priori and peer relevance also varies from peer to peer.

K.M. Węgrzyn-Wolska and P.S. Szczepaniak (Eds.): Adv. in Intel. Web, ASC 43, pp. 192–197, 2007.
springerlink.com © Springer-Verlag Berlin Heidelberg 2007

In this paper, we address the problem of formation of semantic communities of peers. Each peer stores its information which is going to share in the forms of ontology. Traditional P2P system is accompanied by a certain loss of semantics, so users may not query information they want exactly.

The paper is organized as follows: In section 2, it introduces the foundations of semantic community, such as how to store information in a new way. In section 3, a new algorithm is adopted to solve ontology matching problems. The goal of ontology matching techniques is to find concepts that have a semantic relationship with a target concept. In section 4, it exploits a toolkit named pajek to stimulate the formation of semantic community. Finally, conclusions and future research issues are discussed in Section 5.

2   Foundations of Semantic Community

In a P2P system, each peer acts as an autonomous and independent agent and shares knowledge by submitting discovery queries and by replying with relevant knowledge. In this context, the role of semantic communities of peers is related to the capability of dynamically aggregating nodes with similar interests in efficient structures in order to i) reduce the network load due to overlapping single peer requests and ii) define effective communication mechanisms for sets of nodes which share the same understanding of a domain of interest. We define a semantic community of peers as a set of nodes which show a common interest in a given topic and are organized in a structured way. Semantic communities are autonomously emerging [2], in that they originate from a declaration of interest of a peer and group those peers which spontaneously agree with the declaration, since they have relevant resources for the community. The information is stored as OWL ontology. A P2P system includes a large, scalable, dynamic, autonomous and heterogeneous network, where nodes (peers) can exchange data and services in a completely decentralized and equal manner. P2P networking supports knowledge management in a natural manner by closely adopting the conventions of face-to-face human communications. Knowledge exchange among peers is facilitated without the traditional dependence on servers. The original motivation for the early P2P systems is file sharing. Search methods in P2P networks were surveyed in Wulff and Sakaryan and Tsoumakos and Roussopoulos. Both the network structure and the search algorithm significantly influence the properties of P2P applications.

2.1   The Unstructured P2P System

The unstructured P2P systems mostly employ flooding and random-walk search approaches to locate files. The flooding approach forwards a peer's queries to all its neighbors, which results in traffic problems. The random-walk approach forwards a peer's queries only to randomly selected neighbors. Each selected peer repeats this process until the required data are found. To be effective, a query-routing strategy should forward queries only to peers who are likely to match the queries.

2.2 The Structured P2P Systems

Structured P2P systems, such as Can, Chord, Pastry, and Tapestry, use a distributed hash table (DHT) for routing [3]. DHT-based P2P systems are not suitable for complex queries because they only support keyword-based and exact-match lookup. Issues related to complex queries in DHT-based P2P systems were outlined in Harren etal. Mechanisms for locating data using incomplete information in P2P DHT networks were introduced, where multiple and hierarchical indices were used to help users to locate data even from poor information in the query [4]. The main purpose of the Ontology is to share and manage globally distributed knowledge resources in an efficient and effective way. P2P networks can be adopted as the semantic overlay layer of the Ontology. The scalability and autonomy make the P2P network a promising underlying infrastructure for a scalable Ontology. Integrate and manage heterogeneous knowledge and information in large-scale web resource[5].

3 Ontologies Matching Algorithm

The semantic matching algorithm[6] proposed in this paper is based on heuristics, which means that the similarity of concepts can be determined according to the properties and concepts attached closely to them on the taxonomic structure and other constraints information on the properties. In our semantic matching algorithm, we only consider the relative properties and the property constraints (values of domain and range) without taking the subclass-superclass into consideration. It is necessary for us to use more information for getting a high accuracy. We present the mechanism of our algorithm as follows.

We use the example to explain the mechanism of the algorithm. In the example the two concepts are desktop-computer called concept A and game-computer called concept B. We can export the property lists of the two concepts from the concept-lattice repository. Then we need to confirm all the matching pairs between the two property lists. The principle of confirming the matching pair is that if the two members from the two lists respectively have the same name or the namespace, then they are the matching pair; or the member of one list is the sub-property of the member of the other list, so vice versa. If there isn't any member in one list A that matches with one member of list B, then return the corresponding value of -1.

On the principle we get each matching pair P_i and P_j, now we need to compute the compatibility between them according to the rules showed in Figure1. And apply the rules to the property matching pairs of our example, we get the values of 1,-1, 1, 1, 2/5. Then we need to assign the weights to the returned values according the importance of the property in having impact on the feature of the concept. To be simple, we let the domain expert to give the weights of 0.4, 0.1, 0.1, 0.3, and 0.1. Or we could generate the weights on computing the frequency of the name of property through scanning the OWL documents. Then we get a value of 0.74 through (1*0.4+(-1*0.1)+1*0.1+1*0.3+0.4*0.1). Finally we can get the result that desktop-computer and game-computer are semantically similar because 0.74 lies between 0.5 and 1.

$$compatibility = \begin{cases} -1 & P\_i \ mismatch \ P\_j \\ -1 & range\_i \cap range\_j = \emptyset \\ 1 & range\_i \subseteq range\_j \\ |range\_i \cap range\_j| \ / \ |range\_i| & otherwise \end{cases}$$

Fig. 1. The rules of computing the compatibility

In order to put in evidence the role of ontology matching in the semantic community formation, we consider an example and we discuss the role of the matching algorithm. In our semantic community, peers are represented together with a portion of their peer ontologies described with this ontology matching algorithm. The peer ontology is relative to the computer domain .when the founder initializes a semantic community, the invited peers with resources stored in ontology compare with the founder's topic or interest ontology. Each receiving peer invokes the ontology matching algorithm with matching model to evaluate the semantic affinity between the incoming peer ID and its respective peer ontology. According to the threshold t = 0.5 , P_i and P_j can match with the same topic. The reply to the founder, and will be a member of this semantic community in computer domain.

4   Running Example

Pajek is employed to stimulate the formation of semantic community. Pajek is a very decent network visualization tool. It allows you to visualize large network in a relatively static state. The Pajek toolkit labels itself as a windows-based toolkit that can analyze and visualize large networks containing nodes having millions of relationships.

In Figure 2, dashed lines represent random P2P connections and the path followed by the invitation message (continuous line) defines a tree structure where the root is identified by the community founder and the leaf are represented by the invited peer with TTL = 0. Each invited peer negotiates its participation in the community directly with the community founder. Once it is admitted, the peer exploits the tree structure and communicates within the community through its community neighbors. We define the community neighbors of a community member W as the peer that invited W in the community (i.e., W predecessor) and the peers that W invited in the community (i.e., W successors). An invited peer not interested in the community or discarded by the founder is to be pruned from the tree structure of the community. For this reason, after the approval phase, each community member W notifies to its predecessor P_p of its presence in the community.

As an example, consider peer I_3 in Figure 2. The community member peer A_5 and peer A_6 notify peer I_3 of their participation. Peer I_3 has not joined the community and is to be pruned from the community tree. Then, peer I_3 forwards the notification to peer A_1 and notifies peer A_5 and peer A_6 that peer A_1 is their new predecessor.

For the sake of convenience, W represents community founder, A_1 to A_6 represent accepted member, I_1 to I_4 represent invited member, and N_1, N_2 represents not invited member. The lines represent invitation message and P2P communication path. W wants to initiative a semantic community with respect to pop music. So it send invitation messages to its neighbors (A_1, I_1, I_2) with TTL=2. They send the invitation messages to their neighbors. If they would like to attend this community, they'll return a message to the founder. Or send a refused message to founder. At last, the founder public a member list to its all member. The member in this community can share music among them.

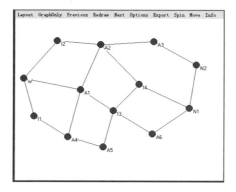

Fig. 2. Example of formation of a semantic community

5 Conclusion

A peer-to-peer system is a natural and extensible underlying layer for ontology because of its autonomy, self-organization, and scalability. As a solution to query answering at the semantic overlay of the scalable ontology, this work proposes: a semantic community, a semantic-based peer similarity measurement for query routing, and a peer schema mapping approach for query reformulation. The results from theoretical analysis and simulations show that the proposed approach is effective.

The main concerns of this work are: (1) the notion of ontology used in a P2P network; (2) the use of semantic structure similarity to measure the similarity between peers to improve the effectiveness and efficiency of query routing.

Ongoing work includes three aspects: (1) semantically clustering relevant peers to formal hierarchical semantic community to improve the performance of the proposed semantic community model; (2) incorporating query reformulation optimization into the proposed approach; (3) incorporating our approach with others, such as those indexing on actual data to improve the effectiveness of queryrouting; (4) adopting a more reasonable ontology matching algorithm to improve the peer ontology matching efficiency; (5) Exploiting a new network stimulator to build the semantic community dynamically; and, (6) How to manage the semantic community. We have a long way to go!

Acknowledgement

Sponsored by the Natural Science Foundation of China (No. 60472093).

References

1. Castano S and Montanelli S (2005).Semantic self-formation of communities of peers. In Proceedings of the ESWC Workshop on Ontologies in Peer-to-Peer Communities, Heraklion, Greece.
2. Castano S,Ferrara A,Montanelli S,Racca G (2004).Semantic Information Interoperability in Open Networked Systems. In Proc. of the Int. Conference on Semantics of a Networked World ,Paris,France.
3. TAND C,XU Z,MAHALINGAM M (2003). Psearch: Information retrieval in structured overlays. ACM SIGCOMM Computer Communications Review, 33(1):89–94.
4. TATARINOV I,HALEVY A (2004). Efficient query reformulation in peer-data management systems. In Proceedings of ACM SIGMOD, Paris, France, 539–550.
5. Crespo A and Garcia-Molina H (2003). Semantic overlay networks for p2p systems. Technical report, Stanford University, January .
6. Li Wen,Yu Xueli (2005) Research on Semantic Interoperability among Heterogeneous Ontology in the Semantic Web. In Proc. of the ISTP, 432–436.

Semantic Resource Management for Research Community

Yanyan Li[1] and Xiaochun Cheng[2]

[1] Knowledge Science & Engineering Institute, Beijing Normal University, 100875, Beijing, China
   liyy1114@gmail.com
[2] School of Computing Science, Middlesex University, London, UK

Summary. In order to provide diverse users an integrated, scalable and easy-to-use interface to manage and exploit the dynamically growing learning resources on the Web, this paper proposes a semantic resource management framework to support the effective organizing and accessing of learning re-sources in an open environment. By exploiting the semantic relationships that characterize learning resources and user profile, the framework supports both semantic querying and conceptual navigation of learning resources pertaining to specific topics. A system has been developed and deployed within an academic setting to improve the resource management and exploitation for cooperative research.

1 Introduction

More and more instructors are developing multimedia learning materials, such as lec-ture notes, software simulations, and videos, to support distance learning. As a result, the amount of resources available to users is increasing continuously. Users spend a great deal of time on the Web searching and browsing for information to "amplify" their intelligence [1]. They try to gather enough information about a topic to be able to answer a question or complete a task, but the acquired knowledge is often disor-dered, disconnected, and not effectively integrated to address their learning needs. This use of the Web is ubiquitous and yet has not been supported adequately by ex-isting web-based learning systems. Therefore, the wealth of resources presents a great challenge: how to provide a coherent, structured, shareable collection of resources to cater for users' specific needs. Some systems have been proposed intending to effec-tively support resources searching and exploitation [2], [3], [6]. However, they are still in infancy and have two weaknesses. One is that the terminology used by differ-ent sources is often inconsistent and there is no common overarching context for the available resources, so maintaining huge weakly structured resources is a difficult and time-consuming activity. The other is that existing keyword-based search without semantic annotations retrieves much irrelevant information, so navigation through a large set of independent sources often leads to users' being lost.

K.M. Węgrzyn-Wolska and P.S. Szczepaniak (Eds.): Adv. in Intel. Web, ASC 43, pp. 198–203, 2007.
springerlink.com   © Springer-Verlag Berlin Heidelberg 2007

2   A Conceptual Model for Semantic Resources Organization

Fig. 1 illustrates the semantic-based resource organization model. As the fig-ure shows, RC-SLN deploying above the repository is a structured semantic link network (SLN) [4] where the node represents the learning objects anno-tated with properties and the arc represent the semantic relationship between learning object. RC-SLN represents the various semantic relationship between learning resources for different research communities (RC), and the base types of semantic links include such as the Cause-effective Link, the Implication Link, the Similar-to Link, the Sequential Link, and the Reference Link, etc [7]. A Hyper-SLN herein only comprises the complex nodes, which depicts the seman-tic relationships between the SLNs of underlying dis-tributed sources, and thus gives an overall view of the underlying sources. The map-ping rules represent the mapping from the SLNs to the hyper-SLN along with the in-formation and constraints of the underlying sources. This can help organize and manage re-sources across different SLN networks with a universal view. Reference ontology indicates the common terminology with respect to the specific domains.

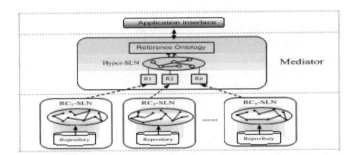

Fig. 1. Conceptual Model for Semantic Resources Organization

In this way, by explicitly defining the semantics of the resources and sepa-rating the knowledge structure from the learning materials, the disordered and heterogeneous learning resources can be adaptively organized and structured in a more flexible, interoperable and coherent manner, enabling effective resources searching, maintenance and semantics interoperability.

3   Process Flow to Search for Semantic-Associated Resources

For the given query proposed by the user, the search is conducted by following the three-step process.

Step 1: Query Analysis. For a given query, the first step is to process the query and map the search terms to the entities in the resource space. This might locate the matching concepts in the ontology, or the entities in the semantic link network. As for the case of multiple matches because of linguistics ambiguity (e.g. synonym), users' profiles are taken as the reference to select the proper annotation. Additionally, the search context and the popularity of the term as measured by its frequency of occurrence in a text corpus can also give hints for selecting the proper annotation for the search terms.

Step 2: Target Entities Selecting. If the target entity is the concept in the resource space, then select all the linking learning objects in the SLN. Afterwards, taken the matching entities as the anchor ones, this step is to find semantic relevant entities. The simple approach for selecting the target entities for the one matching entity, purely based on the structure of the SLN graph, is to collect the first N triples originated from the anchor entity, where N is the pre-defined traversal constraints. As for the case of two matching entities corresponding to the query, it is the key problem to find all the semantic association paths between the two entities so as to select the relevant instances on the paths. The basic idea of the algorithm is to traverse the graph in a breadth-first order starting from the two entities.

Step 3: Semantic Ranking. This step is to rank the target entities in terms of their association weights. Our approach to ranking the semantic associated results is primarily based on capturing the interests of a user. This can be accomplished by enabling a user to browse the ontology and mark a region (sub-graph) of nodes and/or properties of interest. If the discovery process finds some semantic association paths passing through these regions then they are considered relevant, while other associations are ranked lower or discarded. For example, consider a sub-graph of an SLN graph representing a researcher who leads a project and who also is fond of music. If the user is just interested in the "research" domain, the semantic associations involving music related information can be regarded as irrelevant (or ranked lower). In sum, the ranking of an entity in an association path can be computed in terms of the following aspects.

Path weight. It is possible to capture a user's interest through a context specification. So, using the context specified, we adopt the approach in [5] to rank a path according to its relevance with a user's domain of interest

$$P_w^i = \frac{1}{|c|}((\sum_{i=1}^{\#regionsPisIn}(r_i \times (\sum c \in R_i))) \times (1 - \frac{\#c \notin R}{|c|}))$$

Entity location. The location of entities in a path also affects the relevance weight of entities. The shorter the path length from the anchor entity to the candidate entities, the more relevance the entities are. The weight of the ith entity's location is computed with equation $E_l^i = \frac{L_i}{|C|}$, Where L_i denotes the path length from the anchor entity to the ith entity (i.e. the total number of entities in the path and includes the start and end entities), $|C|$ denotes the total number of entities in the path, and $E_l^i \in \hat{E}$ [0, 1].

Entity Frequency. The entities with higher appearance frequency in paths are taken as more relevant to the search terms. The weight of the ith entity's frequency is computed with equation $E_f^i = \frac{F_i}{|E|}$, Where F_i denotes the appearance frequency of the ith entity, $|E|$ denotes that total number of candidate entities in the paths, and $E_f^i \in \hat{E}$ [0, 1].

Furthermore, different weights can be given to different criterion according to users' preferences. So, the overall weight of the ith candidate entity is computed with formula $W^i = k_1 * \sum_{t=1}^{n} (P_w^t * (1 - E_l^i)) + k_2 * E_f^i$, Where n denotes the number of the selected association paths, k_i add up to 1 and are intended to allow fine-tuning of the different ranking criteria.

4 Implementation

We have developed the scientific research platform with java language. An authoring tool has been developed for defining the SLN and uploading the materials with metadata annotation.

Fig. 2. The interface for defining an SLN

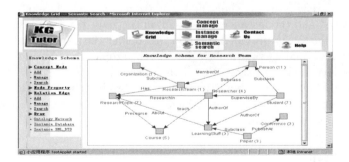

Fig. 3. The graphical view of an SLN

Fig. 2 shows the user interface for defining the SLN. Multiple operations are provided for users to constructs their own SLN for a scientific research community. Additionally, the authoring tool provides a graphical view of the constructed ontology, as shown in Fig. 3. The number attached to each concept indicates the

Fig. 4. The interface for managing the instances

Fig. 5. Semantic Search results

amount of instances that belong to the concept, and users can click the number to view the list of instances. This intuitive display manner offers the conceptual navigation in a map-view pattern for users.

Fig. 4 shows the interface to manage the instances. The scalable knowledge category hierarchy is arranged on the left portion of the interface offering a tree-view for users, and the instances are listed on the right portion of the interface. A reference button is provided for users to define the related instances to a specific student, and users can click the reference topics to view and specify the topics in which he has interest. Fig. 5 shows the searching results. In addition to the general information about the item (e.g. introduction, link-address), more related resources are given for users' reference, and thus users can click any topic to get more focused information.

5   Conclusion

The proposed semantic resource management framework for research community has the following advantages. First, the knowledge structure is separated from the media resources, which enables effective resource organization and reusing. Second, users benefit from the thematic search that discovers and provides semantic associated resources to address the users' focused and personalized learning needs. Third, the adaptive learning resources delivered in a coherent structure empower users to efficiently explore the learning resources at their own paces. We have developed a system and deployed it within an academic setting to improve the resource management and exploitation for cooperative research.

Acknowledgement. This paper is supported by the China Natural Science Fundation (No. 60402016).

References

1. Andersson, M., Person plus Web – samples from everyday life. Word Conference on e-Learning (e-Learn), Phoenix, AZ, 2003.
2. Goecks, J. and Cosley, D., NuggetMine: Intelligent groupware for opportunistically shar-ing information nuggets, Proceedings of the Intelligent User Interfaces Conference, pp. 87-94, 2002.
3. Jari, K., Stroulia, E., EduNuggets: an intelligent environment for managing and delivering multimedia education content, In Proceedings of the 8th international conference on Intel-ligent user interfaces, Miami, Florida, USA, pp. 303 - 306, 2003, ACM Press.
4. Zhug, H., Li. Y, Learning with Active E-Course in Knowledge Grid Environment, Con-currency and Computation: Practice and Experience, 18(3), 2006, 333-356.
5. Meza, B.A., et al. Context-aware semantic association ranking. Semantic Web and Data-bases Workshop Proceedings, Belin, September, 2003.
6. Young, R.L., Kant, E., and Akers, L.A., A knowledge-based electronic information and documentation system, Proceedings of the Intelligent User Interfaces Conference, pp. 280-285, 2000.
7. Zhuge, H., Active Document Framework ADF. In Proceedings of the 5th Asian-Pacific Web Conference (APWeb 2003), Xi'an, China, September 2002.

Approximation Property of Fuzzy Finite Automata*

Yongming Li

College of Computer Science, Shaanxi Normal University, Xi'an, 710062, China
liyongm@snnu.edu.cn

Summary. We have shown that *nondeterministic fuzzy finite automata* (or NFFAs, for short) and *deterministic fuzzy finite automata* (or DFFAs, for short) are not necessary to be equivalent in the previous work. We continue to study the approximation of fuzzy finite automata in this paper. In particular, we show that we can approximate an NFFA by some DFFA with any given accuracy even the NFFA is not equivalent to any DFFA, related construction is also presented. Some characterizations of NFFA and DFFA are given.

1 Introduction

Much of the information which resides in the Web-and especially in the domain of world knowledge-is imprecise, uncertain and partially true. Existing bivalent-logic-based methods of knowledge representation and deduction are of limited effectiveness in dealing with information which is imprecise or partially true. To deal with such information, bivalence must be abandoned and new tools, such as fuzzy logic or fuzzy inference, should be employed [10].

We can use different fuzzy inference method in constructing a fuzzy system [9, 6], such as fuzzy web intelligence. The commonly used fuzzy inference methods are known as compositional rule of inference (or CRI) and its generalization max-$\star$ compositional inference for some t-norm $\star$. Usually, we can use different kinds of t-norms when we implement max-$\star$ compositional inference in the practical processes. However, they are equivalent in the approximate sense. That is to say, for any practical process, if we can approximately describe the process by a fuzzy system with max-min compositional inference, we can always approximately describe it by a fuzzy system with max-$\star$ compositional inference for any other t-norm $\star$. In fact, the universal approximation property of fuzzy systems has been demonstrated. For the detailed explantation of universal approximation theory of fuzzy systems, we refer to [9, 6]. On the other hand, fuzzy finite automaton can be seen as one kind of discrete fuzzy system [8, 6]. We can use fuzzy finite automata to describe practical processes. For example, we can use fuzzy automata to model discrete event systems with fuzzy uncertainty, as done

---

* This work is supported by National Science Foundation of China (Grant No. 10571112) and National 973 Foundation Research Program (Grant No. 2002CB312200).

in [7] using a *nondeterministic fuzzy finite automaton* (or NFFA, for short) with max-product compositional inference, while [2] using an NFFA with max-min compositional inference, and we can also describe recurrent fuzzy systems by a fuzzy finite automaton with max-$\star$ compositional inference for some t-norm $\star$ in [1]. Fuzzy finite automata are also the potential models for web intelligence, especially for fuzzy web intelligence. Theoretically, once there is a t-norm $\star$, there will be corresponding to one type of fuzzy automaton using max-$\star$ compositional inference that could be used in modeling practical processes. Therefore, there will be many types of fuzzy finite automata that can be implemented in practical processes. A crucial theoretical problem arisen is whether these fuzzy finite automata using max-$\star$ compositional inference for different kinds of t-norm $\star$ are equivalent in the power of modeling the practical processes. However, contrary to ordinary case, we have shown that a nondeterministic fuzzy finite automaton (NFFA) under max-$\star$ compositional inference is in general not equivalent to a *deterministic fuzzy finite automaton* (or DFFA, for short) [5] in the power of processing fuzzy languages. Some necessary and sufficient conditions for the equivalence between NFFA and DFFA are given in [5]. So it is necessary to require the proposed models of NFFAs using max-$\star$ compositional inference for different kinds of t-norm $\star$ to be equivalent in the approximate sense. We express this problem explicitly as follows: for any discrete practical process, if we can approximately describe the process by a fuzzy finite automaton with max-min compositional inference, whether can we always approximately model it by a fuzzy finite automaton with max-$\star$ compositional inference for any other t-norm $\star$? We have proved that DFFA and NFFA under max-min compositional inference are equivalent in the power of processing fuzzy languages. In order to discuss whether NFFAs using max-$\star$ compositional inference for different kinds of t-norm $\star$ are equivalent in the approximate sense, we only need to study the approximation property of NFFA by DFFA. This forms the main topic of this study.

2   Approximation of Nondeterministic Fuzzy Finite Automata by Deterministic Fuzzy Finite Automata

A *nondeterministic fuzzy finite automaton* (NFFA) is a 6-tuple $A = (Q, \Sigma, \delta, \sigma_0, \sigma_1, \star)$ such that Q is a finite nonempty state set, Σ is a finite input alphabet, and δ is a fuzzy subset of $Q \times \Sigma \times Q$, which represents fuzzy transition function, and σ_0 and σ_1 denote the fuzzy initial state and the fuzzy final state as a fuzzy subset of Q, $\star$ is a specified t-norm which is used to define the fuzzy language accepted by an NFFA. Where a t-norm is a binary operation $\star$ on the unit interval $[0, 1]$ such that $\star$ is commutative, associative, nondecreasing in both arguments and $\star$ has the unit 1. For example, the minimum operation "$\wedge$", the product operation, the Lukasiewicz t-norm and nilpotent minimum are t-norms on the unit interval $[0, 1]$. We omit $\star$ in the definition of an NFFA if $\star$ is understood in the context.

For an NFFA $A = (Q, \Sigma, \delta, \sigma_0, \sigma_1, \star)$, if δ is a crisp function from $Q \times \Sigma$ to Q and σ_0 is an element of Q, that is, $\sigma_0 = q_0 \in Q$, which denotes the crisp initial state of A, then we call A a *deterministic fuzzy finite automaton* (DFFA).

For a fuzzy transition function $\delta : Q \times \Sigma \times Q \to [0,1]$, we can extend δ onto $Q \times \Sigma^* \times Q$ inductively as follows, where Σ^* denotes all these finite strings over Σ including empty string ε, (1) $\delta^*(q, \varepsilon, q) = 1$ and $\delta^*(q, \varepsilon, p) = 0$ for any distinct elements q and p in Q; (2)for any input string $\omega \in \Sigma^*$ and input symbol $x \in \Sigma$, $\delta^*(q, \omega x, p) = \bigvee\{\delta^*(q, \omega, r) \star \delta(r, x, p) : r \in Q\}$, where we use the symbol "$\bigvee$" to represent the supremum of real numbers. Then we define an extension $\delta^* : Q \times \Sigma^* \times Q \to [0,1]$.

The *fuzzy language* accepted by or recognized by an NFFA A is defined as a fuzzy subset of Σ^*, denoted f_A, for any input string $\omega \in \Sigma^*$,

$f_A(\omega) = \bigvee\{\sigma_0(q) \star \delta^*(q, \omega, p) \star \sigma_1(p) : q, p \in Q\}$.

In particular, if A is a DFFA, then $f_A(\omega) = \sigma_1(\delta^*(q_0, \omega))$ for any $\omega \in \Sigma^*$.

We need the following theorem to characterize the fuzzy languages accepted by DFFAs.

Some notations need to be introduced here. Let X be a set. For a fuzzy subset $f : X \to [0,1]$, let $R(f) = \{f(x) : x \in X, f(x) > 0\}$. For any $a \in [0,1]$, two subsets f_a and $f_{[a]}$ of X are defined as, $f_a = \{x : x \in X$ and $f(x) \geq a\}$ and $f_{[a]} = \{x : x \in X$ and $f(x) = a\}$. The support of f is a subset of X defined as $supp(f) = \{x : x \in X, f(x) > 0\}$.

Theorem 1. *Let $f : \Sigma^* \to [0,1]$ be a fuzzy language, the following statements are equivalent for f.*

1) f is accepted by a DFFA.

2) f is accepted by an NFFA with t-norm chosen as minimum operation $\wedge$.

3) $R(f)$ is finite, and f_a is regular for any $a \in R(f)$.

4) $R(f)$ is finite, and $f_{[a]}$ is regular for any $a \in R(f)$.

Remark 1. As implied by Theorem 1, fuzzy finite automaton using max-min compositional inference can be realized by a simpler model: DFFA. Then, the problem to approximate an NFFA under max-$\star$ compositional inference by an NFFA under max-min compositional inference is equivalent to the problem to approximate the NFFA by a DFFA.

To study the further properties of NFFA and DFFA, we need to introduce some notations here. Let $\star$ be a t-norm. For any $a \in [0,1]$, we can inductively define the power of a as follows: $a^0 = 1$, $a^1 = a$, and $a^{n+1} = a^n = a^n \star a$ for any non-negative integer n. For any subset D of $[0,1]$, the subalgebra of $([0,1], \star)$ generated by D, denoted $S(D)$, is defined as follows:

$S(D) = \{a_1^{l_1} \star \cdots \star a_k^{l_k} : a_1, \cdots, a_k \in D$ and $l_1, \cdots, l_k$ are nonnegative integers$\}$.

For any $a \in (0,1]$, we use $S_a(D)$ to represent a subset of $S(D)$ consisting of elements of $S(D)$ that are larger than or equal to a, i.e.

$S_a(D) = \{b : b \in S(D)$ and $b \geq a\}$.

We have two conditions imposed on t-norm $\star$.

Finite generated condition (FGC, for short): *For any finite subset D of $[0,1]$, $S(D)$ is finite.*

Weakly FGC (WFGC, for short): *For any finite subset D of $[0,1]$ and any $a \in (0,1]$, $S_a(D)$ is finite.*

Obviously, if a t-norm $\star$ satisfies FGC, then it also satisfies WFGC. For example, minimum, Lukasiewicz and nilpotent minimum t-norms satisfy FGC, and product t-norm satisfies WFGC but not FGC.

Theorem 2. *Let $\star$ be a t-norm satisfying WFGC. If a fuzzy language f can be recognized by an NFFA, then for any $a \in (0,1]$, f_a is a regular language.*

Theorem 3. *Let $\star$ be a t-norm satisfying WFGC. If a fuzzy language f can be recognized by an NFFA A, then for any $a \in (0,1]$, the set*

$$R_a(f) = \{f(\omega) : \omega \in \Sigma^*, f(\omega) \geq a\}$$

is always finite and there exists an DFFA B such that $f_B = f \vee a$.

Obviously, the DFFA constructed in Theorem 3 satisfies the condition $|f_B(\omega) - f(\omega)| \leq a$ for any $\omega \in \Sigma^*$. Therefore, an NFFA can be approximated by some DFFA with the prescribed accuracy. We present it in the following theorem.

Theorem 4. *Let $\star$ be a t-norm satisfying WFGC. For any NFFA A, and any small positive number ϵ, there always exists a DFFA B such that the following inequality holds for any input $\omega \in \Sigma^*$*

$$|f_B(\omega) - f_A(\omega)| \leq \epsilon.$$

Remark 2. Because of the above results about the approximation of NFFA by DFFA, for any approximation accuracy ϵ, we can construct a class of NFFAs in which all the fuzzy transitions, fuzzy initial states and fuzzy final states take values in a fixed finite subset, written as D, of unit interval $[0,1]$, and each NFFA can be approximated by a DFFA in this specified class with given accuracy ϵ. Here we say that an NFFA A can be approximated by an NFFA B with accuracy ϵ, if the condition $|f_A(\omega) - f_B(\omega)| \leq \epsilon$ holds for any input string ω.

In order to do this, let us assume that $\star$ is any t-norm satisfying WFGC, ϵ is a given approximation error, and Σ is a given input alphabet. Take a positive integer n such that $\frac{1}{n} \leq \epsilon$, and let

$$D = \{0, \frac{1}{n}, \frac{2}{n}, \cdots, \frac{n-1}{n}, 1\}.$$

We construct a class of NFFAs with input alphabet Σ, denoted $NFFA_n$, such that $NFFA_n$ satisfies the following property:

For any NFFA A with input alphabet Σ, there is a DFFA B in the class $NFFA_n$ such that A can be approximated by B with accuracy ϵ, i.e., $|f_A(\omega) - f_B(\omega)| \leq \epsilon$ for any input string $\omega \in \Sigma^*$.

The construction of $NFFA_n$ is based on Theorem 3 and Theorem 4. The element B in $NFFA_n$ satisfies the following condition:

B is an NFFA, write B as $(Q, \Sigma, \delta, \sigma_0, \sigma_1)$, then the fuzzy sets $\delta, \sigma_0, \sigma_1$ only take values in the given finite subset D of the unit interval $[0,1]$.

We show that the class $NFFA_n$ so constructed satisfies the above mentioned approximation property.

Take any NFFA A with input alphabet Σ. By Theorem 3, there is a DFFA C such that $f_C = f_A \vee \frac{1}{n}$. Write $C = (Q, \Sigma, \delta, q_0, \sigma)$. Constructing a new DFFA B from C as follows, where $B = (Q, \Sigma, \delta, q_0, \tau)$, the only difference between B and C lies in the definitions of fuzzy final state σ and τ, where τ is defined in the following manner: For any state q in Q, $\tau(q) = \frac{i}{n}$ for some nonnegative integer $i \leq n$ if $\frac{i}{n} \leq \sigma(q) < \frac{i+1}{n}$. It is clear that τ is defined well, and $|\tau(q) - \sigma(q)| \leq \frac{1}{n} \leq \epsilon$ holds for any state $q \in Q$. Evidently, B is in the class $NFFA_n$ and B is a DFFA. For any input string $\omega \in \Sigma^*$, let $q = \delta^*(q_0, \omega)$, then $f_C(\omega) = \sigma(q)$ and $f_B(\omega) = \tau(q)$. Furthermore, if $f_A(\omega) < \frac{1}{n}$, then $f_C(\omega) = f_A(\omega) \vee \frac{1}{n} = \frac{1}{n} = \sigma(q)$. By the definition of τ, $f_B(\omega) = \tau(q) = \frac{1}{n}$. Hence $|f_A(\omega) - f_B(\omega)| = \frac{1}{n} - f_A(\omega) \leq \frac{1}{n} \leq \epsilon$. If $f_A(\omega) \geq \frac{1}{n}$, then $f_C(\omega) = f_A(\omega) \vee \frac{1}{n} = f_A(\omega)$, i.e., $f_C(\omega) = \sigma(q) = f_A(\omega)$. By the definition of τ, it follows that $|f_A(\omega) - f_B(\omega)| = |f_C(\omega) - f_B(\omega)| = |\sigma(q) - \tau(q)| \leq \epsilon$. Therefore, B is a DFFA in the class $NFFA_n$ which can approximate A with accuracy ϵ.

Remark 3. By Theorem 1, Remark 1, Theorem 4 and the constructions given in Remark 2, for a t-norm $\star$ satisfying WFGC, we know that NFFAs under max-$\star$ compositional inference and NFFAs under max-min compositional inference are equivalent in the approximate sense. Therefore, for different kinds of t-norms satisfying WFGC, NFFAs under max-$\star$ compositional inference are equivalent in the approximate sense. This gives an affirmative answer to the mentioned problem in the introduction section, and NFFAs under different compositional inference mode have the ability to describe discrete event systems with fuzzy uncertainty and fuzzy discrete recurrent systems. That is to say, just as fuzzy systems using different compositional reasoning methods are universal approximators to the general fuzzy control systems [9, 6], NFFAs with different compositional inference are also universal to discrete fuzzy systems. This is just the significance of the consequence of Theorem 4. Furthermore, since a continuous t-norm can be uniformly approximated by a sequence of Archimedean t-norms [4], the results presented in this paper can also be applied to any continuous t-norm.

3 Conclusion

In this work, we study the relationship between DFFA and NFFA under max-$\star$ compositional inference for some t-norm satisfying weakly finite generated condition. We show that DFFA and NFFA are not equivalent in general. However, we can approximate an NFFA by some DFFA with any given accuracy, the related construction is also presented in Remark 2. The significance of this approximation is that it shows the approximation equivalence between NFFAs using max-$\star$ compositional inference for different kinds of t-norms. Therefore, as the models of discrete event systems with fuzzy uncertainty, fuzzy finite automata can be universally used. Some general results are obtained which will be potentially useful for the application of fuzzy finite automata.

References

1. Adamy J, Kempf R (2003) Regularity and chaos in recurrent fuzzy systems. Fuzzy Sets and Systems 140: 259-284
2. Cao Y, Ying M S (2006) Observability and decentralized control of fuzzy discrete-event systems. IEEE Trans. Fuzzy Systems. 14(2): 202- 216
3. Hájek P (1998) Metamathematics of Fuzzy Logic. Kluwer Academic Publisher, Dordrecht
4. Klement E P, Pap E (2000) Triangular Norms. Kluwer Academic Publisher, Dordrecht
5. Li Y M, Pedrycz W (2005) Fuzzy finite automata and fuzzy regular expressions with membership values in lattice-ordered monoids. Fuzzy Sets and Systems 156: 68-92
6. Li Y M (2005) Analysis of Fuzzy Systems(in Chinese). Science Press, Beijing
7. Lin F, Ying H (2002) Modeling and control of fuzzy discrete event systems. IEEE Trans. Systems, Man and Cybernectics-part B: Cybernectics. 32(4): 408-415
8. Mordeson J N, Malik D S (2002) Fuzzy Automata and Languages: Theory and Applications. Chapman & Hall/CRC, Boca Raton, London
9. Wang L X (1997) A Course in Fuzzy Systems and Control. Prentice-Hall PTR, Englewood Cliffs, NJ
10. Zadeh L A (2004) A note on web intelligence, world knowledge and fuzzy logic. Data and Knowledge Engineering 50:291-304

Context Aware System for Future Homes*

Tae-Hun Lim[1], Jin-Heung Lee[2], and Sang-Uk Shin[1]

[1] Department of Interdisciplinary Program of Mechatronics Engineering, Pukyong National University,
599-1, Daeyeon3-Dong, Nam-Gu, Busan 608-737, Republic of Korea
lth1553@hanmail.net, shinsu@pknu.ac.kr
[2] MOBILIZONE Inc. 916, 21C Century City B/D, Daeyon3-dong Nam-Gu, Busan, Republic of Korea
jinhung@seoul.com

Summary. With the rapid development of ubiquitous computing technology in the home and community, users can access information anytime and anywhere via personal devices such as PDA and internet mobile device. Ubiquitous application will need flexible access control mechanisms and more suitable access control decisions. Also context awareness throughout intelligent system is emphasized in order to provide automatic services in future home system. In this paper, we use a variety of device for collecting data from events for users in the home. This data is stored in the form of cases. This case form of data is used as training data for learning with a neural network algorithm, and the result is applied to an intelligent context aware system for future home automation. This system provides the intelligence required to "right situations do right things" for users. Consequently, the proposed system is an intelligent one.

Keywords: Context-aware, Neural Network, CBR.

1 Introduction

As computers become more common in the home and broadband technology is introduced into residential communities, new applications will allow a wide range of human activities (*e.g.*, education, entertainment, social and community gatherings, *etc.*) to be conducted over the internet. Such applications often will use information about the residents of smart home, as well as various resources inside the home. Furthermore, these applications will access this sensitive information from many different locations. Therefore, the protection of private information about each smart home's resources and residents is a critical concern that must be addressed before such applications can be successfully deployed. Also the proliferation of smart gadgets, mobile devices, PDA and sensors has enabled the construction of pervasive computing environments, transforming regular physical spaces into intelligent spaces [1]. Also intelligent spaces provide services and

---

* This research was supported by the Program for the Training of Graduate Students in Regional Innovation which was conducted by the Ministry of Commerce Industry and Energy of the Korean Government.

K.M. Węgrzyn-Wolska and P.S. Szczepaniak (Eds.): Adv. in Intel. Web, ASC 43, pp. 210–216, 2007.
springerlink.com

resources that users can access and interact with via personal portable devices such as PDAs using short range wireless communications such as Bluetooth or IEEE 802.11. The resulting anytime,anywhere access infrastructure is enabling a new generation of applications that can leverage this home environment information by continuously managing, adapting and optimizing it. The home system will effectively integrate the previously separated communication and computing networks in the home and incorporate the core tenets of ubiquitous computing. The future home system requires dynamical adaptation according to the user's activities and environments. Many researchers have focused on context-aware architectures and context-aware applications [2]. Automatically collecting the context information and reacting in ways that fit in with the environment are the main design goals of context-aware systems. Machine learning, data mining, and intelligent decision algorithms with context information are the key technologies required to implement context-awareness in the home environment.

Until now, neural networks have been successfully applied to many classification problems in data mining. The complex structure of neural networks provides the powerful performance required to solve difficult problems, but also make it difficult to construct a proper architecture. For efficient and effective data mining tools to be developed, there are several requirements and challenges to be met[3]. These include performance, user interaction, and smart architecture. Performance includes accuracy and efficiency. User interactivity allows users to perform customized functions. Smart architecture implies the ability of self-adaptation to the environment.

In this paper we demonstrate how to use CBR (case based reasoning) with neural network algorithm as context awareness solution in future home environment. The context content and case's organization format are discussed respectively. For case retrieving, multi-level similarity is suggested. A framework was developed the novel system based on CBR with a neural network algorithm. According to the action to be performed, a different solution to the case adaptation is defined. In future homes, a specific solution for case revising is suggested.

The rest of this paper is organized as follows: Section 2 presents the background and related work. Section 3 presents the proposed novel system based on CBR with a neural network algorithm for future homes. Section 4 presents the conclusion and future works.

2   Background and Related Work

Nowadays, a lot of smart home projects have been initiated by many research institutes such as the Georgia Tech Aware Home [4], MIT Intelligent Room [5] and Neural Network Hose [6] at the University of Colorado at Boulder. Context-awareness is one of the primary characteristics of future homes. Collecting information from sensors, reasoning based on information from knowlege databases, and thus adopting corresponding activities are the main steps of context-aware applications. Much work has been done on how to use reasoning algorithms to achieve context awareness.

A. Ranganathan et al. proposed a context-aware model based on first order predicate calculus [7]. They used a first order model to express complex rules, which involve contexts. This knowledge expression enables automated inductive and deductive reasoning to be easily performed on contextual information. M. Wallace et al. developed a context aware clustering algorithm for mining a user's consumption interests of multimedia documents, based on the user history [8].

CBR is a problem solving technique that reuses previous cases and experiences to find a solution to current problems. L.D. Xu et al. discussed the advantages and process of CBR and provided an application that uses CBR to judge the AIDS (Acquired Immune Deficiency Syndrome)[9]. W. J. Yin et al. used a combination of genetic learning approach and case based learning to solve job-shop scheduling problems [10].

In general in the case of category classifications with multilayer feed-forward neural networks, the hidden units are learned to maximize the useful information from the input pattern and the output units are learned to discriminate the information obtained from the hidden units [11]. Therefore, it is reasonable to provide more information to the output units in order to improve the discrimination power in category classification. A back-propagation neural network offers a framework suitable for reusing the output values of the network during training.

3   Context in Future Home

3.1   Scenarios

To illustrate the motivation of our research, let us discuss an example application that could be used by a pervasive computing infrastructure in a future home. The home has several rooms including bedrooms, a dining room, kitchen, living room, bathroom and garage. Sensors in the home can capture, process and store a variety of information about the home, the users and their activities.

The future home sometimes means an automated home. One possible scenario is as follows: "At 19:00, Miss Kim enters the garage; the garage's door opens and the garage's light switches on automatically. After parking her car, Miss Kim leaves the garage; the light turns itself off and the door is closed. Next, Miss Kim enters the living room; the room temperature is 0 °C, the heater will automatically turn itself on to increase the temperature. At the same time, the TV is turned on and tuned to the entertainment channel is tuned."

"At 23:30, Miss Kim leaves the living room and enters the bedroom, the heater and TV in the living room are turned off. The light in the bedroom is turn on, and the brightness is set to low." Although the above scenarios seem to be relatively simple, it would be challenging to achieve these "simple" scenarios in the real home environment. The purpose of context awareness is "to do the right in the right situations". The basic principle of the commonsense reasoning and context awareness is the understanding of the current state. However, there are many situations in which the TV, heater and light might be encountered. In terms of adjusting the TV channel: some people prefer the news, while others prefer entertainment. Sometimes TV stations provide comedy, while sometimes they provide sitcoms. Also, some programs fit the living room, while other programs fit the bedroom, and so on.

In term of switching on the heater: some people like the room to warm while others like it to be cool, some like to turn on the heater while they are sleeping while others do not, and so on. It almost means impossible for the system designer to envision all possible contexts before the deployment of the system. The future home system will inevitably perform in unexpected and undesirable ways on occasion and inevitably perform in unexpected and undesirable ways on occasion and, thus, disappoint the home occupant. A common learning algorithm also cannot solve this problem, because the training set will not contain examples of appropriate decisions for all possible contextual situations.

3.2 Content of Context

Context is usually classified into three categories: the environment, the user's activity, and the user's physiological states. Each category has its own subcategories. In the beginning, our context information model will be simple and idiographic. We don't deal with abstract concepts. We assume that the context information can be simplified into a collection of discrete facts and events with numeric parameters.

According to the above analysis, there may be different rooms. The occupant's habits cas also vary. So,the context in smart home can be classified into three dimensions: (1) time, (2) environment, and (3) person. In each category, there are several entities as shown in Table 1. Since the occupants deal with a large amount of information, the context information is modeled hierarchically.

As an analogical reasoning system, CBR with a neural network algorithm bases itself on the concept that "similar problems have similar solutions", finds a solution by analyzing the information from previous cases and effects the necessary modifications to adapt to new circumstances.

In the proposed system, a database is used to store the case information. This case information is organized into related tables. The target is the development of a solution in order to find the best matched set from the existent cases. Then, we then support the "do the right thing in the right situation" paradigm. The stored data learned using the neural network algorithm is used to find the best match.

Table 1. Context categories and entities in future home

Time	Time	*Minute/Hour/Day/Week/Month/Season*
	Time sequences	*Event occurring sequence*
	Location	*Bedroom/Bathroom/Kitchen/Dining room/Living room/Garage*
Envi-	**Status**	*Leaving/Staying/Entering*
ronment	**Temperature**	*Environment's temperature*
Person	**ID**	*Person's ID*
	Profile	*Name/Gender/Age*
	Habit	*Sports/News/Entertainment/Warm/cool*

Our learning process can be described as follows:

(1) Initialization
- All interconnected weight values have a small random value [0,1]
(2) Application of training data
- The stored data is used as training data for the input node
(3) Forward computation
- The training data should be forward passed through the hidden layer between the input layer and output layer
- In this step, computing errors are signaled by comparing the target value to the input value
(4) Backward computation
- Compute the local error value of the network by proceeding backward, layer by layer.
- This signal was passing backward.
(5) Iteration
- In the case where new training data is available, reiterate steps (2) to (4) until that average square error for all of the training data is sufficiently low or approximately equal to the target value.

In the case of the neural network algorithm, there is no theory which cas be used to identity the neural network architecture. Therefore, it is very difficult to decide the learning rate value η, the momentum constant value α, and how many hidden layers there should be. Therefore, in our paper, we select the value based on the experimental result.

To select the number of hidden nodes, we carry out the experiment as following experiment: set the learning rate value η to 0.7 and the momentum constant value α to 0.3, increase the number of hidden nodes from 12 to 26 in steps of one. Then, the experiment is iterated 20,000 times. As a result of the test, we show that most of the cases have over 98% accuracy and the difference is below 0.25%. Therefore, we select the best case: the number of hidden nodes is 15.

Next, in order to select the learning value η and the momentum constant value α, we conduct a similar test as follows: set the number of hidden nodes to 15 and change the learning rate value η and the momentum constant value α as table 2. Then, the experiment is iterated 20,000 times. As a result of the test, table 2 shows that the average squared error had lowest value when the learning rate η is 0.7 and the momentum constant $\alpha$is 0.3. Also, accuracy for adequate service had the highest value when learning rate η: 0.7 and momentum constant α: 0.5. Thus, we select the following values: learning rate η: 0.7, momentum constant α: 0.3, the number of hidden layers: 2 and the numbers of hidden nodes: 15.

The proposed neural network model therefore has input layer, hidden layers and output layer. Its elements are as follow:

- $X_1, \cdots, X_n$: Input data (e. g., time(X_1), time sequences(X_2), location(X_3), status(X_4), temperature(X_5), ID(X_6), profile(X_7), habit(X_n))
- $P_1, \cdots, P_n$: Output data (adequate service(P_x))

Table 2. Performance comparison of the back-propagation neural network with the learning rate and momentum constant

Learning Rate	Momentum Constant	Average Squared Error	Accuracy for adequate service(%)
0.3	0.1	0.0199	98.67
0.3	0.3	0.0185	98.86
0.3	0.5	0.0255	98.76
0.3	0.7	0.0174	98.98
0.3	0.9	0.0178	98.98
0.5	0.1	0.0214	98.83
0.5	0.3	0.0175	98.98
0.5	0.5	0.0186	98.98
0.5	0.7	0.0167	99.03
0.5	0.9	0.0339	98.35
0.7	0.1	0.0176	98.89
0.7	0.3	**0.0161**	99.05
0.7	0.5	0.0170	**99.08**
0.7	0.7	0.0183	99.05
0.7	0.9	0.1189	81.98

4 Conclusion

Intelligence is the core of the future home and context-awareness plays an important role in providing it. With context-awareness, systems can decide which services should be provided. Building a case based expert system is a feasible solution to this problem. Firstly, the future home is a complex system. Secondly, it is impossible to enumerate all of the possible scenarios that may appear before the context awareness system is deployed.

In the proposed system, the stored data learned using a neural network algorithm is used to find the best match. This system supports the "do the right thing in the right situation" paradigm. Thus, CBR with a neural network algorithm is a suitable solution to use in the future home. There are a lot of context contents in the future home environment and more case features will be added to the case tables within this scheme in a future study.

References

1. Campbel M, Al-Muhtadi J, Ranganathan A and Mickunas M. D (2002) "A flexible, privacy-preservering authentication framework for ubiquitous computing environments," *In International Workshop on Smart Appliances and Wearable Computing*, Vienna, Austria.
2. Abowd G, Ebling M, Hunt G, Lei H and Gellersen H (2002) "'Context-Aware Computing," *In PERVASIVE Computing*, JULY-SEPTEMBER, pp. 22-23.
3. Sohn S and Dagli C (2004) "Ensemble of Evolving Neural Network in Classification," *Neural Processing Letters*, pp.191-203.

4. Helal S, Winkler B and Lee C, "Enabling Location-Aware Pervasive Applications for the Elderly," *In Proc. of 1st IEEE Int. Conf. PERVASIVE Computing and Communications (PerCom'03)*, pp.531-538.
5. Roy A, "Location Aware Resource Management in Smart Homes," *In Proc. of 1st IEEE Int. Conf. PERVASIVE Computing and Communications (PerCom'03)*, pp.481-488.
6. Cook D, "MavHome: An Agent-Based Smart Home," *In Proc. of 1st IEEE Int. Conf. PERVASIVE Computing and Communications (PerCom'03)*, pp.521-524.
7. Ranganathan A and Campbell R (2003) "An Infrastructure for Context-awareness based on First Order Logic," *Personal & Ubiquitous Computing*, Vol.7, pp.353-364.
8. Wallace M and Stamou G (2002) "Towards a context aware mining of user interests for consumption of multimedia documents," *In Proc. Of the IEEE Multimedia and Expo*, Vol.1, pp. 733-736.
9. Xu L (1993) "Case based reasoning for AIDS Initial Assessment," *In Proc. Of the IEEE Int. Conf. Systems science and systems engineering*.
10. Yin W and Liu M, "A genetic learning approach with case based memory for job-shop scheduling problems," *In Proc. Of the 1st Int. Conf. Machine learning and cybemetics, Beijing(2002)*, pp. 1683-1687.
11. Sethi I (1990) "Entropy Nets: From Decision Trees to Neural Networks," *In Proceedings of the IEEE*, Vol.78, pp. 1605-1613.

Document Clustering Based on Semantic Smoothing Approach

Yubao Liu[1,2], Jiarong Cai[1], Jian Yin[1], and Zhilan Huang[1]

[1] Department of Computer Science of Sun Yat-Sen University, Guangzhou, 510275, China
[2] Guangdong Province Key Laboratory of Information Security, Sun Yat-sen University, Guangzhou, 510275, China

Summary. Clustering of text documents is an important data mining issue and has wide application fields. However, many clustering approaches fail to yield high clustering quality because of the complex document semantics. Recently, semantic smoothing, which has been widely studied in the field of Information Retrieval, is proposed as an efficient solution. However, the existing semantic smoothing methods are not effective for partitional clustering. In this paper, based on the principle of TF*IDF schema, we propose an improved semantic smoothing method which is suitable for both agglomerative and partitional clustering. The experimental results show our method is more effective than the previous methods in terms of cluster quality.

1 Introduction

Clustering of text documents is an important data mining issue and has wide application fields. A volume of previous work has been focused on this area [1][6]. However, many clustering approaches fail to yield high clustering quality because of the complex document semantics. In general, agglomerative and partitional clustering are two main approaches for document clustering. Then how to improve the cluster quality is a critical problem for these two approaches.

A document is often full of class-independent general words and short of class-specific core words, which leads to the difficulty of document clustering. Recently, the researchers show that the semantic smoothing, which has been widely studied in the field of information retrieval (IR) [2][3], is an efficient solution for such problems [1]. The key idea of semantic smoothing approach is to discount general words and assign reasonable counts to unseen core words. The experimental results in [1] show the semantic smoothing methods are suitable for agglomerative clustering. However, the experiments also show the existing semantic smoothing methods can not efficiently reduce the general words and are not effective for partitional clustering.

In this paper, inspired by the principle of TF*IDF schema, we propose an improved semantic smoothing method which is suitable for both agglomerative and partitional clustering. The experimental results also show our methods are more effective than the previous methods in terms of cluster quality.

K.M. Węgrzyn-Wolska and P.S. Szczepaniak (Eds.): Adv. in Intel. Web, ASC 43, pp. 217–222, 2007.
springerlink.com

2 The Existing Clustering Based on Semantic Smoothing

The existing context-sensitive smoothing approach includes three steps: (1) multiword phrase extraction and translation, (2) building the semantic smoothing model, (3) clustering based on the semantic smoothing model.

Many previous approaches use word extraction method and single word vector as the document features. However, they suffer from the context-insensitivity problem. The terms in these models may have ambiguous meanings. To solve this problem, [1] uses the multiword phrases as topic signatures (document features). They index all documents in a given collection with terms (individual words) and topic signatures (phrases). Then the training process, which determines the probability $p(w|t_k)$ of translating the given multiword phrase t_k to term w in the vocabulary, is used. After training, we can estimate the semantic smoothing model. There are two kinds of context-sensitive semantic smoothing models, one is the document model-based smoothing, and the other is cluster model-based smoothing.

The document model-based smoothing approach is a kind of context-sensitive semantic smoothing approach that is used for agglomerative clustering. For each word in the vocabulary, its likelihood generated by a given document is obtained using the formulas (1)(2)(3).

$$p_{bt}(w|d) = (1 - \lambda)p_b(w|d) + \lambda p_t(w|d) \tag{1}$$

$$p_b(w|d) = (1 - \alpha)p_{ml}(w|d) + \alpha p(w|C) \tag{2}$$

$$p_t(w|d) = \sum_k p(w|t_k)p_{ml}(t_k|d) \tag{3}$$

The document model with semantic smoothing (i.e. formula (1)) is a mixture model with two components, that is, a simple language model and a topic signature (multiword phrase) translation model. The influence of two components is controlled by the translation coefficient (λ) in the mixture model. The first component is the simple language model (i.e. formula (2)), which can be obtained using the maximum likelihood estimator (MLE) document model $p_{ml}(w|d)$ together with background smoothing model $p(w|C)$ [2] with the controlling coefficient α. The second component of document model is the topic signature (multiword phrase) translation model (i.e. formula (3)). Here, the probability $p(w|t_k)$ of translating t_k to term w (individual word) is estimated in the training process.

The cluster model-based smoothing approach is a kind of context-sensitive semantic smoothing approach that is used for partitional clustering. The estimation of the cluster model $p(w|c_j)$ is very similar to the document model by replacing the document d in the above document model (i.e. formulas (1)(2)(3)) with the cluster c_j.

For agglomerative clustering, it is essential to measure the distance of two clusters, which is further reduced to the calculation of pairwise document distance by the complete linkage criterion. Give two document models $p(w|d_1)$ and $p(w|d_2)$, the KL-divergence distance [5] can be used to evaluate their distance, which is defined as,

$$\Delta(d_1, d_2) \equiv \sum_{w \in V} p(w|d_1) \log \frac{p(w|d_1)}{p(w|d_2)} \tag{4}$$

where V is the vocabulary of the corpus. Since KL-divergence is not a symmetric metric. Thus, the distance of two documents is defined as the minimum of two KL-divergence distances:

$$dist(d_1, d_2) \equiv \min\{\Delta(d_1, d_2), \Delta(d_2, d_1)\} \tag{5}$$

For partitional clustering, the problem is how to estimate the likelihood of a document d generated by a cluster. Based on the multinomial model [4], the log likelihood of document d generated by the j-th multinomial cluster model $p(w|c_j)$ is described in formula (6).

$$\log p(d|c_j) = \sum_{w \in V} c(w, d) \log p(w|c_j) \tag{6}$$

where $c(w, d)$ denotes the frequency count of word w in document d and V denotes the vocabulary.

3 The Improved Semantic Smoothing Model

3.1 The Improvement with IDF Factors

The experimental results in [1] show the semantic smoothing models are suitable for agglomerative clustering. However, the experiments also show the semantic smoothing models are not effective for partitional clustering. The key reason is that the existing semantic smoothing models are effective for handling the sparsity of core words that is the major problem affecting the agglomerative clustering. On the other hand, the density of general words but not the sparsity of core words is critical in partitional clustering, and the existing semantic smoothing models are not enough effective for discounting general words. Actually, Based on the principle of TF*IDF [6], we propose an improved semantic smoothing model. The principle of TF*IDF scheme contains two aspects. That is, if a term appears frequently in a given document, then it is important for the document, and on the other hand, if a term appears in many documents in the corpus, it is less important to the document. Generally, the terms appearing in many documents are probable to be the general words. The existing semantic smoothing models just consider the frequency counts of the words and topic signatures, and that corresponds to the TF factor of TF*IDF schema. In our improved method, we use the IDF factors to enhance the ability of discounting general words. In detail, we define the IDF factors as follows.

Definition 1. *The δ factor of term w_i is defined as $\log_2(N/n_i + 0.01)$, where N is the total number of document in the corpus and n_i is the number of document contains term w_i.*

Definition 2. *The φ factor of topic signature (multiword phrase) t_i is defined as $\log_2(N/n_i + 0.01)$, where N is the total number of document in the corpus and n_i is the number of document contains topic signature t_i.*

Based on the above two factors, we define the improved document model-based semantic smoothing model as the formulas (7)(8)(9).

$$p'_{bt}(w|d) = (1 - \lambda)p'_b(w|d) + \lambda p'_t(w|d) \tag{7}$$

$$p'_b(w|d) = (1 - \alpha)p_{ml}(w|d)\delta + \alpha p(w|C) \tag{8}$$

$$p'_t(w|d) = \sum_k p(w|t_k)\delta p_{ml}(t_k|d)\varphi \tag{9}$$

Similarly, we can get improved cluster model-based smoothing model by replacing the document d in the improved document model (i.e. formulas (7)(8)(9)) with the cluster c_j.

3.2   The Improvement with Equal Background

From the above formulas, we can also know that the background model $p(w|C)$ is computed by the frequency count of word w in background corpus $(c(w, C))$. Actually, the background model $p(w|C)$ largely influences the likelihood of general word generated by a given document or cluster. Since the general words often have high frequency in the background corpus. So, we introduce the equal background model as another improvement. In equal background model, the general words would have the same probability as the other words. We use the average frequency count to compute the model estimation value. The estimation of the equal background model is given in formula (10).

$$p(w|C) = \frac{avg_{w_i \in C}(c(w_i, C))}{\sum_{w_i \in C} c(w_i, C)} \tag{10}$$

In (10), $avg_{w_i \in C}(c(w_i, C))$ denotes the average frequency count of the words in the background corpus. By replacing the original background model in the previous formulas with the equal background model, we can obtain another improved smoothing model.

4   The Experimental Results

In the experimental studies, we evaluate both agglomerative and partitional clustering with different semantic smoothing methods. We denote the original context-sensitive semantic smoothing method as *O-smoothing*, denote the improved context-sensitive semantic smoothing with IDF factors as *I-smoothing*, and denote the improved context-sensitive semantic smoothing with both IDF factors and the equal background model as *IE-smoothing*. Similar to [1], the evaluation functions, *FScore*, *Purity* and normalized mutual information (*NMI*), are used to evaluate the cluster quality, and 20ng-newsgroups (*20NG*) is used as the test dataset.

In our experiments, we use the complete linkage criterion and document model-based semantic smoothing model for agglomerative clustering, and we use K-Means and cluster model-based semantic smoothing model for partitional clustering. Since the converging of the agglomerative clustering is slow, we only test it with the small corpus. We randomly pick 100 random documents from each selected class of *20NG* dataset to form the small corpus (2000 documents). The Large dataset contains all 20 classes (19660 documents) of *20NG* dataset.

The experimental results with different kinds of smoothing are in Table 1, 2, 3. For all smoothing methods except TF*IDF, we run with different translation coefficient(λ) and choose the best evaluation function values to report and the value in the bracket corresponds to the translation coefficient value for the best values.

Table 1. Agglomerative clustering with different kinds of smoothing

Evaluation Function	Vector Cosine	KL-Divergence		
	TF*IDF	O-smoothing	I-smoothing	IE-smoothing
FScore	0.092	0.206 (λ=0.6)	**0.217 (λ=0.7)**	0.193 (λ=0.4)
NMI	0.132	0.198 (λ=0.6)	**0.258 (λ=0.4)**	0.246 (λ=0.8)
Purity	0.098	0.220 (λ=0.6)	**0.255 (λ=0.9)**	0.249 (λ=0.9)

Table 2. Partitional clustering with different kinds of smoothing (small corpus)

Evaluation Function	Standard K-Means	Model-based K-Means		
	TF*IDF	O-smoothing	I-smoothing	IE-smoothing
FScore	0.368	0.298 (λ=0.6)	0.320 (λ=0.2)	**0.463 (λ=0.4)**
NMI	0.377	0.318 (λ=0.9)	0.352 (λ=0.8)	**0.481 (λ=0.8)**
Purity	0.402	0.319 (λ=0.5)	0.350 (λ=0.2)	**0.484 (λ=0.4)**

Table 3. Partitional clustering with different kinds of smoothing (large corpus)

Evaluation Function	Standard K-Means	Model-based K-Means		
	TF*IDF	O-smoothing	I-smoothing	IE-smoothing
FScore	0.438	0.455 (λ=0.4)	0.536 (λ=0.6)	**0.551 (λ=0.6)**
NMI	0.508	0.472 (λ=0.6)	0.537 (λ=0.6)	**0.567 (λ=0.4)**
Purity	0.485	0.463 (λ=0.4)	**0.576 (λ=0.6)**	0.533 (λ=0.4)

5   Conclusions

In this paper, based on the IDF factors and the equal background strategy, we propose an improved context-sensitive semantic smoothing for model-based document clustering. The improved smoothing model has much stronger ability of discounting general words and is more suitable for document clustering. The experimental results also confirm our method is more effective than the previous methods in terms of cluster quality.

Acknowledgments

This paper is supported by the National Natural Science Foundation of China (60573097), Natural Science Foundation of Guangdong Province (05200302, 04300462), Research Foundation of National Science and Technology Plan Project (2004BA721A02), Research Foundation of Science and Technology Plan Project in Guangdong Province (2005B10101032), Research Foundation of Disciplines Leading to Doctorate degree of Chinese Universities (20050558017) and Opening Foundation of Guangdong Province Key Laboratory of Information Security.

References

1. Xiaodan, Z., Xiaohua Zhou, Xiaohua Hu. Semantic Smoothing for Model-based Document Clustering. In:Proc. IEEE ICDM 2006, pp.1193-1198.
2. Zhai, C. and Lafferty, J. A Study of Smoothing Methods for Language Models Applied to Ad hoc Information Retrieval. In: Proc. ACM SIGIR 2001, pp.334-342.
3. Zhou, X., Hu, X., Zhang, X., Lin, X., and Song, I.-Y. Context-Sensitive Semantic Smoothing for the Language Modeling Approach to Genomic IR. In: Proc. ACM SIGIR 2006, pp.170-177.
4. Zhong, S. and Ghosh, J. Generative model-based document clustering: A comparative study. Knowledge and Information Systems, 8(3): 374-384, 2005.
5. Kullback, S. and Leibler, R. A. On information and sufficiency. Annals of Mathematical Statistics, 22(1): 79-86, March 1951.
6. Steinbach M., Karypis G., Kumar V. A Comparison of Document Clustering Techniques. In: Proc. of Text Mining Workshop, KDD 2000, pp.1-20.

WIR-A Graph-Based Algorithm for Friend Recommendation

Shuchuan Lo and Chingching Lin

National Taipei University of Technology, Taipei, Taiwan 106 R.O.C.
sclo@ntut.edu.tw

1 Introduction

Content-based [6] and collaborative-based methods [7] [8] [9] are two main underlying techniques used in recommendation system. Some researchers proposed hybrid recommendation method to compromise the advantages and disadvantages of content-based and collaborative recommendation approaches [4] [10]. These three approaches share the common goal of assisting in the user's search for items of interest not recommending themselves to other users. For friend recommendation, we proposed a graph-based method, in this research, named Weighted Information Ratio (WIR) borrowed idea from information theory [1] [2]. We compare the method WIR with our prior algorithm in friend recommendation named Minimum-message Ratio (WMR) [3]. The precision and recall of WMR method are 15% and 8% for a target member with 15 recommendations in the testing prediction, respectively. This result is acceptable compared to a hybrid recommender system for digital library [4], where the testing precision and recall are 3% and 14% for 100 customers. Both recommendation algorithms generate a limited, ordered and personalized friend lists by the real communication number among web users. Communication number is more representative than most of the population variables because they are lack of diversity [5].

There are three indexes in the quality evaluation of recommender system, that is, recall, precision and F1 metric [9] [11]. **Recall** is the ratio of recommendation lists to the real future happenings of a target user. **Precision** is the ratio of correct guess in all the recommendations. In order to average recall and precision, we use the standard **F1 metric** that gives an equal weight to both measures and is computed as $F_1 = \frac{2*recall*precision}{recall+precision}$.

The article is organized as follows. Our main theory of WIR and WMR are given in Section 2. Section 3 is the comparison of WIR and WMR by an experimental real life data from a community website. Section 4 is our concluding remarks.

2 Research Method

We chose the minimum interaction number as the strength of relationship in WMR. But some people may be enthusiasm if they issued many letters out. We

K.M. Węgrzyn-Wolska and P.S. Szczepaniak (Eds.): Adv. in Intel. Web, ASC 43, pp. 223–229, 2007.
springerlink.com   © Springer-Verlag Berlin Heidelberg 2007

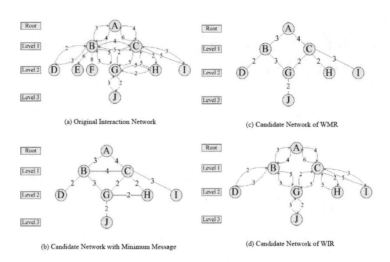

(a) Original Interaction Network

(c) Candidate Network of WMR

(b) Candidate Network with Minimum Message

(d) Candidate Network of WIR

Fig. 1. Interaction Network

consider both information volume and minimum message number in WIR. From the famous research of Milgram in the social network [12], proposed that two Americans acquainted each other only through the most 6-level intermediaries. It makes no sense to get recommendations from a big and complex network in our case study. Therefore, we used the target user as a root to expend 5 levels forming a candidate network. An example is shown in Figure 1. In Figure 1(a), we can observe some members with unbalanced interaction.The minimum-message candidate network is shown as Figure 1 (b). In Figure 1 (b), the member A can acquaint member D directly through member B, noted as P(D→A)={A, B, D}. But it also can be through the path P(D→A)={A, C, B, D}. It seems to be redundant and unnecessary by adding an extra intermediary C. Therefore, we eliminate the interactions between same-level nodes as shown in Figure 1 (c). Figure 1 (d) is two-way interactions of Figure 1 (c). There is an assumption in WMR and WIR algorithm, that the longer the path, the lower the recommendation.

The Weighted Minimum-message Ratio-WMR Recommend Algorithm. The WMR algorithm is based on the minimum message candidate network without interactions on the same level. All the examples are based on the network in Figure 1 (c). Notation defined as

C_{ij}: the minimum messages between the parent node i and child node j.
$L(h)$: all the nodes belong to the level h in the candidate network.
$Pth_k(j)$: the kth path stared from root (the target user) to any node j.
$Pth_k(j)$.CSum: the total messages of path $Pth_k(j)$.
$L(h)$.CSum: the total messages of $L(h)$, for example, $L(1)$.CSum= $C_{AB} + C_{AC}$ = 3 + 4 = 7.
$C(i, j)$: the proportion of message C_{ij} and total messages of the level h where node $j \in L(h)$ and $i \in L(h-1)$, i is the parent node of node j, for example, C(A,

B) $= C_{AB} / L(1).CSum = \frac{3}{7}$.

$WMR_o(j)$: the recommendation score of node j to node o, in WMR, all the recommendations being aimed at the root node (the target user) and $j \notin L(1)$.

Table 1. Ordered Recommendation Lists to Node A in WMR

Recommend Order	Member	Recommend Score
1	G	6.00
2	I	4.00
3	H	3.43
4	D	2.14
5	J	1.62

We assume that the more messages the stronger the relationship between members (nodes). We propose a friend recommended ratio based on the strength of interaction. The recommendation is

$$WMR_o(j) = \Sigma_k[Pth_k(j).CSum * \Pi_i C(S_{i-1}, S_i)] \tag{1}$$

$S_i \in Pth_k(j) \cap L(i), i = 1, ..., h-1$, where $j \in L(h)$ and $h \neq 1$. After we calculated all recommendation scores for all nodes except the nodes in $L(1)$ because they are friends of the root node. The ordered recommendation lists are in Table 1.

The Weighted Information Ratio-WIR Recommend Algorithm. All the examples are based on the network in Figure 1 (d). Notation defined as

$n(i \to j)$: the number of message issued from node i to node j.

$n(i \leftrightarrow j)$: the number of message between node i and node j.

$n(i)$: the number of message issued by node i, $n(i) = \Sigma_j n(i \to j)$.

N: the total messages in this network.

$P(i)$: the message probability of node i, that is, the messages issued from node i divided by all the messages, for example, $P(A)=n(A)/N =8/57=0.140$.

$Pth_k(j)$: the kth path stared from root (the target user) to any node j.

C_{ij}: the minimum messages between the parent node i and child node j.

$Pth_k(j).CSum$: the total minimum messages of path $Pth_k(j)$, for example, $Pth_1(G).CSum = C_{AB} + C_{BG} = 3 + 3 = 6$.

$IR(i,j)$: the information ratio between node i and node j defined by

$$IR(i,j) = (p(i) * \frac{n(i \to j)}{n(i \leftrightarrow j)} + p(j) * \frac{n(j \to i)}{n(i \leftrightarrow j)}) \tag{2}$$

For example, $IR(A,B) =0.140*4/7+0.193*3/7=0.163$.

$L(h)$: all the nodes belong to the level h in the candidate network, for example, $L(1) = \{B,C\}$.

$WIR_o(j)$: the recommended score of node j to node o, in WIR, all the recommendations being aimed at the root node (the target user) and $j \notin L(1)$.

$$WIR_o(j) = \Sigma_k(Pth_k.CSum * log\frac{\Pi_{t \in Pth_k}IR(o,t)}{P(o) * P(j)}) \tag{3}$$

Table 2. Message Possibility Issued by Each Member

	Possibility
P(A)	0.140
P(B)	0.193
P(C)	0.281
P(D)	0.035
P(G)	0.123
P(H)	0.088
P(I)	0.088
P(J)	0.052
Total	1.0

Table 3. Information Ratio between Two Connected

	Information Ratio
IR(A,B)	0.163
IR(A,C)	0.225
IR(B,D)	0.130
IR(B,G)	0.167
IR(C,G)	0.236
IR(C,H)	0.143
IR(C,I)	0.160
IR(G,J)	0.080

Table 4. Ordered Recommendation Lists to Node A in WIR

Recommend Order	Member	Recommend Score
1	G	4.127
2	I	3.260
3	D	3.180
4	H	2.051
5	J	-6.065

For example $WIR_A(I) = (4+3) * log \frac{0.225*0.160}{0.140*0.088} = 3.26$. The probability for each member in Figure 1(d) is listed in Table 2. The information ratio between two connected nodes is listed in Table 3. The recommendation list is shown in Table 4.

3 Experimental Study

Experimental Environment and Data Process. The experimental data come from a community website in Taiwan. The period of data spans from May 1, 2004 to July 7, 2004 with total 317,171 messages issued by 7,620 distinct members. We divided the whole messages into two parts; one part is from May 1, 2004 to May 30, 2004 to be the training data, the other is from June 1, 2004 to July 7, 2004 to be the testing data. The former has 114,969 messages and the latter has 202,202 messages. Lo used rescaled consuming-behavior variables: recency, frequency and monetary to segment 16,094 customers into five clusters by K-means method into five clusters [5]. Pool the actives and potentials and re-clustered by K-means again. There are 678 of potential customers had message records in May and June. We randomly sampled 30 customers from these 678 potential customers being our testing members.

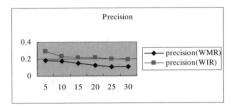

Fig. 2. Precisions Prediction

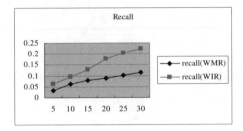

Fig. 3. Recall Prediction

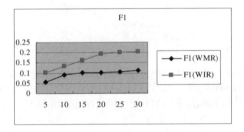

Fig. 4. F1 Metrics Prediction

Experimental Result and Evaluation. For each testing sample, we made 5 to 30 recommendations. Some of 30 recommendation lists for one of our 30 testing customers. All these lists were predictions because we did not recommend these lists in practice. We just compared the lists with the testing data. The recall, precision and F1 metric for WMR and WIR are shown in Table 5 and Table 6, respectively. The comparison of these two method for recall, precision and F1 metric are exhibited in Figure 2, 3 and 4, respectively. The F1 metric of WMR is about the same after 15 recommendations in Figure 4. We chose 15 recommendations as our quantity of recommendation in order to balance the recall and precision where precision = 14.7% and recall = 8%. The F1 metric of WIR is about the same after 20 recommendations in Figure 4. We chose 20 recommendations as our quantity of recommendation in order to balance the recall and precision where precision = 21.8% and recall = 17.6%. The experimental results of WIR are much better than those of WMR in different recommendation quantity from Figure 2 to Figure 4. In order to give statistical evidence

Table 5. Recommend Evaluation Indexes by WMR

# of Recom.	5	10	15	20	25	30
Precision	0.186	0.173	0.147	0.123	0.111	0.107
Recall	0.03	0.06	0.08	0.09	0.1	0.116
F1	0.056	0.09	0.104	0.104	0.105	0.112

Table 6. Recommend Evaluation Indexes by WIR

# of Recom.	5	10	15	20	25	30
Precision	0.293	0.233	0.216	0.218	0.201	0.190
Recall	0.063	0.094	0.128	0.176	0.203	0.223
F1	0.103	0.134	0.161	0.195	0.202	0.205

Table 7. Tests of Precision in WMR and WIR

# of Recom.	Z-Statistics	
	Precision	Recall
5	2.16**	3.73***
10	1.82**	3.24***
15	2.68***	4.08***
20	4.37***	6.94***
25	4.83***	7.37***
30	4.89***	7.31***

about the improvement of WIR comparison with WMR in precision and recall, we made two hypotheses as 1. H0: Precision of WIR=Precision of WMR vs. H1: Precision of WIR>Precision of WMR 2. H0: Recall of WIR=Recall of WMR vs. H1: Recall of WIR>Recall of WMR. The Z-statistics are exhibited in Table 7. The test results show that acceptance H1 for both hypotheses and the proportion differences are supported by statistical significant level equal to at least 5% for precision and 1% for recall from 5 to 30 recommendations.

4   Conclusion and Future Research

There is little research on friend recommendation. In this study, we proposed a graph-based friend recommendation algorithm, WIR. We developed our algorithm and interface under Visual Basic 6.0 and MSSQL 2000. The average precision is around 22% and average recall is 18% for our 20 recommendations in our experimental prediction. Due to no practical recommendations in this experiment, we can not expect the true precision and recall. However the testing precision and recall of WIR are much better than our former graph based algorithm WMR with 15% precision and 8% recall. On the experimental limit, we only provide the predicting precision and recall. However, the true effect of recommendation still requires further validation.

References

1. Chow C.K. and Liu C.N. (1968) Approximating discrete probability distributions with dependence trees. IEEE Transactions on Information Theory, 14:462–467.
2. Cheng J., Bell D.A. and Liu W. (1997) An Algorithm for Bayesian Belief Network Construction from Data. In Proc. of AI and START'97:83–90.
3. Lo S.C. and Lin C.C. (2006) WMR-Graph-based Algorithm for Friend Recommendation. In Proc. of 2006 IEEE/WIC/ACM International Conference on Web Intelligence, December 18 - 22, Hong Kong.
4. Huang Z., Chung W., Ong T. and Chen H. (2002) A Graph-based Recommender System for Digital Library. In Proceedings of JCDL'02, July 13-17, Portland, Uregon, USA.
5. Lo S. (2004) Online Customer Segment Based on Two-stage K-means. Technique Report of E-commerce Technology Laboratory, National Taipei University of Technology, Taipei, Taiwan.
6. Buckley C. and Salton G. (1995) Optimization of relevance feedback weights. In Proceedings of the 18th Annual International ACM SIGIR Conference on Research and Development in Information Retrieval, July, Seattle.
7. Herlocker J.L., Konstan, J.A. and Riedl J. (2000) Explaining Collaborative Filtering Recommendations. In Proceeding of ACM 2000 Conference on Computer Supported Cooperative Work.
8. Sarwar B.M., Karypis G., Konstan J.A. and Riedl J. (2001) Item-Based Collaborative Filtering Recommendation Algorithms. In Proceedings of the 10th international World Wide Web Conference:285–295.
9. Sarwar B.M., Karypis G., Konstan J.A. and Riedl J. (2000) Analysis of Recommendation Algorithms for E-Commerce. In Proceedings of the ACM EC'00 Conference:158–167, Minneapolis, MN.
10. Herlocker J.L. and Konstan J.A. (2001) Content-Independent Task-Focused Recommendation. IEEE Educational Activities Department, 5(6):40–47.
11. Kowalski G. (1997) Information Retrieval Systems: Theory and Implementation. Kluwer, Academic Publishers, Norwell, MA.
12. Milgram S. (1967) The small-world problem. Psychology Today, 2:60–67.

Automatic and Dynamic Composition of Web Services Using Ontologies

Nicola Mazzocca[1], Rosa Anna Micillo[2], and Salvatore Venticinque[2]

[1] Department of Computer Science - University Federico II of Naples
nicola.mazzocca@unina.it
[2] Department of Information Engineering - Second University of Naples
rosaanna.micillo@unina.it, salvatore.venticinque@unina2.it

Summary. Ontology-driven Web Services composition represents a novel approach to provide value added services or to improve services availability. Here we describe how ontology can be exploited to express the composition criteria, which can be automatically processed to obtain different workflows with the same semantic. Each workflow can be used to generate an orchestrated or a chorographical execution plan, whose actor contributes to provide the final service. We present a simple implementation as proof of concept using an OWL description of the composition criteria and the JENA APIs to process it.

1 Introduction

The Service Oriented Architecture (SOA) is the most accepted model for developing distributed systems, where heterogeneous applications cooperate to satisfy user's requests. This model is based on software agents, which interact by assuming the role of Service Requestor or Service Provider, and a Service Registry, which stores services' description files. Web Services (WSs) are the wide accepted standard technology for building SOAs. In the last years a lot of efforts in research activities have focused on WSs composition. Much of the work on WSs composition adopt a human-driven approach [5, 6, 7]. The developer generates, manually, a composition plan that will be published in a public register as a new service and will be executed by a centralized engine. When the composed WS is requested, the engine executes the plan and invokes the WSs mentioned in the plan. This kind of composition is not scalable when the number of WSs increases. Furthermore the developer needs to have low-level knowledge about adopted technologies. The execution will fail, if at least a service mentioned in the plan is not available. In [3] a semi-automated approach is used. The developer exploits a framework that, by a semantic analysis of the available services, is able to help him to select the suited components during the composition process. Step by step the developer chooses a service from a list. The framework makes a matching, based on functional and not-functional properties, between the chosen service and the component requirements. The new composed service is

K.M. Węgrzyn-Wolska and P.S. Szczepaniak (Eds.): Adv. in Intel. Web, ASC 43, pp. 230–235, 2007.
springerlink.com

stored like a DAML-S CompositeProcess. This approach is still unscalable. The last approach is completely automatic [2]. It must be supported by a semantic description of the provided service capabilities in order to recognize the similarities between the provided capabilities and the requested functionalities. The service description must provide the information about service requirements, interaction protocol and low-level mechanisms used for invocation, too. It means that automatic composition requires full semantic support, not only syntactic interoperability. Some standards [6, 7] describe how multiple WSs could be composed together to provide a more complex WS, but they only address composition at syntactic level. Here we propose an ontology-based approach and an architecture to support automatic and intelligent composition of WSs. Composition will be exploited in order to provide value added services and to improve availability of existing ones. Our approach is divided into three phases [1]. Furthermore we present a framework, that we have partially implemented, that supports planning of composite services and choreographic execution by mobile agents. In the following we focus on composition. About a preliminary work on mobile agents based execution the reader can refer to [4]. In the section 2 we present our approach to the composition of WSs and a proof of concept implementation. In the section 3 we present an example of use. Finally conclusions and future works are due.

2   Framework Overview

The conceived approach aims at supporting automatic and intelligent WSs composition granting semantic compliance of composed services using OWL-S. OWL-S [10], which is proposed by W3C, defines a set of OWL ontologies to describe WSs in a more expressive way than the one allowed by WSDL [8]. In particular these ontologies describe what a WS does, how it works and how to access it. Processing OWL-S description of services it is possible to find relationships among services such as equivalence and dependencies [3], which will be used to find new equivalent workflows to improve services availability or to build new services categories. In this section we introduce those components, and their prototypal implementation. In Figure 1 we show how our framework has been conceived to be exploited in a distributed system according to the SOA standard model. It means that the client will act as a standard Service Requestor that is able to discover the available services in a public Service Register and to invoke them according to their WSDL descriptions. The framework will act as a standard Service Provider able to handle client's requests. In the same way Service Providers, which aim at deliver their services exploiting the framework functionalities, will have to publish their services in a private Service Register together their OWL-S descriptions. Using the inner Execution Engine our framework will compose and invoke services, that are mentioned in the composition plan. When a new service is published the framework will compare its

OWL-S description with the other ones already stored in the private Service Registry, extracting the composition criteria and updating the knowledge base. Subsequently, the framework will verify if the new service can be exploited in some composition plans to provide new alternative solutions or to consider new service categories. The building process of a composed Service is divided into

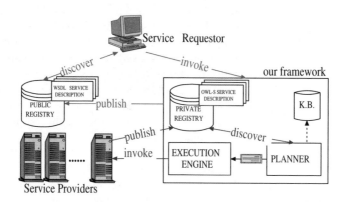

Fig. 1. Proposed Architecture

three phases: Planning, Discovery and Execution. The requestor can ask for a service. The Planner receives the service request and makes an inference on the knowledge base, described by our *Workflow* ontology, in order to evaluate all composition plans, which are semantically equivalent to the invoked service. Furthermore for each composition plan the availability of the components needs to be assured, so the Planner looks for their implementations in the private Service Registry, and chooses the first composition plan for which all required component services are available. The composition plan is sent to the Execution Engine, which interprets and performs it and then it sends an answer to the requestor.

2.1   Workflow Ontology

The description of services and composition criteria has been provided by the Workflow ontology, which is written in OWL [9]. At the state of the art, the Workflow ontology describes only the composition rules of services, but it will be replaced with OWL-S. In Figure 2(a) we can see a graphical representation about how the Workflow ontology defines the composition rules. It is composed of three classes. The Workflow class represents all possible composition plans. The ServiceType class represents an available implementation of that can be found in the Service Registry. The Composition class represents a combination of two *Workflow* instances to which it is connected by the relation isComposed. The ontology defines also six properties:

- *isEquivalent: ServiceType → Composition*
  It is a symmetric property, which expresses the equivalence between a *ServiceType* instance and a *Composition* instance. Its minimum cardinality is 1.
- *isComposed: Composition → Workflow*
  It associates an instance of the *Composition* class to two instance of the *Workflow* class. In fact a *Composition* is modelled as a composition of two *Workflow* instances.
- *isComposedBySeries: Composition → Workflow*
- *isComposedByParallel: Composition → Workflow*
  They are subproperties of *isComposed* and specify that the *Workflow* components have to be executed in sequential order or in parallel.
- *isAntecedent: Composition → Workflow*
- *isConsequent: Composition → Workflow*
  They are subproperties of *isComposedBySeries* and define the particular order of composition for the sequence because it is semantically relevant. The cardinality of these properties is 1.

2.2   The Planner

We have used the platform JADE and the JENA API to implement the Planner. At first the Planner establishes if there is a composition, which is equivalent to the requested service (an instance of the *ServiceType* class). If this instance of the *ServiceType* class exists, it will consider both the equivalent instance of the *Composition* class and an implementation of the requested service. If the considered entity is not composed, the inference will give back the category name that identifies the requested service in the Service Registry. This is done for all the entities of the ontology, which are visited while the plan is going to be built. For each composition the Planner identifies the operator and the entities to be composed. Each entity can be an instance of the *ServiceType* class or an instance of the *Composition* class. So the inference algorithm is recursive and supplies for every requested service, all possible compositions of WSs, which realize it. Figure 2(b) shows the algorithm. The result is a list of feasible plans expressed which are expressed by a simple language that support, at the moment, only two control constructs:

- **Sequence (#):** Two services composed by the operator ' # ' are executed sequentially. For example, A#B means 'perform service A and then perform service B'.

- **Parallel (//):** Two services composed by the operator ' // ' are executed in parallel. For example, A//B means that services A and B are executed concurrently.

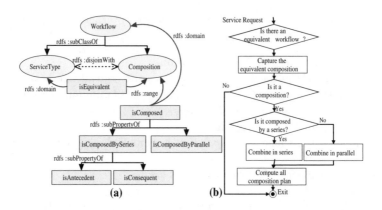

Fig. 2. (a) Workflow Ontology and (b) Inference Algorithm

3 An Example

In Figure 3 we will show an application of our framework that uses Imicitation WS (http://webservices.imacination.com/) and ViaMichelin WSs (http://ws.viamichelin.com/). Protege tool is used to describe their composition criteria using Workflow ontology. The Zip Distance Calculator Service allows to calculate the distance between two points of U.S. that have different Zip code. The Geocoding WS allows to enter an address and obtain an ordered list of locations with an address description and the associated WGS84 encoded geographic coordinates. The DistanceCalculation WS allows to obtain the distance 'as the

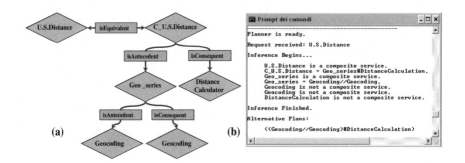

Fig. 3. (a) Planner's inference and (b) log

crow flies' between two points with different coordinates. When requestor asks for Zip Distance Calculator code service, called U.S.Distance, the planner makes an inference on the ontology and obtains all compositions of published services, which are semantically equivalent to the requested one. In this case we can use Geocoding service to detect the two points and then DistanceCalculation service to calculate the distance, rather than the Zip Distance Calculator service. After that, the planner looks for available implementations in the Service Registry. If it does not find necessary component to execute the current plan it will consider the next equivalent one. Logically if at least one component for each plan is not available the service request can not be satisfied.

4 Conclusions and Future Works

We introduced an approach for the automatic and intelligent composition of distributed services to increase availability in SOAs and to build new value added services. We described a prototypal implementation with some limitations due to the early stage of our research. Actually service categories are identified by name and equivalence among a service and the composition of N other services is written explicitly in the ontology by developer. In future work the equivalence between two services will be automatically evaluated by processing their OWL-S descriptions in order to provide intelligence to our system for automatic recognition of candidates of composition plans. Furthermore we are going to integrate our framework in a platform for service delivery to heterogeneous client devices in order to estimate service performance and availability provided by our approach.

References

1. K. Sycara, M. Paolucci, A. Ankolekar, N. Srinivasan. Automated Discovery, Interaction and Composition of Semantic Web Services. Journal of Web Semantics, Volume 1, Issue 1, December 2003.
2. S. A. McIlraith, T. C. Son, H. Zeng. Semantic Web Services. IEEE Intelligent Systems. Special Issue on the Semantic Web. 16(2):46-53. 2001.
3. E. Sirin, J. Hendler, B. Parsia. Semi-automatic Composition of Web Services using Semantic Descriptions. In WSMAI-2003, 2003.
4. R. Aversa, B. Di Martino, N. Mazzocca, S. Venticinque , Web Services composition and delivery using a Mobile Agents based Infrastructure, Proceedings of the 5th International Symposium on Parallel and Distributed Computing (ISPDC06), Timisoara, July, 2006, pp.337-344
5. IBM Alphaworks BPWS4J http://www.alphaworks.ibm.com/tech/bpws4j
6. F. Curbera, Y. Goland, J. Klein, F. Leymann, D. Roller, S. Thatte, S. Weerawarana. BPEL4WS white paper. 2002
7. A. Arkin, S. Askary, S. Fordin, W. Jekeli, K. Kawaguchi, D. Orchard, S. Pogliani, K. Riemer, S. Struble, P. Takacsi-Nagy, L. Trickovic, S. Zimek. Web Services Choreography Interface (WSCI) 1.0. 2002
8. WSDL - Web Services Description Language 1.1 . W3C Note 15 March 2001. http://www.w3.org/TR/wsdl
9. M. K. Smith, C. Welty, D. L. McGuinness. OWL Web Ontology Language Guide. W3C Recommendation 10 February 2004
10. OWL Service Coalition. OWL-S: Semantic Markup for Web Services. W3C Member Submission 22 November 2004.

Service Retrieval for Distributed Environment: An Approach Based Service-Ontologies Mapping

Nacima Mellal and Richard Dapoigny

Université de Savoie, ESIA Laboratoire d'Informatique, Sytèmes, Traitement de l'Information et de la Connaissance B.P. 80439, 74944 ANNECY-Le-Vieux Cedex, France
Nacima.Mellal@univ-savoie.fr

Summary. In the Web service community, services may be exchanged among distributed environments. Actually, web services are described in terms of ontologies. When ontologies are distributed, arises the problem of achieving sematic interoperability. This problem is undertaken by a process which defines rules to relate relevant parts of different ontologies, called "Ontology Mapping". The present paper describes a methodology for automatic and semantic mapping of ontologies, basing our approach on a mathematical model called Information FlowModel and denoted IF Model.

1 Introduction

Interaction and exchange of information between distributed systems is a crucial topic. Systems usually need to communicate and interoperate between them, they also need to understand what they exchange. This introduces the notion of *Semantic Interoperability* denoted SI. F.Vernadat has determines the objectives of semantic interoperability in [13]. It is the ability to exchange services and data among systems that make sense (common Şmeaning\~T). Ontologies present a good way to reach the SI. From most proposed definitions for the philosophical term "Ontology", generally is seen as an explicit specification of a conceptualization [2], [1]. It should give an explicit definition of concepts and relations between them.

In order to exchange information between systems modeled by ontologies, it is necessary to relate these ontologies. This practice is called "Ontologies's Mapping". Recent advances have spurred the developpment of some techniques using ontologies in order to achieve SI. In [5], the authors proposed an approach which is mainly built on the IF-Map method to map ontologies in the domain of computer science departments from five UK universities. Other approach is MAFRA (MApping FRAmework for distributed ontologies). It supports the interactive, incremental and dynamic ontology mapping process [7]. RDFT is a mapping meta-ontology for mapping XML DTDs to/and RDF schemas targeted towards business integration task, where each enterprise is represented as a Web service specified in WSDL language [11]. C-OWL (Context-OWL) is another approach on ontology mapping, which is a language that extends the ontology language OWL both syntactically and semantically in order to allow for the

K.M. Węgrzyn-Wolska and P.S. Szczepaniak (Eds.): Adv. in Intel. Web, ASC 43, pp. 236–241, 2007.
springerlink.com

representation of contextual ontologies [1]QOM (Quick Ontology Mapping) is a (Semi-)automatic mapping of ontologies, and a core task to achieve interoperability when two agents or services use different ontologies [8].

In [12], authors made a step towards semantic integration by proposing a mathematically sound application of channel theory to enable semantic interoperability of separate ontologies. The formalization of the coordination process between ontologies is described, based on exchange that captures progressive partial semantic integration, using the Barwise-Seligman theory.

The different cited works propose semi automatic mappings to reach the SI. Thus, it is necessary to develop automatic techniques for mapping ontologies. Our approach shares the idea in [5], which uses of IF Model to solve semantics coordination of ontologies in distributed systems. We propose a methodology which allows an automatic mapping between distributed ontologies, basing on IF [4]. Following what R.Kent said in [6] "Information Flow is the logical design of distributed systems, provides a general theory of regularity that applies to the distributed information inherent in both the natural world of biological and physical systems".

This article is divided into two mains sections. In the first one, we give the definition of service ontologies and introduce the IF model. In the second section, the ontology mapping in distributed environment is described.

2 Preliminaries

2.1 Ontologies

Ontologies promise a shared and common understanding of some domain that can be communicated and interoperate across people and computers. Ontologies find applicability in many domains of application, in system engineering, knowledge management, interoperability between systems..etc. An ontology is generally seen as a formal explicit specification of a shared conceptualization [2], [1], which is a description of the concepts and relationships between them.

In our context, ontologies will represent web services, where the relationship among them expresses dependency between them. We call ontologies describing services "Service Ontologies".

We associate to the web service two main concepts in our approach, "Web Service Type WSTp" and "Web Service Token WSTk". When we say a service of booking a room in hotel is a generic service, so we classify it as a WSTp. But, when we specify the name of hotel, the service becomes WSTk. This last is seen as an instance of WSTp.

The Relations between Ontology Concepts
Modeled systems are in general expressed by a set of formal rules, so by logics. As ontologies in origin model systems, logics define formally its attributes (concept

---

[1] The term contextual ontology refers to the fact that the contents of other ontologies via explicit mappings to allow for a controlled form of global visibility.

relations...). Our aim is to define an ontology as a Gentzen system, which is a deduction system introduced by Gerhard Gentzen in 1934, and expressed by first order logic. The notion of sequent is central in Gentzen system. Given a set S, a sequent of S is a pair $\langle X, Y \rangle$ of subsets of S.

A binary relation $\vdash$ between subsets of S is called a consequence relation on S. We suppose that, $X = \{x_1, x_2, ..x_n\}$, and $Y = \{y_1, y_2, ..y_n\}$. The syntax of the Gentzen sequent is :

$$x_1, x_2, ..x_n \vdash y_1, y_2, ..y_n,$$

It is interpreted by its symbol $\vdash$ as an implication, the comma on the left is interpreted like a conjunction, the comma on the right like a disjunction. Coming back to our purpose. Gentzen sequents will represent ontology concepts, such as SWTp. The consequence relation will express the functional dependency between these concepts (services).

Definition 1. *A Service Ontology is described by an oriented graph $< C, R >$, where: C is the set of nodes: $C = \{c_1, c_2, ..., c_n\}$, where c_i is an ontology concept, and R is the set of edges: $R = \{r_1, r_2, ..., r_m\}$, where r_j links two nodes if the corresponding ontology concepts are related by the consequence relation $\vdash$.*

The defined service ontology describes a set of concepts, service types, and the relations between them. This set can be seen as complex service which we call a global service and its elements (ontology concepts) as sub-services. So, we deduce that a global service is associated with a service ontology. It is a particular service token which have an action verb and a context token.

Example. We propose a distributed system which employs ticket agents. Each agent is situated in a sub-system. Ticket agents attempt to achieve some important services for distributing, selling, buying, booking tickets from a range of sources. Agents may communicate, interoperate and exchange services on the Web. We suppose that an agent attempts to achieve a service of buying a ticket to go from Annecy to Fontainebleau. Once the depart and the destination are identified, agent may obtain the itinerary. The service Web of buying is considered as a global service. Services of identifying the depart, the destination and the itinerary are defined as WSTp.

2.2   IF Model

For Barwise and Seligman, dynamic and distributed systems are the result of the fusions of their local components. Each local component is described by a an *IF Classification*. This last is a very simple mathematical structure. As it is defined in [4], it consists of a set of objects to be classified, called tokens and a set of objects used to classify the tokens. In our context, IF classifications are used to model systems. Their types represent all the characteristics of a given system and the tokens represent all the instances of the system. That is, we are interested to services provided by systems, classified as WSTp and WSTk. Classifications are linked by applications called *Infomorphisms*. Infomorphisms provide a way to move information back and forth between systems. In our approach, the utility of infomorphisms is not to link classifications of the same system, but to link those of a distributed system, because we need to map between distributed service ontologies.

The information flow in a distributed system is expressed in terms of an *IF theory* of this system, that is a set of laws describing the system. These laws are expressed by a set of types. The theory is specified by a set of sequents, so by a set of types and the relation between them ($\vdash$).

The overall "Classification" and "IF theory" constitute what is called a *local logic*. That is, this system has its own logic expressed by its types (for us WSTp). *Information Channel* is the key for modeling information flow in distributed systems. It is the main step in the process of mapping. The IF theory in the information channel describes how the different types from different classifications are logically related to each other.

Adapting this theory to service ontologies, entities of the IF theory are WSTp. To formalize this mapping, let us recall important notions of the IF theory. On one hand, IF theory is based on a "Classification" of entities handled in the system. On the other hand, it is related to an information channel which retrieves corresponding entities in different systems. The information channel is concretized through a set of infomorphisms, reflecting the whole relationships between the system and its parts.

3   Method of Service-Ontologies Mapping Using IF Model

Systems usually need to exchange services and interoperate. When systems are described by services and modeled by ontologies, the links between them express functional dependencies between their services. Therefore, relating systems deals with mapping between their correspondent service-ontologies.

Let's recall the example cited before. We propose software ticket agents managing services on the Web. They are situated in distributed environments. Each agent has its own knowledge base describing the possible services that it may reach in term of ontologies. Agents attempt to achieve some important services. Agents may communicate, interoperate and exchange services on the web. In the present paper, we treat a case where two agents are communicating and exchanging services $Agent_1$ and $Agent_2$. $Agent_1$ provides the Web service token s_1 and $Agent_2$ provides the Web service token s_2. Relating s_1 and s_2 means that the achievement of one Web service depends on the achievement of the other. The process of mapping may be summarized into four steps.

1. The first step consists in identifying the type of the Web service token s_1
2. In this step, relevant Web services types related by the $\vdash$ relation are found. According to the IF model, types are related in terms of IF theory. So, the first thing is that the type of s_1 is related with other types (we suppose type of s_2).
3. From the deduced type of Web service, the corresponding Web service token is retrieved,
4. Finally, the Web services s_1 and s_2 are related.

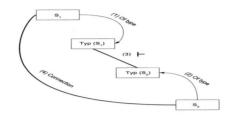

Fig. 1. The connextion between Web Service Tokens

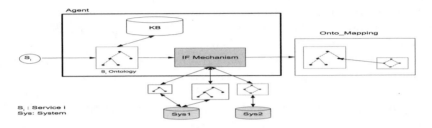

Fig. 2. The Process of Service-Ontologies Mapping

The agent proceeds as in following:

1. The agent gets the service ontology corresponding to its Web service S_i.
2. Agent examines the service ontology. Then by applying the IF mechanism, it detects the distant corresponding services from distributed resources.
3. Agent maps between the services and finally achieves the initial service.

This process is well detailed in our previous works [10], [9].

4 Conclusion

In this paper, we have presented a formal mechanism for mapping distributed service ontologies in a sound and automatic manner, basing on IF model. Besides the advantages of our approach, we plan to implement a multi-agent system, applying the IF model on the Web. The notion of dependency between agents is a challenging problem. Some authors have proposed a graph structure to formalize the relationships between agents [3]. In this work, the IF-based approach tackles the problem of building these dependencies from distributed logics. Investigating these possibilities will be the next work.

References

[1] Richard Fikes. Ontologies: What are they, and where's the research? In *KR'96, the Fifth International Conference on Principles of Knowledge Representation and Reasoning*, Cambridge, Massachusetts, November 1996.

[2] T. R. Gruber. A translation approach to portable ontology specifications. *Knowledge Acquisition*, 5(2):199–220, 1993.

[3] Rosaria Conte Jaime Simão Sichman, São Paulo. Multi-agent dependence by dependence graphs. In *International Conference on Autonomous Agents*, pages 483–490, Italy, 2002.

[4] Jerry Seligman Jon Barwise. *Information Flow: The Logic of Distributed Systems*. Number 44. Cambridge University Press, 1997.

[5] Y. Kalfoglou and M. Schorlemmer. If-map: an ontology mapping method based on information flow theory. *Journal on Data Semantics*, 1(1):98–127, October 2003.

[6] Robert Kent. The information flow foundation for conceptual knowledge organization. In *Proceedings of the Sixth International ISKO Conference. Advances in Knowledge Organization 7*, pages 111–117, Ergon Verlag, Würzburg, 2000.

[7] Alexander Maedche, Boris Motik, Nuno Silva, and Raphael Volz. MAFRA - a mapping framework for distributed ontologies. In *EKAW '02: Proceedings of the 13th International Conference on Knowledge Engineering and Knowledge Management. Ontologies and the Semantic Web*, pages 235–250, London, UK, 2002. Springer-Verlag.

[8] York Sure Marc Ehrig. Ontology mapping - an integrated approach.

[9] Dapoigny Richard Mellal Nacima. A multi-agent specification for the goal-ontology mapping in distributed complex systems. In *7th International Baltic Conference on Databases and Information Systems*, pages 235–243, Lithuanie, July 2006.

[10] Laurent Foulloy Mellal Nacima, Dapoigny Richard. The fusion process of goal ontologies using intelligent agents in distributed systems. In *3rd International IEEE Conference on Intelligent Systems*, pages 42–47, London, UK, September 2006.

[11] B. Omelayenko. Rdft: A mapping meta-ontology for business integration. In *Proceedings of the Workshop on Knowledge Transformation for the Semantic for the Semantic Web at the 15th European Conference on Artificial Intelligence (KTSW2002)*, pages 77–84, Lyon, France, 2002.

[12] M. Schorlemmer and Y. Kalfoglou. Using information-flow theory to enable semantic interoperability. In *6th Catalan Conference on Artificial Intelligence (CCIA'03), Palma de Majorca, Spain*, pages 421–432, October 2003. Volume 100 of Frontiers in Artificial Intelligence and Applications, IOS Press, ISBN:1-58603-378-6.

[13] F.B. Vernadat. Interoperable enterprise systems: Architectures, methods and metrics. Technical report, LGIPM, Université de Metz, France, 2007.

Issues in Semantic File Sharing

Claude Moulin[1] and Cristian Lai[2]

[1] Compiègne University of Technology
Heudiasyc, CNRS, France
`claude.moulin@utc.fr`

[2] Center for Advanced Studies in Sardinia, Italy
`clai@crs4.it`

Summary. This paper presents some issues inherent to the semantic indexing of resources in decentralized peer to peer networks. We first present a scenario that fixes the objectives of the indexing. We also explain the way to construct indexing keys in relation with ontologies. We also describe the way to distribute the global index through the nodes of the network and the way to retrieve resources. Then, we discuss some problems relative to the semantic indexing of personal resources respecting the previous requirements.

1 Introduction

We introduce the paper with a use case in relation with the e-learning domain, but all the considerations could be easily applied to other domains. Teachers are generally interested by resources written or created by other teachers in order to update their courses. They also agree to give a free access to their own material, but they do not want to manage heavy systems or depend on centralized repositories. The context is a file sharing system but not based on document titles.

Types of resources can be very different: texts, images, program source, software, etc. Content domains may also be very large. As resources are not just textual, the indexing can not be done automatically; for some of them it could be done semi-automatically, but the number of resources for each people is not very important and attaching natural language processing procedures to systems would be very cumbersome. More over, automatic indexing can not add information not included in documents and however such information is very useful in our scenario (for example information about point of view on resources, or type of addressees or difficulty level).

We can consider two ways of indexing: key word indexing and semantic indexing i.e. the indexing on concepts of knowledge representation, generally ontology. The second type of indexing preserves the possibility of reasoning on the structure of the underlying knowledge representation when searching for resources.

These considerations lead to choose the following solution in order to achieve the task: a completely decentralized solution, easy to install on a personal

K.M. Węgrzyn-Wolska and P.S. Szczepaniak (Eds.): Adv. in Intel. Web, ASC 43, pp. 242–247, 2007.
springerlink.com

computer, not requiring many time and where all types of resources could be indexed. A peer to peer (P2P) network solution is completely adapted because the current technology is completely reliable. The success of such networks introducing semantics were considered as a challenge in [2]. A manual semantic indexing based on ontologies freely selected by users is sufficient.

2   Semantic Indexing

2.1   Index

In the traditional file sharing, titles of resources, like songs or films are used both for indexing and searching. In this case all needed information shared by users is contained in these titles. At each title is associated the reference to the computers that physically own the corresponding resources.

We also face a file sharing situation. The semantic information is not contained in the title of the resource but in a key that annotate it. For example, when we want to express that a resource is treating of the concept of grammar in the domain of the theory of languages, and that this resource is a difficult exercise, we have to build a key that contain all this information. Some criterion can be objective (grammar, exercise) but others present a point of view on the resource (difficulty level).

An index can be seen as a table associating values to keys. The keys are generally encrypted for enhancing the search in the table. For indexing a specific resource, it is necessary to provide for a key containing all the needed elements that describe its contents and that allow the further discovery of the resource.

2.2   Ontological Elements for Indexing

For a manual semantic indexing we have designed a tool. The user is requested first to select the ontologies used for building the indexing keys. A key may contain concepts belonging to different ontologies. In the previous example two ontologies has to be found, one for the theoretical domain (theory of language) and another for the description of the resources (an ontology based on the Learning Object Metadata (LOM) [5] for example). Generally several ontologies are required [1] Studying the previous example and analyzing the required ontologies, we are led to the following conclusions when indexing resources:

1. a user can select a concept: the concept of grammar is described in the knowledge domain ontology.
2. a user can select an instance of a concept occurring in an ontology: Exercise is an instance of the concept of Learning Resource Type and Difficult is an instance of the concept Difficulty Level in the second ontology.
3. extending the scenario it could be interesting to index using an instance of a concept, instance not included in an ontology. For example indexing a document that is treating of Paris requires Paris as an instance of the concept of Town. A document treating of the concept of City in general is different from a document treated of a city in particular.

In an ontology, an element is completely defined by its unique Unified Resource Identifier (URI)[12] composed of the name space of the ontology followed by the identifier of the element in the ontology. Thus, it is enough to insert the URIs of the ontological elements that characterize a resource in the key that annotates it. From this information any software agent can discover the type of the element inside a key (concept or individual) and can decide to build new queries if some do not give satisfying results. The way to consider the third case is explained in the following section.

2.3   Distributed Knowledge Base

A problem occurs with semantic indexing when one wants to refer to an instance that is not in an ontology. The idea we want to follow is to insert a URI in a key like in the previous section. We create a new instance of the concept as if it was defined in the ontology itself. The expression created by the user has to be transformed in a string representing the local identifier of this instance. We decided to associate to this identifier the namespace of the ontology the instance refers to. Obviously two users could insert two different identifiers for representing the same instance. The only way to solve this problem is to normalize the transformation algorithm for producing the identifiers and to ask users to follow some guidelines.

Inserting this new URI in a key is not enough for a software agent, because it can not find the type of the instance with only its id. In order to solve this problem and to maintain a unique structure, we insert in a key a couple consisting in the concept and instance identifiers. In the case of indexing on a concept, we consider that this concept is instance of its meta concept. In OWL ontologies, this meta concept is referred by owl:Class.

2.4   Examples of Keys

A key may be simple or be composed of simple keys. For the previous example we obtain the following XML index entry:

```
<index>
<key>
<concept>http://www.owl-ontologies.../p2peducational#
Difficulty</concept>
<instance>http://www.owl-ontologies.com/2006/10/p2p
educational#veryDifficult</instance>
</key>
<key>
<concept>http://www.owl-ontologies.../p2peducational#
LearningResourceType</concept>
<instance>http://www.owl-ontologies.../p2peducational#
Exercise</instance>
</key>
<key>
```

```
<concept>http://www.w3.org/2002/07/owl#Class</concept>
<instance>http://www.owl-ontologies.com/2006/10/p2plt.
owl#grammar</instance>
</key>
</index>
```

Obviously the indexing must be supported by appropriate tools.

3 Discovery of Resources

3.1 Type of Network

Resources are published in a P2P network, i.e. a set of nodes, in which the index of semantically annotated resources is distributed on the nodes. The data structure that has been considered suitable for such aim is called Distributed Hash Table (DHT) [10, 6, 11, 9]. The index is composed of entries that are pairs of data (key, value). Each node of the network contains a portion of the index as part of the whole data structure.

The publication or the indexing is the operation of insertion of a resource inside the DHT. More exactly it is the operation of inserting a new index entry in the index, so within the DHT. The discovery is the operation which allows to find some resources in the network as soon as they correspond to a research key.

For building the network that has to contain the resources, we choose the Kademlia [7] type of network, a basic approach of many P2P systems. Each participating computer to the network houses a node and is attributed a node identifier (its ID, for brevity). The set of encrypted keys contains both the keys used to index the resources and the encrypted node IDs. The key space is the set of all the keys that can be constructed. Its volume depends on the number of bits used to encrypt the keys. This allows to compare nodes ids and encrypted keys in order to determine the nodes containing the part of the DHT in which a key could be located.

3.2 Comparison

For publishing a resource a user builds a key with an appropriate tool and selects the name of the file that contains the resource. These elements are then published in a network (the file name is completed with the computer identifier). The algorithms used to insert and retrieve resources encrypt the keys and comparisons are made on the encrypted values. So, for discovering resources in the network, it is necessary to produce a research key exactly equal to a key generated during the publication.

Publishing a resource with several keys corresponds to a conjunction of criteria. During a query the research can be made on one or several criteria. It is thus necessary to create several entries to the index when a resource is indexed on more than one element. For example with three criteria, it is necessary to create seven entries containing one, two and three simple keys.

The other issue to take into consideration is the order of the keys in a composed key because two simple keys in inverse order will be considered different even if they are logically equal. Composed keys can be created according to the lexicographic order of their instance URIs.

3.3   Multi-lingual Strings

Indexing using a URI and not keywords eliminates problem due to lingual comparisons. However, one issue remains with multi-lingual strings i.e. strings corresponding to the same element but having a different representation according to the language. For example, if a resource is treating of cities in general, there is no problem; but, there could be difficulties treating of the city Paris, (the third case of indexing in Section 2.2) which relies to the indexing on a virtual instance.

We proposed to build a key from the word entered by the user but this one might be different (Paris in English and French, Parigi in Italian). With a unique key, it is not possible to search for documents written in different languages. The solution we propose is to modify the XML format (see Section 2.4) for writing keys with multi-lingual strings. It is possible to insert an xml:lang attribute in the instance tag of the key element. The value is the corresponding language. Strings representing languages are normalized which prevents any problem when writing the language string itself.

4   Main Issues and Conclusion

Semantic indexing relies on ontology availability. Ontologies are a shared representation [3, 4] and obviously, it is necessary for a user to select ontologies that a community largely accepts, else the resources could not be accessible by other people. The difficulty is to find ontologies that can be easily reused. Indexing resources in such networks requires tools for building the keys used during the publication and the discovery of resources and for that purpose we built an application that allows the three types of indexing summarized in Section 2.2 from several OWL [8] ontologies that a user can freely choose and has to install on a computer.

We also performed some experiment for publication and retrieval of resources on a virtual network and have demonstrated that our system can be used to implement the initial scenario. The experiments intended to demonstrate that it was possible to use keys long enough to contain several concept ids (several hundreds of characters) and to index on the same key a huge amount of documents. Nevertheless, even if the indexing uses a few of ontology structures, the user has to be familiar with intologies and has to spend time to discover ontology content with appropriate tools in order to decide if an ontology is interesting or not.

References

1. A. Abel, D. Lenne, C. Moulin, and A. Benayache. Using two ontologies to index e-learning resources. In *WI '04: Proceedings of the Web Intelligence, IEEE/WIC/ACM International Conference on (WI'04)*, pages 549–552, Washington, DC, USA, 2004. IEEE Computer Society.

2. J. Davies, D. Fensel, and F. van Harmelen. *Towards the Semantic Web: Ontology-Driven Knowledge Management.* John Wiley & Sons, January 2003.

3. A. Gomez-Perez, O. Corcho, and M. Fernandez-Lopez. *Ontological Engineering: with examples from the areas of Knowledge Management, e-Commerce and the Semantic Web. First Edition (Advanced Information and Knowledge Processing).* Springer, July 2004.

4. T. R. Gruber. Towards Principles for the Design of Ontologies Used for Knowledge Sharing. In N. Guarino and R. Poli, editors, *Formal Ontology in Conceptual Analysis and Knowledge Representation*, Deventer, The Netherlands, 1993. Kluwer Academic Publishers.

5. Lom ieee 1484.12.1, http://ieeeltsc.org/wg12lom/

6. D. Malkhi, M. Naor, and D. Ratajczak. Viceroy: A scalable and dynamic emulation of the butterfly. In *Proceedings of the 21st annual ACM symposium on Principles of distributed computing.* ACM Press, 2002.

7. P. Maymounkov and D. Mazieres. Kademlia: A peer-to-peer information system based on the xor metric. In *IPTPS02, Cambridge, USA, March 2002*, 2002.

8. Web ontology language overview http://www.w3.org/tr/owl-features/.

9. S. Ratnasamy, P. Francis, M. Handley, R. Karp, and S. Schenker. A scalable content-addressable network. In *SIGCOMM '01: Proceedings of the 2001 conference on Applications, technologies, architectures, and protocols for computer communications*, pages 161–172, New York, NY, USA, 2001. ACM Press.

10. A. Rowstron and P. Druschel. Pastry: Scalable, decentralized object location and routing for large-scale peer-to-peer systems. In *IFIP/ACM International Conference on Distributed Systems Platforms (Middleware)*, pages 329–350, Nov. 2001.

11. I. Stoica, R. Morris, D. Karger, M. F. Kaashoek, and H. Balakrishnan. Chord: A scalable peer-to-peer lookup service for internet applications. In *Proceedings of the ACM SIGCOMM '01 Conference*, San Diego, California, August 2001.

12. Uniform ressource identifier, http://www.gbiv.com/protocols/uri/rfc/rfc3986.html.

A Unified Firewall Model for Web Security

Grzegorz J. Nalepa

Institute of Automatics, AGH University of Science and Technology,
Al. Mickiewicza 30,30-059 Kraków, Poland
gjn@agh.edu.pl

Summary. The paper presents a new formalization for firewall systems, called the *Unified Firewall Model* (UFM). It offers an abstraction over firewall implementations, and uses formal concepts of Rule-Based Systems to describe firewall syntax and semantics. It is backed by the XTT/ARD design methods. It allows for improving system quality, by introducing a formal verification during the design stage.

1 Introduction

Security issues play an important role in the development of real life web systems. These issues include problems of: access control, privacy, system monitoring, and data protection. The focus of this paper is on access control, provided by the firewall systems. The design of such systems remains a challenge, due to complex security requirements, and number of different incompatible implementations. There are no standard design approaches, or methods. Quality issues are even more complex than the design.[1]

This paper continues the line of research started in [3]. In that paper an idea of using Rule-Based Systems (RBS) formalism, to the design and implementation of firewall systems was put forward. Strong logical foundations [2] of RBS allow for introducing a *formal* design and analysis into the firewall design process. These ideas were presented in [5], with the application of the XTT knowledge representation method. Since then, the design process has been extended by the conceptual design method ARD [4], which allows for formalization of RBS requirements. Using this approach, this paper presents a new formalization for firewall systems. It is called the *Unified Firewall Model*, since it offers an abstraction over common firewall implementations. It uses formal RBS concepts in order to describe the syntax and semantics of the most common firewall systems. It is backed by the XTT and ARD.

The paper is organized as follows: in Sect. 2 the architecture of web firewalls is discussed; next, in Sect. 3, the formal XTT design process is briefly discussed; then in Sect. 4 the UFM is introduced. A practical example is presented in Sect. 5. Directions for future work are presented in the Sect. 6.

---

[1] The paper is supported by the *HEKATE* Research Project.

K.M. Węgrzyn-Wolska and P.S. Szczepaniak (Eds.): Adv. in Intel. Web, ASC 43, pp. 248–253, 2007.
springerlink.com © Springer-Verlag Berlin Heidelberg 2007

2   Firewalls for Web Security

The practical definition of *the firewall system* has been changing, along with the changes in both technology and security requirements. The first firewalls were just simple network packet filters, working on the IP protocol level. Later on, the statefull inspection, involving the analysis of the TCP sessions, and network translation technology has been introduced. Recently, the technology has been extended by the application-level firewalls at the HTTP level.

The general idea behind this paper is to consider a hybrid network and web application level firewall model. In this model a statefull network firewall serves as a gateway to a demilitarized zone (DMZ), where the main web server is located. The web server itself is integrated an the application-level firewall, in this approach an open implementation, called *ModSecurity* (www.modsecurity.org), available for the well known opensource *Apache2* (httpd.apache.org) web server is considered. It provides protection from a range of attacks and intrusions against web applications (including server-side solutions, such as PHP). It allows for HTTP traffic monitoring and analysis.

Ultimately, the application of the UFM/XTT approach presented in this paper should develop into an integrated design methodology, combining both network-level and application-level firewall. However, in this short paper only the network-level statefull firewall design using the UFM/XTT is discussed.

3   Application of the XTT-Based Process

The UFM has been developed, with XTT, the design methodology for RBS, in mind. In this case, the design process of a RBS (the firewall system) consists of several stages [4, 2]: 1) attribute specification, with explicit domains, 2) the conceptual design with *Attribute-Relationship Diagrams* (ARD) that model functional dependencies between system attributes [4], 3) the logical system design with *EXtended Tabular Trees* (XTT), that includes full firewall rule specification, 4) on-line formal analysis, using a dynamically generated Prolog-based description of the XTT logical structure [6], 5) the physical design, in this case providing the rule translation to particular firewall target languages.

4   The Unified Firewall Model

In the solution proposed herein, the *Unified Firewall Model*[2] (UFM) [1] is introduced as a middle layer in firewall design process, enabling logical analysis of the created firewall. It is a formal, implementation-free firewall model, build on top of the XTT methodology, providing a unified attribute specification for representing network firewalls. Generation of target language code for specific firewall implementation is achieved by defining translation rules transforming abstract firewall model into a specific implementation.

---

[2] The first description of the UFM has been given by Michał Budzowski, in [1].

A full list of conditional firewall attributes, corresponding to the information found in packets header, is specified. In UFM some attributes that are not related to packet header, but rather to firewall design will also be distinguished. Following attributes are considered: Source/Destination IP address, Input/Output interface, Source/Destination group, Protocol, Destination port, Service, ICMP type and error code. The following firewall decision attributes are considered: Accept, Reject, Snat, Dnat. The **nat** decisions refer to the network translation technique present in all advanced firewalls.

Domains for UFM attributes are presented in the Table 1. The specification is given in the Table 2, where each attribute is specified with: *Name, Symbol, Subset* (the position in inference process, specifying whether attribute is *input, output* or its value is defined during inference process – *middle*), *Atomicity* (specifying whether attribute takes only *atomic* values from specified domain or also *sets* or *ranges* of these values). Attributes *aSGR* and *aDGR* define groups, that organize network traffic and facilitate specifying decisions.

Table 1. UFM attribute domains

Domain	Type	Constrains
Ipaddr	integer	$< 0, 2^{32} >$
Port	integer	$< 0, 2^{16} >$
Protocol	enum, symbolic	{tcp, udp, icmp }
Interface	enum, symbolic	int$i, i \in Integer$
Icmptype	enum, symbolic	{echoreq, ... maskrep }
Icmperrcode	enum, symbolic	{net-unr, ... host-unk}
Tcpflags	enum, symbolic	flag/mask, flag, mask $\subset$ {fin, syn, rst, psh, ack}
Service	enum, symbolic	{www, ssh, dns, smtp, icmp, other$i$}, $i \in Integer$
Group	enum, symbolic	{$fwi, neti, neti\_hostj, neti\_hostk$}, $i, j, k \in Integer$
Action	enum, symbolic	{dnat, accept, reject, log, snat}

Table 2. UFM attribute specification

Name	Symbol	Subset	Domain	Atomic
Source/Destination IP	aSIP/aDIP	input	Ipaddr	*set*
Protocol	aPROTO	input	Protocol	*set*
Destination port	aPORT	input	Port	*atomic*
Input/Output interface	aIINT/aOINT	input	Interface	*atomic*
ICMP type	aICMPT	input	Icmptype	*atomic*
ICMP error code	aICMPC	input	Icmperrcode	*atomic*
TCP flags	aTCPF	input	Tcpflags	*atomic*
Service	aSERV	middle	Service	*set*
Source/Destination group	aSGR	middle	Group	*set*
Action	aACT	output	Action	*atomic*

In order to construct practical firewall implementation, it is necessary to provide a formal translation from the unified model to particular implementations.

Two open firewall implementations have been considered: Linux NetFilter [8] and OpenBSD PacketFilter (PF) [7]. Full translation contained in [1] is long and detailed, and is out of scope of this paper.

5   Formal Hierarchical Design Example

The conceptual design of the firewall is conducted using ARD. The logical structure, including rules is based on XTT. The ARD/XTT design, and a formal Prolog-based analysis [6], is supported by the prototype Mirella tool.

In this case a non-trivial network firewall is considered. It is discussed in detail in [1]. It consists of: the firewall host and connections to four networks: the Internet, DMZ1, DMZ2, and LAN. The goal of the system is to monitor the traffic between networks and accept, reject or forward packets between them basing on the characteristics of the packets traversing the networks.

Decision taken by the firewall system can be defined if *source, destination* and *protocol* of the packet are known. This dependency can be denoted with ARD diagram of level 0, where Source, Destination and Protocol are *conceptual* variables. Later in design process these conceptual variables are replaced with one or more *physical* attributes present in firewall rules. Decision taken by the firewall is defined with aACT, attribute also alerting source and destination is possible, so conceptual variable. Decision will be replaced with aACT, aSIP, aDIP, and aPORT UFM attributes. A three-level ARD diagram (Fig. 1) models the functional dependencies between firewall attributes.

Following conceptual design and the ARD diagram, the XTT tables are designed. There are four XTT tables: source table, a *root table* checking source group, destination table checking destination group, service table inferring the type of service, decision table specifying firewall decision.

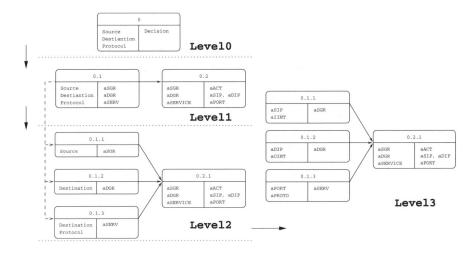

Fig. 1. The ARD diagram for the firewall

Each table is created based on final the ARD diagram. Left part corresponds to Precondition part in XTT table, right part corresponds to Assert/Retract or Decision part. Firewall policy is implemented by filling XTT tables with rules. Ctrl part of XTT table specifies control algorithm. In the `source` table packet is classified to one of the source groups basing on its source IP address and interface it is coming from. Similar operations are conducted in the `destination` table, but the packet is classified to one of the destination groups. The same control algorithm is implemented, but the next table is the `service` table. The `decision` table is the final table where the actual decision is inferred from the values of *middle* attributes: $aSGR$, $aDGR$, $aSERV$. For brevity only the last, `decision` table is shown in Tab. 3

Table 3. XTT decision table

Info	Prec			Retract			Assert			Dec	Ctrl	
I	aSGR	aDGR	aSERV	aSIP	aDIP	aPort	aSIP	aDIP	aPort	aACT	N	E
1	inet	fw_inet	www		f_inet	80		d_3w	8080	dnat	2.1	4.3
2	inet	dmz1_www	www							accept	1.1	4.3
3	dmz1_dns	inet	dns							accept	4.4	4.5
4	dmz1_dns	inet	dns	d_dns			f_inet			snat	1.1	4.5
5	sDNS	dmz1_dns	dns							accept	1.1	4.6
6	employee	dmz1_mail	smtp							accept	1.1	4.7
7	dmz2_proxy	inet	www							accept	4.8	4.9
8	dmz2_proxy	inet	www	d_prox			f_inet			snat	1.1	4.9
9	admin	sMACHINE	ssh							accept	1.1	4.10
10	employee	inet	www	–		80		d_prox	8080	dnat	2.1	4.12
11	employee	dmz2_proxy	forward							accept	1.1	4.12
12	–	–	icmp							accept	1.1	4.13
13	–	–	–							reject	1.1	1.1

Physical design of the system consists of the translation high-level rules in XTT tables to target firewall language of Linux NetFilter (iptables) or OpenBSD PF. Let us show how rules translation is performed for the rule 4.1 from the table `decision`. The firewall rule for the NetFilter is:

```
iptables -t nat -A PREROUTING -s 0/0 -i eth0 -d 83.29.224.129
    -p tcp --dport 80 -j DNAT --to-destination 192.168.2.2:8080
```

The PF form allows for observing differences in the target firewall language:

```
rdr on eth0 proto tcp from any to 83.29.224.129 port 80->192.168.2.2 8080
```

This difference is even more important in case of rule 4.4. While the NetFilter translation requires the use of two rules:

```
iptables -t nat -A POSTROUTING -s 192.168.2.4 -d 0/0 -o eth0
    -p tcp --dport 53 -j SNAT --to-source 89.29.224.129
iptables -t nat -A POSTROUTING -s 192.168.2.4 -d 0/0 -o eth0
    -p udp --dport 53 -j SNAT --to-source 89.29.224.129
```

the PF form is more compact:

```
nat on eth0 proto {tcp,udp} from 192.168.2.4 to any port 53->83.29.224.129
```

The full translation is discussed in [1]. The expressiveness of the UFM is high, so it is closer to the more expressive target language. However, all of the UFM syntactic structures can be translated to any firewall language, provided that the the implementation has the features represented by the UFM.

The XTT approach offers a possibility of automatic, on-line formal analysis of the firewall structure *during* the logical design. The analysis is accomplished by an automatic transformation of the XTT model into a corresponding code in Prolog. Some important features of the firewall system can be analyzed, including completeness of the system, or its determinism. Unfortunately, these issues are out of scope of this short paper. They have been discussed in [6].

6 Future Work

The original contribution of this paper is the new formalization of the firewall model, the *the Unified Firewall Model*. The research is considered a work in progress. Future work includes: a new version of Mirella, using the *Eclipse Modelling Framework*, and UFM application to *ModSecurity* and IDS. The UFM is being developed within the *HEKATE* project. The approach allows for improving system quality, by introducing the formal verification during the design. It offers an abstract layer over common firewall implementations.

References

1. Michał Budzowski. Analysis of rule-based mechanisms in computer security systems. formulation of generalized model for firewall systems. Master's thesis, AGH-UST, 2006. Supervisor: G. J. Nalepa, Ph. D.
2. Antoni Ligęza. *Logical Foundations for Rule-Based Systems*. Springer-Verlag, Berlin, Heidelberg, 2006.
3. Grzegorz J. Nalepa and Antoni Ligęza. Designing reliable web security systems using rule-based systems approach. In Ernestina Menasalvas, Javier Segovia, and Piotr S.Szczepaniak, editors, *Advances in Web Intelligence. AWIC 2003*, volume LNAI 2663, Berlin, Heidelberg, New York, 2003. Springer-Verlag.
4. Grzegorz J. Nalepa and Antoni Ligęza. Conceptual modelling and automated implementation of rule-based systems. In Tomasz Szmuc Krzysztof Zieliński, editor, *Software engineering : evolution and emerging technologies*, volume 130 of *Frontiers in Artificial Intelligence and Applications*. IOS Press, 2005.
5. Grzegorz J. Nalepa and Antoni Ligęza. Security systems design and analysis using an integrated rule-based systems approach. In Piotr Szczepaniak, Janusz Kacprzyk, and Adam Niewiadomski, editors, *Advances in Web Intelligence: AWIC 2005*, volume LNAI 3528. Springer-Verlag, 2005.
6. Grzegorz J. Nalepa and Antoni Ligęza. Prolog-based analysis of tabular rule-based systems with the xtt approach. In Geoffrey C. J. Sutcliffe and Randy G. Goebel, editors, *FLAIRS 2006*, FLAIRS. - Menlo Park, 2006. Florida Artificial Intelligence Research Society, AAAI Press.
7. OpenBSD Project. *PF: The OpenBSD Packet Filter*, 2006.
8. Rusty Russell. *Linux 2.4 Packet Filtering HOWTO*. NetFilter Project, 2002.

Six New Informativeness Indices
of Data Linguistic Summaries

Adam Niewiadomski

Institute of Computer Science, Technical University of Łódź, Poland
Wólczańska 215, 90-924 Łódź, Poland
aniewiadomski@ics.p.lodz.pl

Summary. The study introduces six new quality measures for linguistic summaries of databases. The basic Yager approach [6], with further improvements [1, 2], based on fuzzy logic, is considered. The informativeness of summaries is discussed, and the new measures are applied in the extended version of the algorithm (introduced in [3]) which enables automated generating of textual descriptions of databases, possibly applied in WWW or RSS news, press comments, etc.

1 Linguistic Summarization of Data

Let $\mathcal{Y} = \{y_1, \ldots, y_m\}$ be a set of objects described by a table (view) in a database. Let $\mathcal{D} = \{d_1, \ldots, d_m\}$ be the table (view) the records of which are described with attributes $V_1, \ldots, V_n$ in the $\mathcal{X}_1,\ldots, \mathcal{X}_n$ domains, respectively. Let the value of V_j, $j \leq n$, for y_i, $i \leq m$, be denoted as $V_j(y_i) \in \mathcal{X}_j$. Hence $d_i = \langle V_1(y_i), \ldots, V_n(y_i) \rangle$, and we use $\mu_{S_j}(d_i)$ for $\mu_{S_j}(V_j(y_i))$. "To summarize $\mathcal{D}$ linguistically" means to build a natural language sentence $Q\ P$ are/have $S\ [T]$ in which Q is a linguistic determination of amount (a quantity in agreement), e.g. *few, about 150*. P is the *subject of summary*. S is a feature of interest, the so-called *summarizer*, e.g. *high salary*. Both Q and S are handled by fuzzy logic [7, 8]. $T \in [0, 1]$ is the degree of truth for the summary

$$T = \mu_Q \left(\sum_{i=1}^{m} \mu_{S_j}\big(V_j(y_i)\big)/M \right) \qquad (1)$$

where $M = m$ if Q is relative or $M = 1$, if Q is absolute. The sample summary is *About half of my friends have big houses [0.71]*. The numerator of (1) is an interpretation of the cardinality of the fuzzy set S_j in $\mathcal{X}_j$, i.e. $card(S_j) = \sum_{x \in \mathcal{X}_j} \mu_{S_j}(x)$. The summary is constructed via the first canonical form of a linguistically quantified proposition, Q^I, [8].

The idea has been extended by George and Srikanth [1] with the so-called *composite summarizer* S expressed as the family of fuzzy sets $\{S_1,\ldots, S_n\}$, and

$$\mu_S(d_i) = \min_{j=1,\ldots,n} \left\{\mu_{S_j}\big(V_j(y_i)\big)\right\}, \ i = 1, 2, \ldots, m \qquad (2)$$

Minimum or another t-norm is a model of the "and" connective. T is still (1).

K.M. Węgrzyn-Wolska and P.S. Szczepaniak (Eds.): Adv. in Intel. Web, ASC 43, pp. 254–259, 2007.
springerlink.com

Linguistic summaries in the second canonical form, Q^{II}, are defined in [2], i.e. Q P being w_g are/have S, where w_g is a *query*, represented by a fuzzy set

$$T\left(Q\ P\ \text{being}\ w_g\ \text{are/have}\ S\right) = \mu_Q\left(\frac{\sum_{i=1}^{m}\mu_S(d_i)\ t\ \mu_{w_g}(d_i)}{\sum_{i=1}^{m}\mu_{w_g}(d_i)}\right) \tag{3}$$

The method allows to achieve more interesting and specific summaries, e.g. *Many of my <u>older</u> friends have big houses*, where *older*=w_g.

2 New Quality Measures for Linguistic Summaries

This section presents the quality criteria for linguistic summaries. The $T_1 \div T_5$, measures, i.e. degree of truth, imprecision, covering, appropriateness, and length of summary, respectively, have been defined by Kacprzyk, Yager, and Zadrożny [2]. The measures are applied to finding the optimum summary, see Sec. 2.5.

Six new measures, $T_6 \div T_{11}$, introduced in Sec. 2.1–2.4, are the author original contribution. The issue of finding the optimum summaries for a given set of summarizers, is originally modified. The measures are also implemented in the extended algorithm of news generating, see Sec. 3.

2.1 Quantification Imprecision

The intuition used in constructing the *quantification imprecision* index, denoted as T_6, is very similar to the one that determines T_2: the flatter a fuzzy set, the less precise the label represented. Nevertheless, the T_2 criterion concerns impreciseness of the fuzzy set that represent a summarizer, and the degree of quantification impreciseness concerns the fuzzy set that represents the linguistic quantifier in the summary.

$$T_6 = 1 - in(Q) = 1 - \frac{card\left(supp(Q)\right)}{card(\mathcal{D}(Q))} \tag{4}$$

where $in(Q)$ is the degree of fuzziness of Q, and $supp(Q)$ is the support of Q. Computing the denominator of (4), where $\mathcal{D}(Q)$ is the domain of the Q fuzzy quantifier, is, in fact, limited to two cases: 1) when Q is relative, hence $card(\mathcal{D}(Q)) = 1$ since $\mathcal{D}(Q) = [0, 1]$, or 2) when Q is absolute and $card\left(\mathcal{D}(Q)\right) = m$, i.e. the number of records, see the example.

Example 1. Consider a summary Q P are S, where Q =ABOUT $1/4$ is relative. See Fig. 1: there are two membership functions for Q presented, μ_{Q_1} and μ_{Q_2}. Let, for example, $\frac{card(S)}{m} = 0.3$, where m is a number of records. Hence

$$T_1(Q_1\ P\ S) = \mu_{Q_1}(0.3) \simeq 0.8\ ,\quad T_1(Q_2\ P\ S) = \mu_{Q_2}(0.3) \simeq 0.6 \tag{5}$$

Thus, the summary that uses Q_1 seems to be better, rather than Q_2, according to T_1, see (1), (3). Nevertheless,

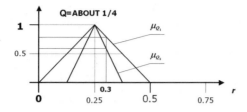

Fig. 1. Two membership functions representing the ABOUT 1/4 fuzzy quantifiers with different T_6 values

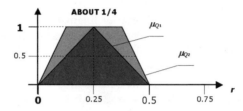

Fig. 2. Two membership functions representing the ABOUT 1/4 fuzzy quantifiers with different T_7 characteristics

$$in(Q_1) = \frac{card\,(supp(Q_1))}{card([0,1])} = 0.5 \ , \ in(Q_2) = \frac{card\,(supp(Q_2))}{card([0,1])} = 0.25 \quad (6)$$

hence

$$T_6(Q_1 \ P \ S) = 1 - in(Q_1) = 0.5 \ , \ T_6(Q_2 \ P \ S) = 1 - in(Q_2) = 0.75 \quad (7)$$

Thus, the summary $Q_2 \ P$ are S is more precise (in sense of the T_6 criterion) since its quantifier is characterized by the lower degree of fuzziness.

2.2 Quantification Cardinality

The intuition of the *quantification cardinality* measure is based, in contrary to the quantification impreciseness T_6, on the cardinality (instead of the support) of a fuzzy set that represents Q

$$T_7 = 1 - card(Q)/N \quad (8)$$

where $N = 1$ if Q is relative, or $N = card\,(D(Q))$ if Q is absolute. T_7 describes the following intuition: the greater the cardinality of Q is, the less precise the model of the quantity pronouncement.

Example 2. Let Q_1, Q_2 be relative fuzzy quantifiers as depicted in Fig. 2. The dark grey area represents graphically the cardinality of Q_1, $card(Q_1)$, and the light grey area – the difference between Q_2 and Q_1, $card(Q_2) - card(Q_1)$. Both quantifiers are characterized by exactly the same $T_6 = 0.5$ value, since cardinalities of their supports are 0.5. Nevertheless, thanks to the T_7 index it is possible to determine which of them represents the quantity more precisely.

$$T_7(Q_1 \ P \ S) = 1 - card(Q_1) = 1 - 0.25 = 0.75$$
$$T_7(Q_2 \ P \ S) = 1 - card(Q_2) = 1 - 0.375 = 0.625 \tag{9}$$

Hence, the Q_1 quantifier describes the quantity more precisely since its quantification cardinality, T_7, is greater. Notice that this subtle distinction is impossible to be obtained via the T_6 index.

2.3 Summarizer Cardinality

This quality measure is related to the cardinalities of the fuzzy sets that represent a summarizer. We denote it T_8. Its meaning is that the greater $card(S_j)$, the smaller precision of S_j. Because of possible several fuzzy sets $S_1,...,S_n$ representing the summarizer, see (2), the form of T_8 is

$$T_8 = 1 - \left(\prod_{j=1}^{n} \frac{card(S_j)}{card(\mathcal{X}_j)} \right)^{1/n} \tag{10}$$

where $\mathcal{X}_j$ is the universe of discourse of the S_j fuzzy set.

2.4 Imprecision, Cardinality, and Length of the w_g Query

Three next indices, denoted as T_9, T_{10}, T_{11} are determined by the degree of imprecision, by the cardinality, and by the length of the w_g query in a summary in the second canonical form, where w_g is represented by a fuzzy set in a $\mathcal{D}(w_g)$ domain

$$T_9 = 1 - \frac{card(supp(w_g))}{card(\mathcal{D}(w_g))} \tag{11}$$

where $in(w_g)$ is the degree of imprecision of the w_g fuzzy set,

$$T_{10} = 1 - \frac{card(w_g)}{card(\mathcal{D}(w_g))} \tag{12}$$

$$T_{11} = 2 \cdot (0.5)^{|w_g|} \tag{13}$$

where $|w_g|$ is the number of sets the query consists of (as the intersection).

2.5 The Optimum Summary

All the presented indices determine a reliable quality measure:

$$T = T(T_1,...,T_{11}; w_1,...,w_{11}) = \sum_{i=1}^{11} w_i \cdot T_i \tag{14}$$

where: $w_1 + ... + w_{11} = 1$. Finding the summary of the highest quality S^* among all possible summaries $\{S\}$ is the optimization task in which $\max_{S^* \in S} T$ is sought [2].

3 The Application: The News Generating Algorithm

The algorithm, in its first version, was introduced in [3]. The version presented in this section is extended with the use of six new measures, $T_6 \div T_{11}$. We assume that k linguistic quantifiers $Q_1,...,Q_k$, and z summarizers of $S_1,\ldots,S_n$, $z \leq n$, are applied.

```
// summaries in the first canonical form Q^I
1. for each non-empty Ŝ ⊆ {S_1, ..., S_z}
1.1. determine μ_Ŝ(d_i) via (2)
1.2. for each quantifier Q_h, h = 1, ..., k
     if (Q_h is absolute)
```
$$T_{1,h} = \mu_{Q_h}\left(\sum_{i=1}^{m} \mu_{\hat{S}}(d_i)\right)$$
$$T_{6,h} = 1 - \frac{card(supp(Q_h))}{m} \text{ via (4)}; \quad T_{7,h} = 1 - \frac{card(Q_h)}{m} \text{ via (8)}$$
```
     else // if Q_h is relative
```
$$T_{1,h} = \mu_{Q_h}\left(\frac{\sum_{i=1}^{m} \mu_{\hat{S}}(d_i)}{m}\right)$$
$$T_{6,h} = 1 - card(supp(Q_h)) \text{ via (4)}; \quad T_{7,h} = 1 - card(Q_h) \text{ via (8)}$$
```
1.3. T_{h_max} = max_{h∈{1,...,k}} {t: t = w_1 T_{1,h} + w_6 T_{6,h} + w_7 T_{7,h}}; remember h_max
```
$$1.4. \text{ compute } T_2 = 1 - \left(\prod_{S_j \in \hat{S}} in(S_j)\right)^{1/card(\hat{S})}$$
```
// T_3, T_9, T_10, T_11 do not appear in Q^I
```
$$1.5. \text{ compute } T_4 = \left|\prod_{S_j \in \hat{S}} \frac{\sum_{d_i \in D} \xi_{supp(S_j)}(d_i)}{m}\right|$$
$$1.6. \text{ compute } T_5 = 2 \cdot (0.5)^{card(\hat{S})}$$
$$1.7. \text{ compute } T_8 = 1 - \left(\prod_{S_j \in \hat{S}} \frac{card(S_j)}{card(\mathcal{X}_j)}\right)^{1/card(\hat{S})} \text{ via (10)}$$
```
1.8. T = T_{h_max} + w_2 · T_2 + w_4 · T_4 + w_5 · T_5 + w_8 · T_8
1.9. generate the summary Q_{h_max} P are Ŝ [T]

//Summaries in the second canonical form Q^II
2. For each non-empty query S_w ⊊ {S_1, ..., S_z}
   and for each non-empty summarizer Ŝ ⊆ {S_1, ..., S_z} \ S_w
2.1. μ_{S_w}(d_i) = min_{S∈S_w} μ_S(d_i)
2.2. D ⊇ D_w = {d_i ∈ D: μ_{S_w}(d_i) > 0}
2.3. for each d_i ∈ D_w determine μ_Ŝ(d_i)
2.4. for each relative Q_h : h ∈ {1, ..., k}
```
$$T_{1,h} = \mu_{Q_h}\left(\frac{\sum_{d_i \in D_w} \min\{\mu_{\hat{S}}(d_i), \mu_{w_g}(d_i)\}}{\sum_{d_i \in D_w} \mu_{S_w}(d_i)}\right) \text{ via (3)};$$
$$T_{6,h} = 1 - card(supp(Q_h)) \text{ via (4)};$$
$$T_{7,h} = 1 - card(Q_h) \text{ via (8)}$$
```
2.5. find T_{h_max} as in 1.3.; remember h_max
2.6. compute T_2 as in 1.4.
```
$$2.7. \text{ compute } T_3 = \frac{\sum_{d_i \in D_w} \xi_{supp(S \cap S_w)}(d_i)}{\sum_{d_i \in D_w} \xi_{supp(S_w)}(d_i)}$$
$$2.8. \text{ compute } T_4 = \left|\prod_{S_j \in \hat{S}} \frac{\sum_{d_i \in D} \xi_{supp(S_j)}(d_i)}{m} - T_3\right|$$
```
2.9. compute T_5, T_8 as in 1.6., 1.7., respectively
2.10. compute T_9, T_10, T_11 via (11),(12),(13)
```
$$2.11. \quad T = T_{h_max} + \sum_{i=2}^{5} w_i \cdot T_i + \sum_{i=8}^{11} w_i \cdot T_i$$
```
2.12. generate the summary Q_{h_max} P being S_w are Ŝ [T]
```

The algorithm considers $k \sum_{i=0}^{z-1} \binom{z}{i} \left(2^{z-i} - 1\right)$ summaries, and for each combination of $S_1, \ldots, S_n$ the Q_h quantifier for which the T in Step 1.8. and/or 2.11. is maximized, is chosen. Finally, the resulting textual message consists of $\sum_{i=0}^{z-1} \binom{z}{i} \left(2^{z-i} - 1\right)$ summaries, which are then processed by a human editor or analyst or etc. For instance, for $z = 2$, we have 5 summaries:

$Q^{(1)}$ P are S_1 $[T^{(1)}]$,
$Q^{(2)}$ P are S_2 $[T^{(2)}]$,
$Q^{(3)}$ P are S_1 and S_2 $[T^{(3)}]$,
$Q^{(4)}$ P being S_1 are S_2 $[T^{(4)}]$,
$Q^{(5)}$ P being S_2 are S_1 $[T^{(5)}]$,

where $Q^{(1)}$ is a Q_h quantifier, $h = 1, \ldots, k$, for which $T(Q_h \ P \ S_1)$ is the greatest (denoted as $T^{(1)}$), and $Q^{(2)} \ldots Q^{(5)}$ – analogously. For examples and further details, see [3, 5].

4   Conclusions and Future Work

Currently, the author works on similar algorithms for interval-valued and for type-2 linguistic summaries. The use of extended fuzzy sets is supposed to process membership functions based on preferences of several experts, possibly with different confidence levels, see [4, 5].

References

1. R. George and R. Srikanth. Data summarization using genetic algorithms and fuzzy logic. In F. Herrera and J.L. Verdegay, editors, *Genetic Algorithms and Soft Computing*, pages 599–611. Physica–Verlag, Heidelberg, 1996.
2. J. Kacprzyk, R. R. Yager, and S. Zadrożny. A fuzzy logic based approach to linguistic summaries of databases. *Int. J. of Appl. Math. and Comp. Sci.*, 10:813–834, 2000.
3. A. Niewiadomski. News generating via fuzzy summarization of databases. *Lecture Notes in Computer Science*, 3831:419–429, 2006.
4. A. Niewiadomski, J. Ochelska, and P. S. Szczepaniak. Interval-valued linguistic summaries of databases. *Control and Cybernetics*, 35(2):415–444, 2006.
5. A. Niewiadomski and P. S. Szczepaniak. News generating based on interval type-2 linguistic summaries of databases. In *Proceedings of IPMU 2006 Conference, July 2–7, 2006, Paris, France*, pages 1324–1331, 2006.
6. R.R. Yager. A new approach to the summarization of data. *Inf. Sciences*, 28:69–86, 1982.
7. L. A. Zadeh. The concept of linguistic variable and its application for approximate reasoning (i). *Inf. Sciences*, 8:199–249, 1975.
8. L. A. Zadeh. A computational approach to fuzzy quantifiers in natural languages. *Computers and Maths with Appl.*, 9:149–184, 1983.

Sufficient Knowledge Omission Error and Redundant Disjoint Relation in Ontology

Wajahat Noshairwan, Muhammad Abdul Qadir, and Muhammad Fahad

Center for Distributed and Semantic Computing Mohammad Ali Jinnah
University, Islamabad, Pakistan
wajahat.nsoahirwan@gmail.com, aqadir@jinnah.edu.pk, mhd.fahad@gmail.com

Summary. Ontology evaluation is as important as its designing and application. Researchers have identified different errors which should be catered in ontology evaluation process and classified them in error's taxonomy. We have found that some important errors are missing in the error's taxonomy. We have identified and defined new errors i.e. sufficient knowledge omis-sion error (SKO) and redundancy disjoint relation error (RDR) and catego-rized them in appropriate category of error's taxonomy.

1 Introduction

Ontology becomes a standard way to describe the concepts more formally [3]. There are different phases in ontology life cycle like ontology designing, its evaluation, mapping and merging. There is a possibility that the ontologists unintentionally make some errors in ontology designing. So evaluation of ontology is as important as the description of ontology because if ontology itself is error prone then the applications dependent on the ontology have to face some critical prob-lems.

For assistance in the evaluation, domain researchers have identified some errors and defined them in error's taxonomy [1]. In error's taxonomy, there are mainly three types of error that are usually encountered by ontologist i.e. inconsistency, incompleteness and redundancy of information [3]. this error's taxonomy becomes a guideline for evaluators to evaluate the ontology in perspective of such errors. If some errors are not defined in error's taxonomy then we can say that the evalua-tors based on the error's taxonomy will not detect such errors. In this considera-tion we have evaluated the error's taxonomy of ontology that it has covered all types of possible errors or not. Surprisingly we identified that some important er-rors are missed in error's taxonomy i.e. sufficient knowledge omission error and redundancy of disjoint relation and categorized them according to their appropri-ate category. We have defined these errors and the situations where they can occur and explained the importance of these errors by different examples or scenarios.

K.M. Węgrzyn-Wolska and P.S. Szczepaniak (Eds.): Adv. in Intel. Web, ASC 43, pp. 260–265, 2007.
springerlink.com

Rest of the paper is organized as follows: section 2 presents classification of errors and our contribution to error's taxonomy; section 3 presents related work to our domain and section 4 concludes the paper and gives insight on future work.

2 Extensions in Error's Taxonomy

Fig. 1 represents the error's taxonomy by Gomez-Perez [1], slightly extended by us. Our contribution to this classification is represented in dotted box i.e. sufficient omission error and redundancy of disjoint relation.

2.1 Sufficient Knowledge Omission Error (SKO)

Ontology represents different types of information for the concept like concept's description, its hierarchal information and relational information. OWL becomes most adopted ontology language. But defining the ontology by using OWL does not mean that ontologist has provided all types of description for the concepts. In general, there are two types of concept's description, called Necessary description and sufficient description [3]. Necessary description only defines the basic criteria by which new concept is formed like its hierarchal information, and sufficient definition elaborates the characteristics of concept like its self description by using intersection, union, complement or restriction axioms in OWL. Sometimes during ontology designing, ontologists define the concepts but don't provide their suffi-cient definitions. We consider such lack of information as an error and according to nature of error we categorized it in incompleteness partition error as shown in Fig. 1.

Consequence of Sufficient Knowledge Omission Error

We describe the importance of this error by defining some scenarios

- **Ambiguity within ontology:** Consider the ontology of AirFlights, which has two main sub types DomesticAirFlight and CommercialAirFligh. A DomesticAirFlight is the sub concept of AirFlights, is a necessary definition of it. A DomesticAirFlight only flies within a country, is a sufficient definition that differentiates it from the other types of AirFlights. Ontology designer sometimes does not define the sufficient definition of DomesticAirFlight. Due to this, machine fails to infer whether the individual of AirFlights belongs to DomesticAirFlight.
- **Merging of Ontology:** The description of concepts is most important during the merging process of ontologies. The merging system finds the mapping between concepts on the basis of concept's description and other information. The concept's description plays an important role to find corresponding among concepts. If concepts have not sufficient information then basically we did not get the advantages of OWL concept's description richness in merging process.

- **Semantic Search Engine:** The component of semantic search engine used ontology for their purpose like indexer semantically indexes the crawled pages by using ontology [12]. Semantic crawler component crawls the pages and finds semantic relevancy with domain by using ontology. Consider the situations where the concept of ontology itself has not sufficient information. This will affect the results of semantic crawler and semantic indexer.

The above scenarios describe the significant importance of the error and show that if we do not consider the error then we have to face some critical problems to achieve the objectives.

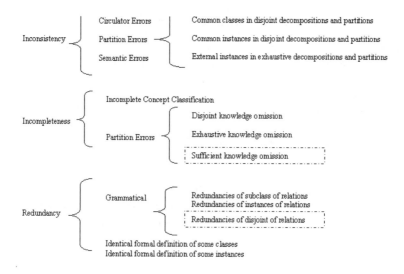

Fig. 1. Extended Error's Taxonomy

Sufficient Omission Checker (SOC)

If concepts have not enough information about itself then warning should be generated against them. We have implemented the evaluation system i.e. sufficient omission checker (SOC). It checks the definition of the concepts and applies the criteria and generates warnings against the concepts, which satisfy all the criteria as shown in Fig 2.

Empirical Results

To prove our concept that the sufficient omission error is usually done by ontologist, we have evaluated some known ontologies [13]. After evaluation we have found that most of the ontologies have sufficient omission error. The summary of evaluation results is shown in Table 1.

```
IF (Concept has parent concept) Then
   // (We ignore the concepts which are basic concept of ontology)
   IF (Concept has no intersection description) Then
      IF (Concept has no Union description) Then
         IF (Concept has not derived by any Restriction) Then
            IF (Concept is not type of enumerated class) Then
               IF (Concept is not complement of any other defined
                  concept) Then
                  IF (Concept has only subclass definition) Then
                     Write: "Warnings, concept has sufficient omission error"
```

Fig. 2. Criteria of Sufficient Omission Checker

Table 1. Summary of Evaluation Results of Ontologies for SKO

Ontology	Concepts have not sufficient definition	Total Concepts
Pizza	19	96
Camera	4	12
Generations	0	18
People-Pets	20	70
Travel	13	35

2.2   Redundancy of Disjoint Relation Error (RDR)

Redundancy of disjoint relation error means that the concept is disjoint with other more than once. We know according to description logic rules [5], if concept is disjoint with any concept then it also disjoint with its sub concepts. The one possi-ble way of occurrence of RDR is that the concept is disjoint with parent concept and also with its child concept. The second possible way is that the parents of con-cepts are already disjoint and their children are also disjoint and third possible way is that parent is already disjoint with the concept and its child is also disjoint with that. We explain such possible ways in Fig. 3. Disjoint relation is shown by dotted line and subclass relation is shown by arrow line.

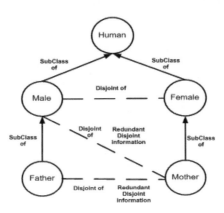

Fig. 3. Example of Redundant Disjoint Relation

The redundant information is due to occurrence of directly disjointness (concepts are directly disjoint) and indirectly disjointness (concept is disjoint with other because its parent is already disjoint with it) at same time [4].

This type of error is not defined in error's taxonomy and the available evaluation systems like Racer [8], Fact [10] and Pellet [11] don't detect it. It will create same problems as other redundant information in ontology like redundancy of subclass relation, so detection of this error is as important as other redundant errors.

3 Related Work

The most related work is presented here in this section.

The main contribution of work in this domain is Gomes [1, 3]. They identified different types of errors and properly categorized them in error's taxonomy. The error's taxonomy becomes a guideline for evaluation of ontology. Several evaluator tools have been developed based on it.

Joachim Baumeister and Dietmar Seipel [9] discuss the evaluation process of ontologies and also identify the new type of errors called design anomalies in on-tology. These new types of error are good contribution in error's taxonomy. They did not only identify the anomalies but also defined the detection method by using prolog and FN-query language. Their identified anomalies help the ontologist to develop consistent ontology.

4 Conclusion

The main contribution of this paper is an extension in error's taxonomy. We have identified two new types of error of different categories first one i.e. sufficient knowledge omission error belongs to incompleteness category and second one i.e. redundancy of disjoint relation belongs to redundant category. We have also de-scribed the importance of detection of sufficient omission error by explaining dif-ferent scenarios and also described the criteria of detection. We evaluated different ontologies and found that the sufficient omission error is present in them. We also described the criteria for detection of second error and brief level implementation details of it. In future work, we will further evaluate error's taxonomy and try to find some other type of errors that are usually encountered by ontologist.

References

1. Gt'omez-Pt'erez, A., et al.: Evaluation of Taxonomic Knowledge on Ontologies and Knowledge-Based Systems. International Workshop on Knowledge Acquisition, Modeling and Management (1999).
2. Antoniou, G., Harmelen, F.V.: A Semantic Web Primer. MIT Press Cambridge2004 ISBN 0-262-01210-3

3. Gomez-Perez,A.,M.Fernandez-Lopez,A.Gsmez-Pirez, O.Corcho-Garcia.: Ontological Engineering:With Examples from the Areas of Knowledge Management, E-Commerce and the. Semantic Web. Published by Springer ISBN:1-85253-55j-3
4. Qadir,M.A.,Noshairwan, W.: Warnings for Disjoint Knowledge Omission in Ontologies. InProc. International Conference of on internet and Web Applications and Services (2007).
5. Nardi,D., et al.:The Description Logic Handbook: Theory, Implementation, and Applications. ISBN: 9780521781763
6. DAML: Available. http://www.daml.org/ (current Jan2007)
7. Web Ontology Language Overview: Available. http://www.w3.org/TR/owl-features/ (current Jan2007)
8. Haarslev,V. and Moller,R.: RACER System Description. In Proceeding of the International Joint Confernce on Automated Reasoning ,IJCARŠ2001, pp 701-705, LNCS, Springer-Verlag,2001.
9. Baumeister,J., Smelly,D.S..: OwlsÜDesign Anomalies in Ontologies. 18th Internaional Florida Artifiical Intelligence Research Society Conference (FLAIRS), pp 251-220, AAAI Press, 2005
10. Horrocks,I.:The FaCT System. International conference. on Analytic Tableaux and Related Methods (TABLEAUX'98), pp 307-312,vol 1397, Springer-Verlag, 1998
11. Pellet.: An OWL Dl Reasoner. Available. www.pelet.owldl.com/ (Current Jan 2007)
12. Ganesh, S.ăă Jayaraj, M.ăă Kalyan, V.ăă SrinivasaMurthyăă Aghila.: Ontology-based Web crawler. International conference of Information technology. ITCC 2004
13. Protégé Ontologies Library. Available: http://protege.cim3.net/cgi-bin/wiki.pl?ProtegeOntologiesLibrary (current Jan 2007).
14. Brank,J., et al.: A Survey of Ontology Evaluation Techniques. Published in multiconference IS 2005, Ljubljana,Slovenia SIKDD 2005.

BiTutor–A Component Based Architecture for Developing Web Based Intelligent Tutoring System

Yaser Nouh[1], Varunkumar Nagarajan[1], R. Nadarajan[1], and Maytham Safar[2]

[1] Mathematics and Computer Applications Department, PSG College of Technology,
Coimbatore 641004, India
yasernouh@yahoo.co.in,
varunkumar_n@yahoo.co.in,
nadarajan_psg@yahoo.co.in
[2] Computer Engineering Department, Kuwait University, Kuwait
maytham@eng.kuniv.edu.kw

Summary. In this paper we present an overview of the architectural design of the *BiTutor* which is a Bayesian Intelligent Tutoring System. This is an on-going research whose final goal is to build an open *Intelligent Tutoring System* (ITS). Every component of the architecture has its own strategies and tools that makes it intelligent. This makes the entire system posses adaptive capabilities. The design includes web-based distributed architecture, AI techniques used and programmer-optimized user interface. It is hypothesized that the completed prototype will be sufficient to prove the concept. A fully developed BiTutor will provide an interactively-rich learning environment for students that will result in increased achievement.

Keywords: Intelligent Tutoring System, Web Adaptive.

1 Introduction

Computers have been playing a vital role in education for the last three decades. Web-based learning systems are becoming increasingly popular providing hyperlinks to allow students to access relevant sites. However, these *HTML* web pages are static and neither interactive nor adaptive to students with different knowledge levels [1]. Applications in this domain are evolving from *Computer Assisted Instruction* (CAI) to *Intelligent Tutoring Systems* (ITS). An ITS is a computer program that makes the system capable of providing the student with personalized, adaptive and effective teaching. Yao and Yao [9] argue that the system should be robust enough to suite various types of learners. Thus, the system is required to be aware of the cognitive state and behavioral skills of a particular student, to diagnose the students' errors and adjust its belief about his current state of knowledge. The goal of our current research is to bring together the recent developments in the fields of ITS, Cognitive Science and AI to construct an efficient intelligent tutor whose eventual target is to identify student ability accurately and also tutor accordingly. One such system proposed by us is BiTutor for

K.M. Węgrzyn-Wolska and P.S. Szczepaniak (Eds.): Adv. in Intel. Web, ASC 43, pp. 266–271, 2007.

tutoring any subject to students. The BiTutor helps a student to navigate through online course materials and recommended learning goals. This paper proposes a component based architecture called BiTutor. It is composed of independent educative modules coordinated by a core that controls the instruction of the student (Open System). The architecture proposes a set of models where each model deals with a particular aspect of the system.

2   BiTutor Architecture

BiTutor is an open system which is a set of autonomous educational components that communicate between themselves following high-level pre-established protocol. It is currently being developed using *JAVA, XML* and *JSP* programs so as to support self-paced learning in *World Wide Web* (WWW) environment. Fig. 1 shows the various elements that compose BiTutor.

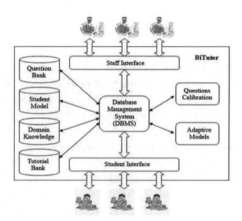

Fig. 1. BiTutor Architecture

2.1   Domain Knowledge

The Domain Knowledge represents the approach that determines how to represent knowledge through different levels of hierarchy [6]. The BiTutor defines five levels of granularity to represent this hierarchy. They can be identified as Subjects (Level 1), Topics (Level 2), Sub-Topics (Level 3) and Concepts (Level 4) and Questions (Level 5) [11]. A Concept is an elementary piece of knowledge, in the sense that it cannot be decomposed into smaller parts. Concepts are considered as basic piece of knowledge. A Sub-Topic is a set of elementary concepts. A Topic is a collection of sub-topics. A Subject is a collection of topics. The Fig. 2 shows the knowledge representation in BiTutor. The link between the nodes is called a relationship and the instructor sets weight to this relationship initially when he creates the network.

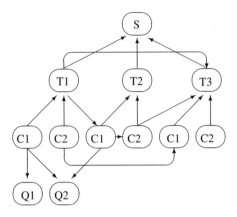

Fig. 2. Knowledge Representation in BiTutor

2.2   Student Model

Here, in student modeling, we adopt a combined approach of integrating belong-to and pre-requisite relationships into one layer improving efficiency of adaptation mechanisms and inference process [10]. This structural model, based on Bayesian Networks and Item Response Theory, allows substantial simplification when specifying parameters (Conditional Probabilities) which measures student ability at different levels of granularity. Previously, much interest has been devoted to the development and use of student models based on Bayesian Networks. In terms of Bayesian Network, the student model will be represented by the network structure and its probabilities and the diagnostic capabilities of Bayesian propagation algorithms. Successful samples can easily be found in research literature: in student modeling [3] [4] [5]; in user profiling for information retrieval [7], for inferring user goals and needs [8], etc. The existing ITS have their student model with relationships like belong-to and pre-requisite in two different layers as in [2]. The complexity in such a model is that the two layers have to be separately managed, by computing individual probabilities for both the layers. Our student model, proposed here, is an open system to develop ITS having both the relationships combined in one layer thus resolving the above mentioned complexity Each element in the hierarchy, as discussed in Knowledge Domain, is represented as a Knowledge Node (KN).

2.3   Question Bank

The Questions Bank contains all the questions that are used by the BiTutor to test the students in various concepts. The questions have two parameters:

- The Discrimination Index (a) denotes how well the question is able to differentiate between students of slightly different abilities.
- The Difficulty Level (b), the value describing how tough the question is.

The question and the answer choices are described in the contents of XML elements. The correct answer for the question is stated in the value of the XML attribute answer. The XML file for the questions on the concept of stack operations under the subject of Data Structures is shown below

```
<Questions>
 <Question qid="Q10" conceptid="C01" answer="B">
   <text>
    Insertion and deletion inside a stack are done by
   </text>
   <Options>
     <Option optid="A">Insert(), Delete()</option>
     <Option optid="B">Push(), Pop()</option>
     <Option optid="C">Append(), Delete()</option>
     <Option optid="D">Add(), Remove()</option>
   </Options>
   <Parameter difficulty="easy"
               discrimination="moderate"/>
 </Question>
</Questions>
```

2.4   Tutorial Bank

The Tutorial Bank contains tutorials for all the concepts. The instructors or the domain experts frame tutorials at the time of adding the concepts or defining the knowledge domain. Whenever a particular concept is selected to be taught to the student or when the student wishes to take up a quick tutorial, this module selects the tutorial depending on the context. They are stored as *HTML* pages with advanced features including *FLASH*, etc. Since these are stored as *HTML* pages, they help a student to navigate through online course materials and recommended learning goals easily.

2.5   Questions Calibration

When a question is added to the Questions Bank by staff, the discrimination index (a), and the difficulty level (b) are given initial values. But as the students are tested using the question, the value of the question parameters may be updated. Here we use a Two-Parameter Logistic Function, which is converted to a Simple Linear Regression by using Least Square Method for computing a and b based on the students' response to the question. If the decision is to assess the student on his mastery of a knowledge node, then each question from the question bank is dynamically connected to the Bayesian network. A question is selected when its difficulty value is marginally higher than the mastery state of the topic attained by the student or if the probability of the student answering the question correctly is marginally less than half.

$$P(x_i = 1/\theta) = \frac{1}{1+e^{-a(\theta-b)}}$$

Where a is the discrimination index, b is the difficulty level, and θ is the student expected mastery value of the concept. Using such an approach, the construction of the network is much easier since not all items need to be included, thus reducing the complexity of BN.

2.6   Adaptive Models

The Adaptive models figure out the next selection based on decision:

- Selection of the next best concept to be taught.
- Selection of the next best question to be asked.
- Conditions on which BiTutor is terminated. The stopping criterion may be due to any one of the following:
  - If the student masters that particular topic.
  - If there exists a pre-defined time-out or the student may choose to exit the tutoring session.
  - If the student is giving atypical responses for a long time (repeated irrelevant or incorrect responses).
  - If the student remains in the same mastery state (no progress).
  - The student is too dependent on the tutorials.

2.7   Staff Interface

The Staff Interface module provides a very user friendly interface for the staff to construct the hierarchy of a topic in the form of a network. This includes the validations such as ensuring the graph is acyclic, absence of no standalone nodes, etc. The parameters for all the concepts and relationships can be entered in the same interface. This enables ease of use. The weights set for the relationships are used to compute the conditional probability table [10] and thus the staff need not undergo the laborious process of entering the values of the Conditional Probability Table (CPT).

2.8   Student Interface

The students use the interface to communicate with BiTutor to select any topic for learning and subsequent testing. This interface provisions for revising the mastered subjects. Quick tutorials of this interface allow the student to explore the subject topics efficiently.

3   Conclusion

In summary, the BiTutor prototype is designed using advanced cognitive science and AI techniques promoting the necessity for on-going research and development in the field of web-based educational tools. The project in progress is based on sound theories and practices used in successful Intelligent Tutoring Systems and draws from the achievements ITS researchers have had in related projects. Furthermore, it is important to the field of Education in both e-learning and traditional settings. BiTutor is still under development, so this presentation is just a first glance of the system. No testing has been carried out yet because it is not fully operational.

References

1. Brusilovsky, P: Adaptive and Intelligent Technologies for Web based Education. Special Issue on Intelligent System and TeleTeaching, Vol. 4. (1999) 19–25.
2. Carmona C., Millan E., Pérez-de-la-Cruz J.L., Trella M., and Conejo R.: Introducing Prerequisite Relations in a Multi-layered Bayesian Student Model. User Modeling 2005, Edinburgh, Scotland, UK (2005) 347-356.
3. Conati C.,Gertner A., and VanLehn K.: Using Bayesian Networks to Manage Uncertainty in Student Modeling. User Modeling and User-Adapted Interaction, Vol. 12(4). (2002) 371–417.
4. Henze N.,and Nedjl W.: Student Modeling for the KBS Hyperbook System using Bayesian Networks. Technical report, University of Hannover, November (1998).
5. Horvitz E.,Breese J., Heckerman D., Hovel D., and Rommelse K.: The Lumiere Project: Bayesian User Modeling for Inferring the Goals and Needs of Software Users. Proceedings of UAI'98. Morgan Kauffman, (1998) 256-265.
6. Joséphine M.P. Tchétagni, Roger Nkambou: Hierarchical Representation and Evaluation of the Student in an Intelligent Tutoring System. Intelligent Tutoring Systems : Proceedings of 6th International Conference, ITS 2002. Biarritz, France and San Sebastian, Spain, (2002).
7. Mayo E., and Mitrovic A.: Using Probabilistic Student Model to Control Problem Difficulty. In LNCS 1839, Springer, (2000) 525-533.
8. Weber G., and Brusilovsky P.: ELM -ART: An adaptive Versatile System for Web-based Instruction. International Journal of Artificial Intelligence in Education 12 (4), Special Issue on Adaptive and Intelligent Web-based Educational Systems, (2001) 351-384.
9. Yao J.T., and Yao Y.Y.: Web-based support system. Proceedings of Second Indian International Conference on Web Intelligence. Canada (2003).
10. Yaser Nouh, Indumathy.R, Karthikeyani.P, Varunkumar Nagarajan, and R. Nadarajan: Bayesian student Modeling in a Distributed Environment. First International Conference on Digital Communications and Computer Applications, DCCA-2007. Jordan (2007) 1132-1140.
11. Yaser Nouh, Karthikeyani. P and R. Nadarajan: Intelligent Tutoring System - Bayesian Student Model. Proceedings of First IEEE International Conference on Digital Information Management, IDCIM-2006. Bangalore, India (2006) 257-262.

A Web-Based Decision Support System for Impacts Assessment of Urban Mobility: WDSS4IA

Hichem Omrani[1,2], Adrien Ogor[1], Luminita Ion-Boussier[2], and Philippe Trigano[1]

[1] Department of Computer Science, Heudiasyc Laboratory(UMR CNRS 6599), Centre de Recherches de Royallieu, B.P. 20529, Compiégne Cedex France
hichem.omrani@utc.fr
[2] EIGSI, 26 rue Vaux de Foletier, 17041 La Rochelle cedex 1, France
luminita.ion@eigsi.fr

Summary. In this paper, we propose a Web Decision Support System for Impacts Assessment of Urban Mobility (WDSS4IA) which is based on a new approach. The latter uses fuzzy set theory for modeling criteria, belief theory for evaluations fusion from various information sources and it is able to handle uncertainty and vagueness. In this paper, the WDSS4IA is presented, focusing on its conception, modules and functionalities.

1 Introduction

Several approaches were used to develop tools of decision-making assistance in the environmental management field and of evaluation of urban mobility such as: LCA (Life Cycle Analysis) [3], CBA (Cost-Benefit Analysis), CEA (Cost-Effectiveness Analysis) [5] and MCDA (Multi Criteria Decision Analysis) [1]. Generally, quantitative assessment methods of the impacts often approach aspects related to only one category like economy, transport or environment [12] and seldom assess all the aspects. Moreover, there is an obvious lack of data or when they exist, they are spoiled with uncertainties and inaccuracies. In this context, we propose a hybrid approach based on fuzzy logic, evidence theory [10](Dempster-Shafer theory(DS)) and multi criteria analysis for the assessment of the environmental and socioeconomic impacts related to the urban mobility and of an assistance tool to web-oriented assessment.

The aim of this paper consists to present a tool based on a new approach for rigorous evaluation of the measure (project) efficiency related to transportation to improve urban mobility in all categories. The development of this tool takes into account the complexity of the evaluation process [8, 9] and interactive aspects. The tool presented is an interactive Web-tool allowing the data collection, information treatment and decision-making aid. In this paper, our major focus is the tool developed within the framework of our research tasks with its conception, modules and functionalities.

K.M. Węgrzyn-Wolska and P.S. Szczepaniak (Eds.): Adv. in Intel. Web, ASC 43, pp. 272–277, 2007.
springerlink.com   © Springer-Verlag Berlin Heidelberg 2007

The paper is structured as follows. The opening section outlines the description of the evaluation process and the tool. This is followed by tool exploitation and ergonomics aspects in order to improve tool usability. The final section offers some general conclusion and discussion.

2   Description of the Evaluation Process and the Tool

2.1   Tool Description

The proposed tool is illustrated in Fig. 1(a). It is entitled "Web Decision Support System For Impacts Assessment of urban mobility" (WDSS4IA). The WDSS4IA is based on client-server architecture that allows the assessment of various impacts of urban mobility via the Internet which enables widespread access by authorized users. With this tool, data are collected from various information sources alike: decision-makers, sensors, studies etc., and then data are stored, fitted and by applying an evidential reasoning approach descried below in order to evaluate a given measure (transport project).

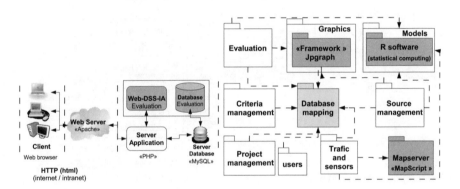

Fig. 1. Tool architecture (a), Package diagram (b)

The tool conception is done with the object formalism UML (Unified Modelling Language) and the navigation is conceived with the Web-ML (Web Modelling Language) formalism. Fig. 1(b) shows the package diagram of the WDSS4IA database and main tools used as environment development. The development environment is composed by open source tools as shown in Fig 1(a). The choice of these tools is justified that they present at my opinion a good compromise to produce an evaluation tool for multi-users, distributed and accessible via Internet tornet (several evaluators, heterogeneous and distant information sources, etc). We used MVC (Model-View-Controller) concept which is a conception model adapted to the creation of software imposing the separation between the data, the processing and the presentation. One of the great advantages of the MVC is due to the fact that it manages the use of various views for a given application. According to MVC concept (see Fig. 2, a), an application is divided into three fundamental components: the Model, the View and the Controller.

Each one of these components plays a well-defined role: the Model describes the data handled by the application and defines the access methods, the View defines the user interface and the presentation and the Controller deals with the management of synchronization events to update the View or the Model.

The tool modules are presented as shown in Fig. 2 (b). In this paper we focus on the evaluation process (data importation, data treatment and analysis, and criteria evaluation).

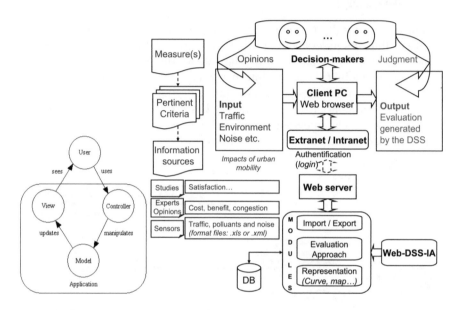

Fig. 2. Model-View-Controller schema (a), A Web-DSS For Impacts Assessment of urban mobility: WDSS4IA (b)

2.2   Evaluation Modules

Specification:

Let $\{C_k, k = 1, \ldots, n\}$ be a criteria set considered relevant for impacts assessment of a fixed measure which is selected previously by the project manager (e.g. traffic flow, noise, environmental data).

Let $\{E_i, i = 1, 2, \ldots, m\}$ be a set of evaluators charged to evaluate a criteria set (for certain criteria, several types of evaluation can be considered, example: acoustic pollution, perception (based on investigations), experimental measurements or modelling from traffic flow).

Data collection: During data collection and data acquisition, the evaluators indicate the reliability degrees of their information sources (α) using a 0-1 scale. Each evaluator can inform about criteria which are not in charge (for example, a questionnaire carrying into the visitors number of a Park & Ride, can bring information on the traffic level). Spatial data relating to the road network (roads, sections, infrastructures, layers etc.) are also integrated in the database, in order to visualize by maps, for example, the intensity of the traffic flow or pollutants concentration and their evolution in time. Fig. 3 presents the interfaces of importation of numerical and spatial data.

Information treatment: To treat possible missing quantitative information, there are various means (e.g. imputation methods etc) [6, 4]. We chose to filter data using an algorithm developed with the software "R", based on the neighborhood principle (version inspired from the k-mean method) but adapted to the nature of the traffic and the environmental data, an example is presented in Fig. 4(b).

Fig. 3. 2 screens of WDSS4IA: Connexion (a), data importation (b)

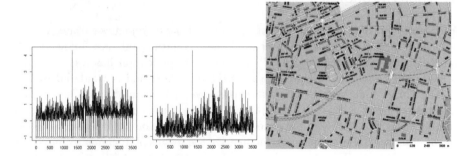

Fig. 4. CO concentration (in mg/m^3), before (a) and after filtering (b)(at the beginning missing data have values equal to "-1") and Traffic Map (c)

In order to follow the evolution of selected criteria, there are two kind of information to utilize.

• Single source information by criterion:

In the case where a single information source is used to evaluate a criterion the results (questionnaire, counting and model) are presented in graphic form (values or semantic evaluation) or cartographic form (map representation) via a GIS (Geographic Information System) such as traffic distribution (e.g. vehicles flow, see Fig. 4, c) or pollutants distribution, (e.g. NO_2)

- Multi-sources information by criterion:

In this case, a common framework is necessary in order to combine quantitative or/and qualitative data. We have proposed an approach [8, 9] for impacts assessment. In this paper, we summarize briefly the proposed approach. It is based on "evidential reasoning" where data are modelled according to fuzzy logic theory. Let Ω be the frame of discernment such as: $\{H_1, H_2, \ldots, H_p\}$ witch represents the evaluation levels (alike Small, Medium and High). We defined p fuzzy functions (triangular or trapezoidal) related to the frame of discernment. The mass assignment is described according to each information sources (experts, questionnaire, sensors), for more details see [8]. The results provided to the project manager are in form numerical, starting from the computation of a utility u_k of the criterion C_k defined by (1):

$$u_k = \sum_{i=1}^{p} u(H_i) \times BetP(H_i) \tag{1}$$

where: $u(H_i)$ represents the utility of an evaluation level H_i, $u(H_{i+1}) \geq u(H_i)$ if H_{i+1} is preferred to H_i and $BetP(H_i)$ represents the "pignistic" probability [11] related to each evaluation level (H_i).

3 Tool Ergonomics

An effort is put on the ergonomics of the application; ergonomics gathers a great number of criteria. The development of the evaluation aid tool follows the 9 heuristic principles suggested in [7], Even if the application has a code which is compatible with the last standards (XHTML, CSS) and is validated by the W3C (World Wide Web Consortium), this does not make of it an "accessible" application, besides this is not the purpose, because the end-users are limited in number. The main aspect of ergonomics is related to the application usability, by the training and appropriation facility. By the means of dynamic technologies such as AJAX (XML and asynchronous Java Script), the application is faster. During the use of the tool, clear and informative error messages are managed. The users can also get information about a functionality given by using an ergonomic "user guide" set up with XML technology.

4 Conclusion

In this paper we have proposed a methodology for the impacts evaluation related to the urban mobility, under the framework of multi criteria analysis, fuzzy logic and belief theory. An interactive Web-tool is presented for data collection, information treatment and decision-making aid. The tool development takes into account two aspects. Firstly, it takes into account the complexity of the evaluation process. In fact, criteria are given numeric or qualitative values, with different degrees of uncertainty and may use several information sources (e.g. experts, sensors, models, questionnaires, etc). Secondly, it has an interactive character, where several evaluators intervene for criteria evaluation. The developed tool has a lot of advantages before described but it is also necessary to mention that it has some weaknesses. The First one is that it can be slow and time consuming when trying to import huge data. To resolve this limitation, we have proposed to optimize the source code. Also, in order to ameliorate the tool interfaces and its usability, the WDSS4IA bas been evaluated with a group of experts and decision makers in transportation and environment respectively from transport service and ATMO (association of monitoring of he air quality) at La Rochelle-France.

Acknowledgment

This work is funded by the conseil général (council general) of La Rochelle, in France. It is under the framework of SUCCESS project which is an European project under the reference (Contract no.: 513785).

References

1. Beinat, Euro. Multi-criteria analysis for environmental management. Journal of Multi-Criteria Decision Analysis, (2001): 51.
2. Kelly G.A, The Psychology of Personal Constructs. New York, Norton, 1955.
3. Guinée J. B, Handbook on Life Cycle Assessment. An operational guide to the ISO standard. London, (2002): 704.
4. Kalton G., and Kasprzyk D., (1986), "The treatment of missing survey data", Survey Methodology, 12, pp. 1-16.
5. Kunreuther H., Grossi P. , Seeber N., Smyth A., A Framework for Evaluating the Cost Effectiveness of Mitigation Measures, Columbia University, USA, (2003).
6. Little, R. J.,Rubin D. B. (1987), Statistical Analysis with Missing Data, New York: Wiley.
7. Nielsen J., Molich R. (1990). Improving a Human-Computer Dialogue. CACM, 33, 3, 338-348.
8. Omrani H., Ion L., Trigano P., An Approach for Environmental Impacts Assessment using Belief Theory, Proceedings of 3rd IEEE Conference On Intelligent Systems (IEEE IS'2006), London, UK, 4-6 September 2006.
9. Omrani H., Ion L., Trigano P., An approach to decision-making for assessment of environmental impacts, Proceedings of the 11th Conference on Information Processing and Management of Uncertainty in Knowledge-Based Systems, (IPMU'2006), Vol II, pages 227-235, Paris, France, 2-7 July 2006.
10. Shafer G., A Mathematical Theory of Evidence, Princeton University Press, Princeton, 1976.
11. Smets P, (2004). Decision Making in the TBM: the Necessity of the Pignistic Transformation.
12. Yang J. B., Xu D. L., On The evidential reasoning approach for Multiple Attribute Decision Analysis Underr Uncertainty, IEEE Transactions on Systems, Man, and Cybernetics, Part A: Systems and Humans, Vol. 32, N0.3, May 2002.l

Privacy Preserving Enhanced Service Mechanism in Mobile RFID Network

Namje Park[1,2], Seungjoo Kim[2], and Dongho Won[2,*]

[1] Electronics and Telecommunications Research Institute (ETRI),
  161 Gajeong-dong, Yuseong-gu, Daejeon, 305-350, Korea
  namjepark@etri.re.kr
[2] School of Information and Communication Engineering,
  Sungkyunkwan University,
  300 Chunchun-dong, Jangan-gu, Suwon-si, Gyeonggi-do, 440-746, Korea
  {njpark,skim,dhwon}@security.re.kr

Summary. Recently, mobile RFID (Radio Frequency Identification) services such as in smart poster or supply chain are rapidly growing up as a newly generating industry. Here, mobile RFID service is defined as a special type of mobile service using RFID tag packaging object and RFID reader attached mobile RFID terminal. This paper is to provide an approach for ensuring RFID net-work security based on security engineering method. To ensure secure mobile RFID service network, we describes an approach which includes security state and security flow. We propose an efficient phone-based middleware platform architecture which is constructed by mobile RFID security mechanism based on WIPI (Wireless Internet Platform for Interoperability). WIPI-based light-weight mobile RFID security platform can be applied to various mobile RFID services that need secure business in mobile environment. So, we will propose the way to protect the personal privacy effectively using privacy preference for secure mobile RFID systems in ubiquitous network on this paper.

1 Introduction

Though the RFID technology is being developed actively and lots of efforts made to generate its market throughout the world, it also is raising fears of its role as a 'Big Brother'. So, it is needed to develop technologies for information and privacy protection as well as promotion of markets (e.g., technologies of tag, reader, middleware, etc.) The current excessive limitations to RFID tags and readers make it impossible to apply present codes and protocols. The technology for information and privacy protection should be developed in terms of general interconnection among elements and their characteristics of RFID in order to such technology that meets the RFID circum-stances.

While common RFID technologies are used in B2B (Business to Business) models like supply channels, distribution, logistics management, mobile RFID technologies are used in the RFID reader attached to an individual owner's

---

* Corresponding author. The research was supported by the University IT Research Center Project funded by the Korean Ministry of Information and Communication.

K.M. Węgrzyn-Wolska and P.S. Szczepaniak (Eds.): Adv. in Intel. Web, ASC 43, pp. 278–283, 2007.
springerlink.com

cellular phone through which the owner can collect and use information of objects by reading their RFID tags; in case of corporations, it has been applied mainly for B2C (Business to Customer) models aiming at their marketing. Though most of current RFID application services are used in fields like the search of movies posters and provision of information in galleries where less security needs are required, they will be expanded to and used more frequently in such fields like purchase, medical cares, electrical drafts, and so on where security and privacy protection are indispensable. Therefore, in this paper we described a privacy preserving enhanced trust building mechanism that extends the extant trust building service mechanisms for mobile RFID network to gain many advantages from its privacy control and dynamic capabilities. This is new technology to mobile RFID will provide a solution to protecting absolute confidentiality from basic tags to user's privacy information.

2   Background on mRFID, and Related Privacy Issues

Networked RFID means an expanded RFID network and communication scope to communicate with a series of networks, inter-networks and globally distributed application systems. So it makes global communication relationships triggered by RFID, for such applications as B2B, B2C, B2B2C, G2C (Government to Customer), etc. Mobile RFID service is defined as to provide personalized secure services such as searching the products information, purchasing, verifying, and paying for the products while on the move through the wireless internet network by building the RFID reader chip into the mobile terminal [1,3]. The service infrastructure required for providing such RFID based mobile service is composed of RFID reader, handset, communication network, network protocol, information protection, application server, RFID code interpretation, and contents development. The service model consists of tag, reader, middleware system, and information server. In the point of view of information protection, the serious problem for the RFID service is a threat of privacy [1,2,5]. Here, the damage of privacy is of exposing the information stored in tag and the leakage of information includes all data of the personal possessing the tag, tagged products and location. The privacy protection on RFID system can be considered in two points of view. One is the privacy protection between the tag and the reader, which takes advantage of ID encryption, prevention of location tracking and the countermeasure of tag being forged. The other is of the exposure of what the information server contains along with tagged items [6]. First of all, we will have a look about the exposure of information caused between tag and reader, and then discuss about the solution proposing on this paper.

3   Privacy-Enhancing Key Approaches

This technology aims at RFID application services like authentication of tag, reader, and owner, privacy protection, and non-traceable payment system where more strict security is needed.

1. Approach of platform level

   This technology for information portal service security in offering various mobile RFID applications consists of application portal gateway, information service server, terminal security application, payment server, and privacy protection server and provides a combined environment to build a mobile RFID security application service easily.

2. Approach of protocol level

   - It assists write and kill passwords provided by EPC (Electronic Product Code) Class1 Gen2 for mobile RFID tag/reader and uses a recording technology prevent-ing the tag tracking.
   - It employs information protection technology solving the security vulner-ability in mobile RFID terminals that accept WIPI as middleware in the mobile RFID reader/application part and provides end-to-end security solutions from the RFID reader to its applications through WIPI based mobile RFID terminal security/code treatment modules.

3. Approach of privacy level

   This technology is intended to solve the infringement of privacy, or random acquisition of personal information by those with RFID reader from those with RFID attached objects in the mobile RFID circumstance except that taking place in companies or retail shops which try to collect personal information. Main assumptions are as follows:

   - Privacy in the mobile RFID circumstance comes into force when a person holds a tag attached thing and both information on his/her personal identity (reference number, name, etc.) and the tag (of commodity) are connected to each other.
   - Privacy protection information are concerned with the tag attached object (its name, value, etc.) and the personal identity (of its owner or reference).
   - When it comes to the level of access authority, the owner can have access to any personal information on his/her object, an authorized person (e.g., pharmacist or doctor in medical care) only to access permitted information, and an unauthorized person to nowhere.

3.1 Overview of Secure Mobile RFID Environment

The mobile RFID is a technology for developing a RFID reader embedded in a mobile terminal and providing various application services over wireless networks. Various security issues - Interdomain security, privacy, authentication, E2E (End-to-End) security, and untraceability etc. - need to be addressed before the widespread use of mobile RFID. Model of mobile RFID service defines additional three entities and two relationships compared to that defined in RFID tag, RFID access network, RFID reader, relation between RFID tag and RFID reader, relation between RFID reader and application server.

Generally, in mobile RFID application such as smart poster, application service provider (ASP) has the ownership of RFID tags. Thus, mobile RFID users have to subscribe to both the ASP for these kinds of RFID services and mobile network opera-tor for mobile communication service. Namely, there exist three potentially distrusted parties: user owned RFID reader, mobile network operator, and ASP. Accordingly, trust relationship among three parties must be established to realize secure mobile RFID service. Especially, when a RFID reader tries to read or change RFID service data stored in tag, the reader needs to get a tag access rights. Additionally, it is important that new tag access rights whenever some readers access a same tag must be different from the already accessed old one.

3.2   Privacy Preserving Enhanced System and Service Mechanism

Widespread deployment of RFID technology may create new threats to privacy due to the automated tracking capability. Especially, in the mobile RFID environment, privacy problem is more serious since RFID reader is contained in handheld device and many application services are based on B2C model. The RPS (RFID user Privacy management Service) provides mobile RFID users with information privacy protection service for personalized tag under mobile RFID environment [4,5]. When a mo-bile RFID user possesses an RFID tagged product, RPS enables the owner to control his backend information connected with the tag such as product information, distribution info, owner's personal information and so on. The proposed network architecture of mobile RFID service is shown in figure 1. Main features of this service mechanism are owner's privacy protection policy establishment and management, access control for information associated with personalized tag by owner's privacy policy, obligation result notification service, and privacy audit service by audit log management. The brief personal privacy protection process using above functions of RPS is as follows.

1. Information of the service system consists of Privacy Reference List (PRL) and Privacy Reference Profile (PRP) as follows:

   - PRL (Privacy Reference List): Information item list (like personal information, product information, distribution information) treated by each service field (like finance, trade, etc.)
   - PRP: Privacy level allotted profile by PRL items, which is decided through its effect estimation by authority concerned (e.g., financial profile, trade profile, medical care profile, etc.)

2. The RPS registration and the generation of default privacy reference profile (DPRP) for company are as follows:

   - Each company may register RPS on RPS registration web site and, upon its registration, has to provide RPS with its OIS (Object Information Service) schema.
   - RPS generates DPRP by matching a privacy level to each attribute of schema of OIS which means an item list treated by company included in a certain service group.

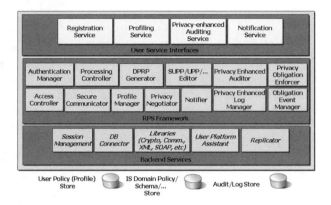

Fig. 1. RPS System Architecture for Mobile RFID Service with Privacy

3. Owner's profile in the RPS system is generated as follows:

- Get access to RPS web server through cellular phone to set an access group profile for generating owner profile and allowing service information access by category groups.
- Each profile is generated in XML (extensible markup language) form by RPS and sent to every service provider.
- When owner sets privacy level (L1-L9) for each item in the service group, OPP (Owner-defined Privacy Policy for information) will be generated, which in turn results in individual OPRP (Owner-defined Privacy Reference Profile) for service providers according to their OIS schema included in the service group.
- Owner is needed to set a single profile input for each service (it is intended to re-duce any burden of owner).
- As owner registers cellular phone numbers of those, who are allowed access, for every access approval level (L1-L5), on the basis of which RPS will generate a security token indicating the approval of access and then PAG (Profile for Access Group) including this token, and send it to every service provider.

4. The following are the mechanism order in RPS applied service network:

- Send purchase events through various ways [for purchase via POS (point of sales management) system, his/her owner information (cellular phone number) is sent to OIS].
- POS sends the confirmation of purchase to OIS, which in turn changes information in connection with the information on user.
- OIS asks RPS its user privacy policy. And, RPS announces a privacy policy and suggests level setting GUI to terminals.
- OPP and PAG are set at user's terminal and sent to RPS.
- RPS sends a protection policy suitable for OIS.

- OIS and user's terminal reflect the protection policy to tag connection information and receive only information satisfying the policy.

4 Conclusion

The mobile RFID technology is being actively researched and developed throughout the world and more efforts are made for the development of related service technologies. Though legal and institutional systems endeavor to protect privacy and encourage protection technologies for the facilitation of services, the science and engineering world also has to develop proper technologies. Seemingly, there are and will be no perfect security / privacy protection technology. Technologies proposed in this paper, however, would contribute to the development of secure and reliable network RFID circumstances and the promotion of the mobile RFID market.

References

1. Jongsuk Chae, Sewon Oh: Information Report on Mobile RFID in Korea. ISO/IEC JTC 1/SC 31/WG 4 N 0922, Information paper, ISO/IEC JTC1 SC31 WG4 SG5 (2005)
2. M. Ohkubo, K. Suzuki and S. Kinoshita: Cryptographic Approach to "Privacy-Friendly" Tags. RFID Privacy Workshop (2003)
3. Wung Park, Byoungnam Lee: Proposal for participating in the Correspondence Group on RFID in ITU-T. Information Paper. ASTAP Forum (2004)
4. Namje Park, Jin Kwak, Seungjoo Kim, Dongho Won, and Howon Kim: WIPI Mobile Plat-form with Secure Service for Mobile RFID Network Environment. Lecture Notes in Com-puter Science, Vol.3842. Springer-Verlag (2006) 741-748
5. Namje Park, Seungjoo Kim, Dongho Won, and Howon Kim: Security Analysis and Imple-mentation leveraging Globally Networked Mobile RFIDs. Lecture Notes in Computer Sci-ence, Vol.4217. Springer-Verlag (2006) 494-505
6. MRF Forum: WIPI Network APIs for Mobile RFID Services (2005)

Enabling Negotiation Between Agents and Semantic Web Services

Saira Parvez Khan Sana Ismaeel[1], H. Farooq Ahmad[2], Hiroki Suguri[2],
and Muhammad Akbar Asim Elahi[3]

[1] National University of Sciences and Technology College of Signals, Pakistan
`sairaparvez_se@yahoo.com`
[2] Communication Technologies 2-15-28 Omachi, Aoba-ku, Sendai, 980-0804 Japan
`{farooq,suguri}@comtec.co.jp`
[3] National University of Sciences and Technology College of Signals, Pakistan
`asimelahi@mcs.edu.pk`

Summary. Emerging Web services standards enable the development of large-scale applications in open environments. In particular, they enable services to be dynamically discovered and invoked. Our research objective is to propose new paradigms for interactions among semantic web services and software agents for problem solving. Agents are autonomous entities capable of acting on behalf of their user while Semantic Web services offer a new potential of automation in e-Work and e-Commerce, where fully open and flexible cooperation can be achieved on-the-fly. We believe that real goal of semantic web can only be realized if the semantic content associated with it can be read and interpreted. Since agent infrastructure has well-established reasoning, decision making and interaction mechanisms, it can contribute exceptionally in this regard. In this paper, we discuss the issue of negotiation between agents and semantic web services and propose architecture that enables negotiation between these heterogeneous entities.

1 Introduction

The aim of the web services (WS) endeavor is to obtain an environment where service customers and service providers can set (negotiate) the terms and conditions of service invocation automatically and then execute the necessary actions according to the prevailing contract. The semantic web adds machine-understandable semantics to data, thus enabling processing on behalf of the human user. Although the new possibilities promised by emerging technologies seem attractive, the Semantic Web with its tools and related technologies like OWL, WSDL, UDDI, SOAP and WS are likely to fall short of realizing an automated interaction and negotiation mechanism [Jennings, 2001]. Many challenges as stated in [20] lie ahead. That is to say, although the Semantic Web promises to make available to programs the meaning of the content of Web pages, these entities alone will not be able to make decisions, interact, and cooperate with other entities. Agent infrastructure, on the other hand has much to offer in this regard. An agent possesses the ability to comprehend and interact with its environment. Because of being context-aware, autonomous and able to interpret

K.M. Węgrzyn-Wolska and P.S. Szczepaniak (Eds.): Adv. in Intel. Web, ASC 43, pp. 284–291, 2007.
springerlink.com   © Springer-Verlag Berlin Heidelberg 2007

semantics with the help of ontological knowledge representation, agents are a necessary complement to web services to realize the vision of semantic web. Several arguments have been made to support the idea of integration of WS and agent infrastructure, including [13,14,20] but perhaps none more evocative than statements made in [21] which clearly expresses the notion that, "software agents are the running programs that drive WS - both to implement them and to access them as computational resources that act on behalf of a person or organization". To enable this integration, several core issues are there out of which bidirectional service discovery, service invocation and negotiation are the most pertinent. Our goal is to take the flexible interaction schemes from the Multi- Agent Systems (MAS) research, and utilize them to enable negotiation among semantic WS, a paradigm that supports rigid and mechanical interaction protocols, and agent infrastructure. In this paper, we propose an abstract architecture for conducting such negotiations. Fig 1 illustrates our vision of an autonomous, flexible and interactive environment.

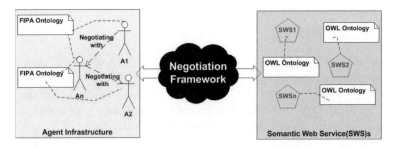

Fig. 1. Block diagrammatic view of an autonomous, flexible and interactive environment

The rest of the paper is organized as follows: Section 2 briefly defines the negotiation process in the light of literature. Role of ontologies in interaction among agents as well as WS has been highlighted in section 3. Section 4 contains a thorough literature review concerning some highly significant issues regarding interoperability of the two paradigms, particularly negotiation, conversation and interaction patterns among the two entities. Section 5 gives the details of the proposed architecture along with UML sequence diagrams for clarity. Section 6 highlights intended future work.

2   Negotiation Process

Negotiation is an iterative communication and decision making process between two or more agents (parties or their representatives) [4] who: (i) cannot achieve their objectives through unilateral actions; (ii) exchange information comprising offers, counter-offers and arguments; (iii) deal with interdependent tasks; and (iv) search for a consensus which is a compromise decision. In this paper, we focus on introducing intelligent interaction patterns between agents and WS aimed at utilizing the semantics associated with the involved parties. In order to achieve this objective, the existence of a common vocabulary (i.e. ontology) is inevitable.

3 Role of Ontologies in Interaction

3.1 Multi Agent Systems

As autonomous problem solvers, agents need to develop model of their environment that allows them to reason on how their actions affect their environment and how those changes lead them to achieve their goals [5]. Ontologies provide the conceptual framework that allows agents to construct such models: ontologies describe the properties of entities that agents encounter, and relations between them. Thus a common vocabulary in the form of ontologies is at the heart of intelligent communication among agents.

3.2 Semantic WS

The semantic web initiative [20, 21] that addresses the problem of XML's lack of semantics by creating a set of XML based languages, also relies on ontologies that explicitly specify the content of the tags. The Web Ontology Language (OWL) is a forthcoming W3C specification for such a language which will supersede the earlier DARPA Agent Markup language (DAML+OIL) [22]. OWL is an extension to XML and the Resource Description Framework (RDF) enabling the creation of ontologies for any domain and the instantiation of these ontologies in the description of resources. The OWL-Services language (OWL-S) [19] is a set of language features arranged in these ontologies to establish a framework within which the WS may be described in the semantic web context.

4 Literature Review

Keeping in mind our long- term objectives, we have conducted a thorough study of the capabilities of OWL-S and the potential of semantic web services [17,18]. With OWL-S markup of services, the information necessary for WS discovery could be specified as computer-interpretable semantic markup at the service Web sites, and a service registry or ontology-enhanced search engine could be used to locate the services automatically [7,15]. Execution of a Web service can be thought of as a collection of remote procedure calls. OWL-S markup of WS provides a declarative, computer-interpretable API that enables automated WS execution [7,11,19]. Given a high-level description of the task by the user, automated composition and interoperation of WS to perform the task is of particular interest to us. With OWL-S, the information necessary to select and compose services would be encoded at the service Web sites [3]. Software agents can be written to manipulate and interpret this markup, together with a specification of the task and thus can be bestowed with the ability to perform the task automatically [7,9,10,11]. [6] combines two recent WS languages, WS-Conversation Language (WSCL) and WS-Agreement to implement Contract Net Protocol (CNP) for negotiation among WS. Yet, the flexibility of negotiation is far-off from that prevalent in the agent infrastructure. FIPA [23] provides detail specifications of Request protocol, Request/ Response protocol, CNP, English Auction, Dutch Auction, Brokering protocol, etc. An exhaustive overview of MAS is beyond the scope of this paper, but essential pointers include [16] and [12]. [7] presents two significant alternatives in empowering WS with agents' properties. One is to implement a wrapper, which turns a current Web service into an agent-like entity. The other alternative is to capture all the functionalities of a Web Service and imbed them into an existing software agent. [2] proposes an architectural model for enabling transparent, automatic connectivity between WS and agent services. A later version of this project [1] uses OWL-S to add semantic aspect to service descriptions. [8] illustrates use of OWL as a content language for ACL messages in MAS and conducts auctions among agents that use OWL ontologies. In this paper, we contend that the degree of flexibility that persists in agent interaction scenarios can never be achieved in negotiations among WS alone. In order to improve flexibility level, negotiation among semantic WS and agents is highly significant.

5   Proposed Architecture

5.1   Assumptions

The following assumptions have been made when designing the proposed architecture: (i) All agents are assumed to be FIPA compliant and capable of communicating with FIPA-ACL encoded messages. (ii) All WS operate using the standard WS stack consisting of WSDL for service descriptions, SOAP for message encoding and UDDI for directory services. (iii) Each web service makes use of OWL-S and exposes its domain ontologies written in OWL, a machine readable form. (iv) The decision engine and a mechanism of web service discovery and invocation from agent infrastructure is assumed to exist. (v) The FIPA specification for CNP as given in fig 2 is followed.

5.2   Major Modules

In order to provide a proof of concept of negotiation between agents and semantic web services, we choose one of the FIPA standard interaction protocols for the agent infrastructure i.e. CNP [23].

According to our proposed architecture as shown in fig 3, an entity called "Ontology Gateway" (OG) acts as a broker to conduct the negotiation between the two heterogeneous entities. Development of this gateway is underway at NUST- Comtec lab. It is an application (neither an agent nor a web service) that operates in a distributed environment. In its preliminary stage, the gateway currently has "Bidirectional OWL-FIPA Translator". We propose another module

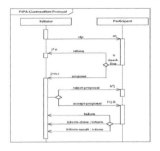

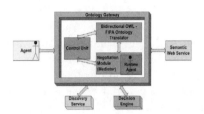

Fig. 2. FIPA Contract Net
Interaction Protocol

Fig. 3. Proposed architecture of
the Ontology Gateway

called "Negotiation Module" (NM) in the OG which supervises the sequence
of messages as they are being exchanged. This would ensure that pre-agreed
protocol is followed and whole process appear transparent to the parties involved
in the negotiation. There are several prerequisite tasks that need to be done
before carrying out the standard CNP. In our architecture, we propose a "Control
Unit" (CU) for handling these initial tasks. Once these have taken place, the
control is shifted to the NM within the OG which then carries out the negotiation
protocol by communicating messages in a sequence. Current implementation of
the Gateway consists only of the bidirectional OWL- FIPA ontology translator.
CU has been added so as to supervise the flow of messages from input to internal
architecture of the gateway as well as of messages from internal architecture to
the output.

This paper discusses only the scenario where an agent asks to conduct negotiation with a semantic web service and not vice versa. Fig3 explains the proposed
architecture of an OG in detail while showing its relation with outside world,
too.

5.3   Detailed Role of Each Module in Ontology Gateway

Control Unit. First of all, the requesting agent who wants to initiate negotiation invokes the service exposed by OG and sends its reference and message. This
message is received by the CU within the OG. The message content contains the
description of the service based on which CU identifies other negotiating party
(i.e. a semantic web service). This is done with the help of a matchmaking service that returns a reference or handle of that service to the CU. This handle
enables the CU fetch the service profile of the service and its ontology, without
which semantic understanding is unattainable.

Bi-directional OWL- FIPA Ontology Translator. Since this ontology is
written in OWL for a semantic web service, which though is allowed as a valid
content language by FIPA but is not as expressive as SL, so there is a need to
translate this ontology from OWL to SL. The CU feeds this ontology to the

OWL to FIPA ontology translator, which returns the FIPA Ontology equivalent of the OWL ontology fed as an input.

Negotiation Module. In order to conduct meaningful negotiation between agent and semantic WS, this module, also known as Mediator, needs the reference of the requesting agent, its FIPA ontology and service profile of the semantic web service. Handles to all of these resources are passed to this module at the time of transfer of control to it by CU. This module shall create an agent at runtime referred to as RuntimeAgent (RA) in the fig 4, which shall be used to query the web service ontology. This agent extracts all negotiable parameters from the profile of the semantic web service with the help of understanding its ontology.

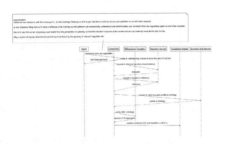

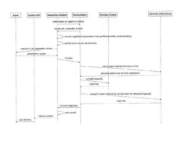

Fig. 4. Sequence diagram listing sequence of message exchanges prior to initiating CNP

Fig. 5. Sequence diagram listing sequence of message exchanges while implementing CNP

5.4   Flow of Control During CNP

Once the prerequisites for the implementation of negotiation protocol are set by the CU, it is now time for the NM to initiate the negotiation process according to a formalized protocol (CNP, in this case). Fig 4 illustrates the flow of control and request- reply pairs involved in CNP as a sequence diagram. The RA acts as Initiator of the protocol and sends a call for proposal to the semantic web service which assumes the role of Participant of CNP. For sending such a request (i.e. a cfp message), NM needs user preferences too. These user preferences can be retrieved with the help of the requesting agent reference that has been passed to NM by the CU.

Along with getting these parameters from the requesting agent, NM needs to have agent's ontology too so as to get a semantic understanding of what these parameters mean based upon a common vocabulary. It can also be accessed with the help of agent's reference that this module has. Such an understanding of semantics will help NM map the information that the requesting agent possesses to the information required by the service method that is going to be invoked as a result of sending a cfp. As a response of this, the participant would return all possible options (proposals) to the RA. Each of this response is the handle of a semantic web service that closely matches the requested service description. These responses are passed to a decision engine. We assume here that the decision engine is an independent component that uses artificial intelligence and semantic deduction rules to choose the best possible option out of many as per the closest match with user preferences. The decision engine sends the chosen option back to RA which invokes the corresponding service. The service is executed as a result of this invocation and a response is sent to the RA indicating whether the service has succeeded or failed. RA forwards this response to NM which stores the results in its knowledge base and returns control to CU. Finally, the requesting agent is informed of the results of negotiation along with all associated details by CU and the negotiation session comes to an end.

6   Future Work

This paper proposes gateway architecture for enabling flexible, autonomous interaction between Semantic WS and agent services. A prototype implementation of this architecture is under development at NUST- Comtec lab. It is intended to improve the proposed design so as to cater more negotiation protocols esp., the auction protocols in future. We expect that this initial effort of conducting negotiation via a Gateway service bridging agents and WS is only a prelude to exploring the immense potential it offers as a means to compose, invoke, and manage heterogeneous service populations.

References

1. Greenwood, D., Calisti, M., Nagy, J.: Semantic Enhancement of a Web Service Integration Gateway, AAMAS 2005 workshop on Service Oriented Computing and Agent Based Engineering (SOCABE) Workshop, Utrecht, Netherlands
2. Greenwood, D., Calisti, M.: An Automatic, Bi-Directional Service Integration Gateway, IEEE Systems, Cybernetics and Man Conference; 10-13 October, 2004, the Hague, Netherlands
3. Laukkanen, M., Helin, H.: Composing workflows of semantic web services. In Proc. of the 1st International Workshop on Web Services and Agent Based Engineering, Sydney, Australia, July 2003.
4. Bichler, M., Kersten, G., Strecker, S.: Towards a Structured Design of Electronic Negotiations. Group Decision and Negotiation, 2003.
5. Sycara, K., Paolucci, M.: Ontologies in Agent Infrastructure. Carnegie Mellon University, USA
6. Paurobally, S., Jennings, N. R. (2005): Protocol engineering for web services conversations. Journal of Engineering Applications of Artificial Intelligence. Special Issue on Agent Oriented Software Engineering, 18(2), March 2005.
7. Tadiou, K. M.: Semantic Web services Interaction
8. Zou, Y., Finin, T., Ding, L., Chen, H.: TAGA: Using Semantic Web Technologies in Multi-Agent Systems, IJCAI-2003
9. Sirin, Hendler, J., Parsia, B.: Semi-automatic Composition of Web Services using Semantic descriptions. 1st Workshop on Web Services: Modeling, Architecture and Infrastructure in conjunction with ICEIS 2003.

10. Laukkanen, M., Helin, H.: Composing workflows of semantic web services. In Proc. of the 1st International Workshop on Web Services and Agent Based Engineering, Sydney, Australia, July 2003.
11. Automated discovery, interaction and composition of semantic web services
12. M. Luck, P. McBurney, C. Preist. Agent Technology: Enabling Next Generation Computing. http://www.agentlink.org/roadmap, 2003.
13. Lyell, M., Rosen, L., Casagni-Simkins, L., Norris, D.: On software agents and web services: Usage and design concepts and issues. In Proc. of the 1st International Workshop on Web Services and Agent Based Engineering, Sydney, Australia, July 2003.
14. Maximilien, E., M., Singh, M., P.: Agent-based architecture for autonomic web service selection. In Proc. of the 1st International Workshop on Web Services and Agent Based Engineering, Sydney, Australia, July 2003.
15. Sycara, K., Paolucci, M., Soundary, J., Srinivasan, N.: Dynamic discovery and coordination of agent-based semantic web services. IEEE Internet Computing, 8(3): 66-73, May 2004.
16. Wooldridge, M.: An Introduction to Multi- Agent System. Wiley and Sons, 2002.
17. McIlraith, S., Son, T., C., Zeng, H.: Semantic Web Service. IEEE Intelligent Systems, 16(2):46-53, 2001.
18. Martin, D., Burstein, M., Lassila, O., Paolucci, M., Payne, T., McIlraith, S.: Describing Web Services using OWL-S and WSDL. http://www.daml.org/services/owl-s/1.1/owl-s-wsdl.html, November 2004.
19. OWL-S Home Page. http://www.daml.org/services/owl-s/, 2003.
20. Berners-Lee, T., Hendler, J., Lassila, O.: The semantic web. Scientific American, 284(5):34-43, 2001.
21. W3C. The Semantic Web http://www.w3.org/2001/sw, 2004.
22. DAML Joint Committee. DAML+OIL (March 2001)Language. http://www.daml.org/2001/03/daml+oil-index.html, 2001.
23. Foundation for Intelligent Physical Agents. FIPA Communicative Act Library Specification. http://www.fipa.org/specs/fipa00037/, June 2002.

Web-Training Resources Construction Principles and Their Interactive Links with Content Distribution Platforms

Jan Piecha[1,2]

[1] Department of Computer Systems, University of Silesia
piecha@us.edu.pl
[2] Department of Transport Informatics, Silesian University of Technology
jan.piecha@polsl.pl

Summary. The paper introduces several aspects of electronic content characteristics with their emphasis into Distance Learning technologies. They all have to be taken under consideration before the application development process starts. They also have to be defined in case the developer wants to run the on-line mode, of distance learning environment. To fulfil the given goals several items of management shell were introduced. The elaborated platform (Multimedia Applications Management Shell - MAMS), for applications development was introduced as well. Both platforms are simplifying the application structure and the unit control processes. The MAMS is supported by Quality Repetitions Unit (QRU) controlling the application progress and application repetition; working in accordance with Kay's [1], [2] interactive model. Finally they both produce the application status, transmitted into the user's personal data record.

1 Introduction

The e-learning units have to work in accordance with well known principles of Programmable Teaching (PT), formulated many years ago, described by Burke in 1982 [1]. The computer courses have to run in an individual schedule of the user's needs, both in the presentation content and in duration of the lesson. The pioneer works for Computer Aided Learning (CAL) systems were defined in several frame assumptions of CAL resources structure (Eberts, 1986 [1] and Piecha, 1989, 1991 [3], [4]); where one can find fundamental solutions for today's e-content units. For many applications an algorithmic way of problems solving is recommended. This multiple choice problems solution makes the computer encouraging for e-learning technology implementation. Many works has recently been done for Internet services development; among them databases organisation, access laws and restrictions, content protection and content distribution principles. Although many efforts have been undertaken, in this area, new challenges one can still observe [5], [6], [7]. Complexity of database content implies an interface structure that makes possible to go through the data via sequence of questions, to extract a specified data part. The author of the paper was trying to draw the complex aspects of e-courses development, with smart validation

K.M. Węgrzyn-Wolska and P.S. Szczepaniak (Eds.): Adv. in Intel. Web, ASC 43, pp. 292–297, 2007.
springerlink.com

method, needed for wide area network services. The distance learning applications developer has to consider many factors of his works, before the computer lesson will be assigned as a satisfactory product. The e-content units have to integrate many interdisciplinary rules to achieve the final result.

2   Some E-Content Characteristics

The computer aided learning products development regulations, allow us making the computer network training very flexible and remarkable smart. For these frame regulations several characteristic features can be distinguished. They define the lesson main goals, by which and what for the lesson is elaborated. Several physical characteristic features were also defined; being pattern solutions for the courseware. They determine the presentation approach into subjects, in accordance with a paper-pencil traditional screenplay. The most often used linear structure of the application is not satisfying the machine algorithmic computing abilities; with good guiding through the databases located at remote servers of the network. Interactivity of the application brings the control unit with various conditions, establishing several possible paths of the lesson selection. These interactivities have to be strictly combined with the questions validation system, controlling the application into the user individual needs. The literature brings several models of interactive lessons structure introduced by Kay's strategy [3]. It produces complex and flexible structure of programmable teaching unit, with several distinctly different levels of available courseware. The e-content is presented on the computer screen in portions, of successive information layers. The data screen, was called a frame; the smallest data unit of the course assigned with an identification index - j (Fig.1.) working as branching switch of the unit.

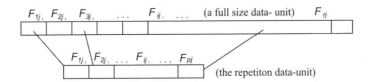

Fig. 1. The data frames of the lesson L_j^k relationship

The highest level of the application track is the shortest one, with modest set of explanations and complex questions. The lowest track of the application is the longest one, with many comments and interactions. Between these boundaries many additional lesson levels (at least one) can be considered.

3   The Application Development Platforms Characteristics

The lesson structure has to allow going through the application under the user knowledge current level estimation. To fulfil this assumption the Multimedia

Applications Management Shell (MAMS), for the application development simplification, was elaborated. The MAMS platform is co-working with additional product: Quality Repetitions Unit (QRU), completing all interactions and calculating the results of interactions. They were provided as applications support for Macromedia - Authorware and Macromedia - Flash environment (Fig. 2.):

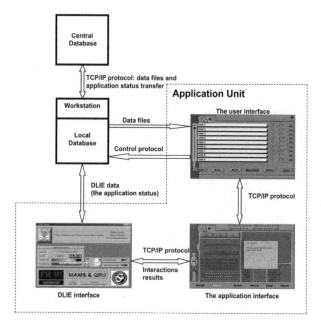

Fig. 2. The distance learning system interfaces and organisation

- DLIE Server, is the Distance Learning Interactions Evaluation machine for QRU unit,
- DLIE library, is a set of functions for client usage that link servers of the network nodes,
- API for DLIE, contains the Application Programs Interface for DLIE server.

Complexity of e-content programming makes troubles with satisfactory courses development. The MAMS shell drives several development factors, as:

- unifying the application structure and development,
- providing interactions quality judgment engine; increasing the interactions quality analysis,
- controlling the application repetition.

The MAMS & QRU platforms were integrated into one structure, made as a client/server application tool for the application frames coordination (indicated in Fig. 2). The client role plays the MAMS presentation platform and the server role plays the QRU - responsible for interactions evaluation. For all data communications the standard TCP/IP packets protocols or http protocols were applied.

DLIE functions

Well done screenplay provides comfortable mechanisms for decision making solutions; in accordance with the user's actions. They provide the user with:

- nonlinear repetition mode, including presentation and question frames of the application,
- multi-level strategy implementation [6], [7] with flexible interfaces through the application.

The application start-up uses default screenplay settings, as a main path of the application execution. After the application is completed the QRU unit judges the interactions results, for controlling the application repetition content.

The nonlinear screenplay

The QRU allows defining the repetition cycle in a flexible way, where presentation frames are put into the selected sequence. What is more every answer can be provided, by its own weight, with limited number of repetitions. The answers are analysed in hierarchical way, with its fill-up format interactions. The judging unit distinguishes subsequent fields in the answer protocol. Each part of the answer is described by an adequate record, assigned by its unique identifier and its local weight. Instead of binary value of the answer (usually applied in well known units) multi-valued measures have been provided. The DLIE server analyses the answers using various measures and algorithms combining the set of answers into a final statement. The QRU engine is used for frames selection to the application repetition as the user interactions results control the data selection process.

Let us consider that the database consists of files-set containing courses (C^k) with several lessons L_j^k in each course.

$$C^k = \bigcup_{j=1}^{m} L_j^k, \quad \text{where}: \ k \in K, K = \{1, 2, ..., m\}, j \in J, J = \{1, 2, ...n\} \quad (1)$$

The ability of an individual path selection, within the application, means that several branches in the data file are available. Each lesson L_j^k of the course k consists of selected frames F_{ij}^k linked into a sequence by the MAMS shell. The i index denotes the frame-number joined into the lesson L_j^k of the course C^k; where j denotes current number of the lesson.

$$L_j^k = \bigcup_{i=1}^{r} F_{ij}^k, \quad \text{where}: \ i \in I, I = \{1, 2, ...r\} \quad (2)$$

This way, the user chooses an own route through the main presentation and repetition sessions. The mentioned above data frames selection controls a question system (QRU) of the application. The assumptions to the database composition are related to programming platforms used for the application development. For simplifying the developOCment process, several programming technologies have been involved into.

4   The Courses Management and Network Security

The MAMS engine produces the application current status. A content distribution platform USE-LMS (University of Silesia - Learning Management System), controls the e-content distribution processes under various restrictions for the registered users. MAMS and USE-LMS are provided with several installation restrictions protecting the training system from applications distribution into eligible users. They are installed on several machines of the computer network, cooperating with learning process management (Fig. 3.). The main machine ŞManagerŢ contains the system main database with a main LMS. The local server (a second layer) works as a local sub-distributing machine (local LMS) of content provided by the main Manager. The local managers are responsible for legal distribution of the content into their node users. The applications are running

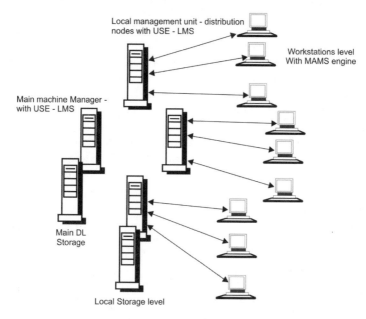

Fig. 3. The Distance Learning computer node management relationship

when the user's workstation is provided with MAMS platform delivered by the local server; as a legal distribution condition of MAMS platforms. The main machine supervisor is able to control the users' legal installations (the license). He is able to interrupt any illegal applications execution. What is more, dismount the installation at the sub machine. The presentation units, developed under Macromedia Authorware and Macromedia Flash or any presentation generators have to be linked into logical information sequences, by the XML manger, added by the application developer. After the application localisation was defined the XML: checks the description code, defines the belonging sub-chapter to the application - bounding the application sequence: chapters and sub-chapters selection.

5   Conclusion

The discussed above solutions show a brief description of recent efforts that have been undertaken at the University of Silesia, Distance Learning Technologies Centre and Department of Computer Systems, in a field of e-learning and distance learning technologies development. The introduced MAMS platform is responsible for smooth controlling of the application. It is also a necessary frame that has to be installed before the application starts running. The elaborated mechanisms protect the network database from its content illegal copying. Every installation has to be confirmed by training system operator, otherwise the application will refuse working. The e-content frames with not satisfying results of interactions are assigned for repetition; defined number of times. The distribution platform (USE-LMS) was provided with handshaking protocols transferring, into the user personal record, many status factors of the application interactions; controlling various conditions of these platforms co-operation. This way several steps, towards the modern learning and teaching system, in Internet environment, were successfully made.

References

1. Eberts R.E. (1986) Learning strategies in CAI design. The International Journal of Applied Engineering Education, no. 2/86, Oxford
2. Piecha J. (1989) The multilevel model for microcomputer ICAI systems. The International Journal of Applied Engineering Education, vol. 5, no. 3/89, Pergammon Press
3. Piecha J. (1991) Remarks to CAL systems designs. Proceedings of the Int. Workshop Computer Aided Learning and Simulation Technologies. Prague, pp.26 - 31
4. Piecha J. (1999) The programmable shell for multimedia applications development. Journal of Applied Computer Science, vol.7, no. 2, pp.31-43, ISSN 1507-0360, Lodz
5. Piecha J. (1999) The Intranet Databases and some Approach Troubles into Multimedia Files. Proc. Int. Conf. Computer Based Learning In Science - CBLIS'99, Enschede, the Netherlands
6. Piecha J., Krol R., Pawelczyk P. (2001). A network node management shell for macromedia applications. Proc. of Conf. KOSYR 2001, ISBN 83-911675-2-6, pp. 493-499
7. Krol R., Piecha J. (2000). The access method to Internet databases developed within the macromedia environment. Journal of Medical Informatics and Technologies. vol.5, Nov. 2000, ISBN 83-909518-2-7, pp. IT 91-98
8. Piecha J., Krol R. (2004). The MAMS an interactive applications management engine. Proc. of Int. Distance Learning Workshop' 04. June 2004, Katowice, Poland ISBN 83-909518-6-X pp. 25-37

On Sequential Bias in Web-Based Reviews

Selwyn Piramuthu

Decision and Information Sciences, University of Florida
Gainesville, Florida 32611-7169, USA
selwyn@ufl.edu

1 Introduction

It is widely recognized that online product reviews by users in sites such as
amazon.com are biased in several dimensions including those due to the sequence
in which the reviews are written as well as the benefit (monetary or otherwise)
that the reviewer gains from writing a review. Although the latter is obvious and
is often practiced by stakeholders wanting to promote their products (e.g., [2],
[3], [4], [7]), the former is not entirely innocuous since subsequent reviews are
not completely independent of preceding reviews by others.

To add to the complexity of.the dynamics at play in this context, the role played
by first impression bias cannot be overestimated (e.g., [1]). Thus, the review that
is first seen by a prospective customer of the product of interest plays a significant role in purchase decisions that follow. These reviews are quite influential
since prospective purchasers of reviewed products rely heavily on these reviews in
making their purchase decisions (e.g., [6]). Piramuthu ([5]) provides an analytical
study of the dynamics of first impression bias and the order sequence of reviews.
In this study, we focus only on the order sequence of reviews and illustrate the
dynamic using an example and analyze the same using induced decision tree.

The sequence in which reviews are written play an appreciable role in how the
reviews that follow later in the sequence are written. For example if a reviewer is
favorable to the product reviewed, she might be biased to write stronger reviews
based on existing negative reviews and *vice versa*. We study this dynamic using
a small synthetic example data set with simplifying assumptions to illustrate
the effects of sequential bias in online reviews. We use the decision tree (J48)
classification method in WEKA ([8]) as a tool to learn the concept of interest.
Preliminary results show that the effects of sequential bias cannot be ignored.

We describe the data used in this study in the following section. This is followed by a brief description of the generated decision tree in Section 3. We discuss
the results and provide possible implications in Section 4.

2 Data Generation

We consider a product that has several attributes. A reviewer of this product
rates several of these attributes before providing an overall recommendation. We
use the following notations:

K.M. Węgrzyn-Wolska and P.S. Szczepaniak (Eds.): Adv. in Intel. Web, ASC 43, pp. 298–303, 2007.
springerlink.com © Springer-Verlag Berlin Heidelberg 2007

i, j - reviewers and attributes of the product reviewed ($i = 1..m; j = 1..n$)

x_{ij} - score derived from reviewer i's review of the jth attribute

$x_{kj}^{+}, (x_{kj}^{-})$ - positive (or negative) score from k's review of jth attribute

y_i - reviewer y_i's overall recommendation of this product.

We calculate the un-biased score for reviewer i's review of the jth attribute as the sum of the score given by reviewer and the bias due to conflict between the reviewer's overall recommendation and previous reviewers' scores as follows:

$$
(\text{unbiased})x_{ij} = \begin{cases}
x_{ij} + \dfrac{\sum\limits_{k=1}^{i-1} x_{kj}^{-}}{i-1} & \text{if } y_i > 0 \\[2em]
x_{ij} + \dfrac{\sum\limits_{k=1}^{i-1} x_{kj}^{+}}{i-1} & \text{if } y_i < 0 \\[2em]
x_{ij} & \text{otherwise}
\end{cases}
$$

Here, the summation of selected scores (only the negative ones from previous reviews if the current overall recommendation is positive and *vice versa*) measures the number of previous conflicting reviews in its denominator and the magnitude of the scores from these reviews in the numerator.

To operationalize this, we use four attributes for the product of interest. For each attribute, the reviewer's score is assigned values -1, 0, and +1 if the review is negative, neutral, or positive respectively. We also consider cases where the review is very positive and very negative, and represent these by +1.25 and -1.25 respectively. We generated a small data set of 100 examples based on these constraints from a Uniform distribution [0,1] with equal probabilities of the score being positive, neutral, and negative. To capture the 'very positive' and 'very negative' cases, we randomly chose 10% of the cases and multiplied the scores by 1.25.

For y-values, we assign a *1* if the sum of the x values is greater than 2, a *-1* if this sum is less than 2, and *0* otherwise. We generated the y-values in this fashion for the data set before removal of sequential bias and kept the same y-values for the other data set. The data set thus randomly generated had 56% positive and 44% negative examples.

Figure 1 shows the first five examples thus generated, and Figure 2 shows the same first five examples after removal of sequential bias. For example, the third line in Figure 2 is generated using information from lines 1-3 in Figure 1 as follows: since the y value is 1, we consider only the negative values in the previous reviews (i.e., lines 1 and 2 in Figure 1). The unbiased x_{3j} values are $x_{31} = 1 + 0 = 1$; $x_{32} = 1.25 + (-1/2) = 0.75$; $x_{33} = 0 + (-1.25/2) = -0.625$; $x_{34} = 1.25 + 0 = 1.25$.

Figures 3 and 4 provide the descriptive statistics for these data sets, corresponding to Figures 1 and 2 respectively.

x_1	x_2	x_3	x_4	class
0	0	-1.25	0	-1
1	-1	0	1	1
1	1.25	0	1.25	1
-1	1.25	-1	-1	-1
0	0	-1	0	-1

Fig. 1. Example (first five) user input data

x_1	x_2	x_3	x_4	class
0	0	-1.25	0	-1
1	-1	-1.25	1	1
1	0.75	-0.625	1.25	1
-0.6667	1.6667	-1	-0.25	-1
-0.25	0.625	-1	0.5625	-1

Fig. 2. Example (first five) user input data *after* removal of sequential bias

	Min.	Max.	Avg.	Std. Dev.
x_1	-1.25	1.25	-0.105	0.884
x_2	-1.25	1.25	0.053	0.891
x_3	-1.25	1.25	0.045	0.855
x_4	-1.25	1.25	-0.018	0.819

Fig. 3. Descriptive statistics of user input data *before* removal of sequential bias

	Min.	Max.	Avg.	Std. Dev.
x_1	-1.679	1.564	-0.169	0.851
x_2	-1.683	1.667	0.009	0.798
x_3	-1.554	1.667	-0.011	0.814
x_4	-1.541	1.5	-0.04	0.782

Fig. 4. Descriptive statistics of user input data *after* removal of sequential bias

3   Decision Trees

We used WEKA ([8]) to generate the decision trees using these data sets - the original data set with the users' reviews and the one after removal of sequential bias. The resulting decision trees are given in Figures 5 and 6 respectively. In addition to the trees themselves, the probability of reaching leaf nodes are provided next to the leaf nodes of the trees. These can be used to estimate the probability of a product being positively or negatively reviewed based on a subset of the attributes.

```
x3 <= -1
|   x2 <= 0: -1 (18.0)
|   x2 > 0
|   |   x2 <= 1: 1 (6.0/1.0)
|   |   x2 > 1: -1 (3.0/1.0)
x3 > -1
|   x1 <= -1
|   |   x2 <= 0: -1 (15.0/2.0)
|   |   x2 > 0
|   |   |   x1 <= -1.25
|   |   |   |   x3 <= 0: -1 (3.0)
|   |   |   |   x3 > 0: 1 (3.0)
|   |   |   x1 > -1.25: 1 (8.0)
|   x1 > -1
|   |   x4 <= 0
|   |   |   x2 <= -1
|   |   |   |   x1 <= 0: -1 (4.0)
|   |   |   |   x1 > 0: 1 (4.0/1.0)
|   |   |   x2 > -1: 1 (17.0/2.0)
|   |   x4 > 0: 1 (19.0)
```

```
Correctly Classified Instances        93              93    %
Incorrectly Classified Instances       7               7    %
```

```
TP Rate   FP Rate   Precision   Recall  F-Measure   Class
 0.909     0.054      0.93      0.909     0.92       -1
 0.946     0.091      0.93      0.946     0.938       1
```

```
=== Confusion Matrix ===

  a   b    <-- classified as
 40   4 |  a = -1
  3  53 |  b = 1
```

Fig. 5. Decision tree using raw input data (before removal of sequential bias)

These tables also include the confusion matrix with the percentages of cases that are correctly or incorrectly classified by the decision tree. Additional information include the true positive (TP Rate) and false positive (FP Rates) rates, precision, and recall. The true positive rate (or, sensitivity), is measured as the ratio between the number of true positive cases and the sum of true positive and false negative cases (i.e., $\frac{TP}{TP+FN}$). The false positive rate is measured as the ratio between the number of false positive cases and the sum of true negative and false negative cases (i.e., $\frac{FP}{TN+FN}$). Precision is measured as the ratio between the true positive cases and the sum of true positive and false positive cases

```
x3 <= 0.95652
|   x3 <= 0.432692
|   |   x3 <= -0.27976
|   |   |   x3 <= -0.53125
|   |   |   |   x3 <= -1.28767: 1 (6.0)
|   |   |   |   x3 > -1.28767: -1 (25.0/4.0)
|   |   |   x3 > -0.53125: 1 (23.0)
|   |   x3 > -0.27976: -1 (15.0/1.0)
|   x3 > 0.432692: 1 (22.0)
x3 > 0.95652: -1 (9.0)
```

```
Correctly Classified Instances          95          95      %
Incorrectly Classified Instances         5           5      %

TP Rate    FP Rate    Precision    Recall    F-Measure    Class
   1        0.089       0.898        1         0.946        -1
 0.911        0           1        0.911       0.953         1

=== Confusion Matrix ===

  a   b    <-- classified as
 44   0 |   a = -1
  5  51 |   b = 1
```

Fig. 6. Decision tree using data after removal of sequential bias

(i.e., $\frac{TP}{TP+FP}$) and recall is measured as the ratio between the true positive cases and the sum of true positive and false negative cases (i.e., $\frac{TP}{TP+FN}$).

4   Discussion

From Figures 5 and 6, it can be seen that the decision trees generated are not the same. The decision tree generated using the raw (before removal of sequential bias) data uses all four attributes for describing the concept of interest, whereas the decision tree generated using the data after removal of sequential bias uses only x_3. The main result of this paper is that the resulting recommendations can be significantly biased by sequential bias. Piramuthu ([5]) analytically shows that the significance of the difference depends on the severity of sequential bias. While the order in which these reviews are written do play a role in introducing this bias, there is some evidence that the review that occurs first plays a significant role in a prospective customer's perception about a product of interest (e.g., [5]). Although we only considered individual attributes when evaluating bias, overall recommendations (the y-values) can also spawn its own dynamic in subsequent reviews. A prospective customer should thus use caution while perusing these reviews for purchasing decisions.

References

1. G. Deffuent and S. Huet. "Collective Reinforcement of First Impression Bias," *First World Congress on Social Simulation*, Kyoto, 2006.
2. C. Dellarocas. "Strategic Manupulation of Internet Opinion Forums: Implications for Consumers and Firms," *Management Science*, 52(10), pp. 1577-1593, October 2006.
3. A. Harmon. "Amazon Glitch Unmasks War of Reviewers," *New York Times*, February 14, 2004.
4. D. Mayzlin. "Promotional Chat on the Internet." *Marketing Science*, 25(2), pp. 157-165, 2006.
5. S. Piramuthu. "Sequential and First Impression Bias in Online Product Reviews," working paper, Decision and Information Sciences Department, University of Florida, 2007.
6. S. Senecal and J. Nantel. "The Influence of Online Product Recommendations on Consumers' Online Choices," *Journal of Retailing*, 80, pp. 159-169, 2004.
7. E. White. "Chatting a Singer Up the Pop Charts," *The Wall Street Journal*, October 5, 1999.
8. I.H. Witten and E. Frank. *Data Mining: Practical Machine Learning Tools and Techniques*, (2nd edition), Morgan Kaufmann, San Francisco, 2005.

Constraint-Based Query Clustering

Carlos Ruiz[1], Ernestina Menasalvas[1], and Myra Spiliopoulou[2]

[1] Facultad de Informatica, Universidad Politecnica, Madrid, Spain
cruiz@cettico.fi.upm.es, emenasalvas@fi.upm.es
[2] Faculty of Computer Science, Otto-von-Guericke-Univ. Magdeburg, Germany
myra@iti.cs.uni-magdeburg.de

Summary. The analysis of query logs over a search engine provides valuable information on the user needs. Query clustering allows for the identification of similar queries and forms the basis for recommendations and query personalization. In this study, we propose constraint-based query clustering to enhance the query grouping process: Prior knowledge about the similarity of given queries is used to guide the formation of clusters, thus resulting in more homogeneous and reliable groups. We apply the constraint-based clustering algorithm C-DBSCAN [10] on query logs and show that constraints improve cluster quality.

1 Introduction

Search engines have become an indispensable tool for information acquisition from the Web or from corporate Intranets. Their popularity also implies high expectations on performance and quality of the results. One method used to enhance the experience of users with a search engine is the formulation of recommendations derived from similar queries. Query clustering [16] forms the algorithmic core for such approaches. As pointed out in [7], query clustering is a challenging problem, mainly due to the diversity of information being retrieved and of user interests to be served. In this study, we propose the improvement of query clustering through the exploitation of background knowledge in the form of *instance-level constraints* [12].

Constraint-based clustering exploits knowledge on the relationships among specific data instances/records: While a clustering algorithm treats all records equally, a constraint may point out that two given records *must* be assigned to the same cluster (or, conversely, they *must* be assigned to different clusters). For example, consider the queries Q_1, Q_2, Q_3, where Q_1 contains the keyword "Basel", Q_2 is on "risk management in banks" and Q_3 mentions "quality control" and "process". Although risk management and Basel (a city in Switzerland) are not related in general, one might know (e.g. by studying the results of the specific queries) that both users were interested in the Basel regulations on bank services, especially with respect to risk management. Basel regulations mention (among else) quality control and certification. However, inspection may reveal that those who launched Q_3 were interested in quality control for industrial processes. Thus we conclude that Q_1 and Q_2 must belong to the same cluster (a

K.M. Węgrzyn-Wolska and P.S. Szczepaniak (Eds.): Adv. in Intel. Web, ASC 43, pp. 304–309, 2007.
springerlink.com

Must-Link constraint between two data instances) and that Q_3 must be assigned to a different cluster than Q_1 (a *Cannot-Link* constraint). Of course, this knowledge is the result of a manual inspection that cannot be afforded for all data. However, research on constraint-based clustering [13, 3, 2] shows that even a small number of constraints can enhance the clustering of the whole dataset.

The incorporation of other forms of background knowledge to query clustering has been studied recently. Wen et al exploit contextual information [15] by checking whether users that issued different queries have clicked on the same documents and whether a set of documents is often retrieved for a given set of queries. In [17], real world events are detected from the Web by analyzing the context and the domain of the click-through data recorded of web search engines based on semantically similar queries but also have similar evolution pattern over time. Wedig and Madani present in [14] a large-scale study on logs of the Yahoo! search engine and elaborate on issues like the variation of users' topical interests and the types of insight that can be gained from query logs. Nettleton et al formulate hypotheses on user types [7]; they enhance the contents of the query sessions with meta-information and use these enriched data to contrast unsupervised and supervised methods for query clustering. However, constraints have not been studied in this context before.

The rest of the paper is organized as follows: Section 2 contains a brief description of C-DBSCAN [10], our constraint-based extension of the density-based DBSCAN [4]. Section 3 describes our constraint-based approach to query clustering and our experiments with query log data. The last section concludes with a summary and an outlook.

2   An Overview of C-DBSCAN

For constraint-based query clustering, we use C-DBSCAN [10], a constraint-based extension of the density-based clustering algorithm DBSCAN [4]. DBSCAN is a very powerful clustering algorithm, even for challenging datasets, but it has been shown to perform poorly if the data have certain properties (e.g. areas of different densities) [11]. In [10], we have shown that C-DBSCAN outperforms DBSCAN for such datasets.

DBSCAN has been recently used for web page and query clustering: In [6], pages accessed by the same users are grouped together using DBSCAN and the users' interests are taken into account for the assignment of each page to a topic/class. In [15], DBSCAN is used to cluster query contents; context information is taken into account, in particular whether users clicked on the same documents for different queries and whether a set of documents is often selected for a set of queries. However, prior knowledge in the form of constraints, as exemplified in section 1, is not considered.

DBSCAN [4] traverses the data and builds *neighbourhoods*. These are spherical dense regions around *core points*, i.e. data records that have at least *MinPts* proximal records within a radius *Eps*. Records in the same neighbourhood are *density-reachable* from each other and may become themselves core points – centers of further, overlapping neighbourhoods. Records in overlapping neighbourhoods are *density-connected*. A cluster is defined as a group of over-lapping neighbourhoods.

C-DBSCAN [10] extends DBSCAN: First, the data are organized into a KD-Tree [1]. Then, Cannot-Link constraints are enforced when building neighbour-hoods and when merging them iteratively towards the final clusters. Must-Link constraints cause the merging of clusters that are not necessarily density-connected. In more detail:

1. *Building Neighbourhoods in KD-Tree Nodes:* The KD-Tree algorithm parti-tions the data space iteratively into cubes, as long as the subcubes contain at least *MinPts* data records. Then, C-DBSCAN builds neighbourhoods for each leaf of the KD-Tree in turn. In the next step, only neighbourhoods within the same node are merged; they form "local clusters".

2. *Creating Local Clusters under Cannot-Link Constraints:* C-DBSCAN en-forces *Cannot-Link constraints* while building and merging neighbourhoods into local clusters. For each KD-Tree leaf node, C-DBSCAN first places each record involved in such a constraint to a singleton group. Then, it scans the remaining records in the node, builds neighbourhoods and puts all records that are density-reachable from the same core point into the same local cluster.

3. *Merging Local Clusters under Must-Link Constraints:* For each two records involved in a Must-Link constraint and belonging to different local clusters, C-DBSCAN merges the clusters into a "core local cluster".

4. *Merging Clusters under Cannot-Link Constraints:* In the last step, C-DBS CAN performs hierarchical agglomerative clustering. It lets core local clus-ters absorb their most proximal clusters. To compute proximity, only dis-tances between density-reachable records of the two clusters are considered. Remaining Cannot-Link constraints are enforced, in the sense that clusters with records involved in such constraints are not merged.

3   Applying C-DBSCAN on Query Log Data

We tested our method on datasets derived from a query log of 69,790 records. This log was recorded at the site of a real-estate agent that operates a local GSA (Google Search Appliance). Next to the query requests, the log also contained information about further activities of the users on the search page, e.g. select-ing among alternatives (locations, price, categories etc) to restrict their query. During data preparation, the log was cleaned and the whole interaction between user and search page was reconstructed. 45,450 "query sessions" were retained, each one composed by a query and its associated user activities.

The queries were further enhanced with help of an ontology. The ontology contained a taxonomy of concepts, which was used to remove stopwords and map keywords to concepts, and a set of further properties describing the objects of the site (e.g. types of real-estate, prices, locations). We used this information to assign the queries into categories/classes. This allowed us to evaluate the query clustering results against labeled data.

For the evaluation, we created 10 subsets á 4,000 records $DS_1, \ldots, DS_{10}$ and performed two experiments on them. In *Experiment 1*, we studied how a set of constraints from a dataset improves the quality of the clustering on this dataset. In *Experiment 2*, we studied how the *same set of constraints* influences the quality of the clustering upon different datasets. For *Experiment 1* we used datasets $DS_1, \ldots, DS_5$ and for *Experiment 2* we used $DS_6, \ldots, DS_{10}$.

We evaluated the clusters derived by DBSCAN and C-DBSCAN against the labeled data. For this, we used the Rand Index [8] on the similarity between the clusters and classes in each of $DS_i, i = 1 \ldots 5$. The Rand Index takes values in $[0, 1]$ and becomes 1 when clusters and classes are identical.

Experiment 1

Goal of the first experiment was to study how knowledge on the label of some queries may influence clustering. For each $DS_i, i = 1 \ldots 5$, we disclosed the label of a number of queries, identified queries that had the same label and combined them into Must-Link constraints. This approach was also used in [12, 5, 10] for constraint generation. For each DS_i, we generated a set of 200 Must-Link constraints $ML_{i,[200]}$ and a superset of 400 Must-Link constraints $ML_{i,[400]}$. We then derived Cannot-Link constraints from them, which reflect that queries with different labels should not belong to the same cluster.

For each DS_i, we repeated the constraint set generation five times and computed the average Rand Index value over the corresponding five runs. The results are depicted in Fig. 1(a). C-DBSCAN outperforms DBSCAN on all datasets and the Rand Index value is higher for the larger number of constraints. Obviously, the impact is different among the datasets, reflecting the differences in the underlying clusterings, but the trend is the same.

Experiment 2

Goal of the second experiment was to study how heuristics can lead to the generation of useful constraints. For this experiment, we have performed an independent data mining run, in which association rules were discovered upon the query log. From the derived associations, we formulated a set of 200 Must-Link constraints $ML_{[200]}$ and a superset of 400 Must-Link constraints $ML_{[400]}$. We applied these two constraint sets on each of $DS_6, \ldots, DS_{10}$. A similar method for the evaluation of constraint impact was used in [12, 3].

The results are shown in Fig. 1(b). Similarly to Fig. 1(a), we see that the constraint set improves the quality of the clustering results and the larger set of constraints leads to clusters of higher quality. The influence of the constraints upon the clustering quality differs among the datasets, being highest for DS_9 and lowest for DS_8. It is also remarkable, that for those two datasets, we observe the largest variations in the Rand Index values: For DS_8, the Rand Index improves by more than 5% when adding 200 constraints, while the Rand Index for DS_9

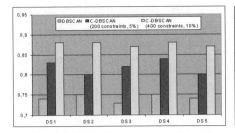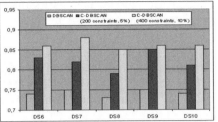

Fig. 1. Comparative results for (a) Experiment 1 and (b) Experiment 2

has almost the same values for $ML_{[200]}$ and $ML_{[400]}$. Those findings are in lieu with the observations of [12], who also found that the same set of constraints can have different influence upon the clustering results of different datasets.

4   Conclusions

We have studied the potential of *instance-level constraints* for query clustering for search engines. Such constraints reflect background knowledge about individual queries, e.g. that they refer to the same subject. We have performed constraint-based query clustering with C-DBSCAN [10]. This algorithm extends the density-based algorithm DBSCAN, which progressively merged groups of records into clusters, by enforcing two types of constraints: If two records/queries are known to be similar, C-DBSCAN merges the clusters to which they are assigned. If two queries are known to be dissimilar, C-DBSCAN prevents the merging of the clusters to which they belong. Our experiments on real query logs show that the quality of query clusters improves over conventional density-based clustering.

Many open issues remain for future work: More experimentation is needed to reveal the semantics of the discovered query clusters. The formulation of constraints requires manual inspection of part of the query log; methods are needed to enhance and possibly automate this process. Finally, a query log is not a static file but a stream; we intend to extend our work on constraint-based clustering for a static query log towards constraint-based clustering for a query log stream, using our work in [9] as a basis.

References

1. J. L. Bentley. Multidimensional Binary Search Trees Used for Associative Searching. *Communications of ACM*, 18(9):509–517, 1975.
2. I. Davidson and S. Basu. Clustering with Constraints: Theory and Practice. In *KDD'06: Tutorial at The Int. Conf. on Knowledge Discovery in Databases and Data Mining*, 2006.
3. I. Davidson and S. S. Ravi. Clustering with Constraints: Feasibility Issues and the k-Means Algorithm. In *SIAM'05: SIAM Int. Conf. on Data Mining*, 2005.

4. M. Ester, H.-P. Kriegel, J. Sander, and X. Xu. A Density-Based Algortihm for Discovering Clusters in Large Spatial Database with Noise. In *KDD'96: Proc. of 2nd Int. Conf. on Knowledge Discovery in Databases and Data Mining*, 1996.
5. M. Halkidi, D. Gunopulos, N. Kumar, M. Vazirgiannis, and C. Domeniconi. A Framework for Semi-Supervised Learning Based on Subjective and Objective Clustering Criteria. In *ICDM'2005: Proc. of IEEE Int. Conf. on Data Mining*, pages 637–640, 2005.
6. X. Liu, P. He, and Q. Yang. Mining user access patterns based on Web logs. *Journal of Electrical and Computer Engineering*, pages 2280– 2283, 2005.
7. D. Nettleton, L. Calderon-Benavides, and R. Baeza-Yates. Analysis of web search engine query sessions. In *WebKDD'06: Proc. of the 8th Int. Workshop on Web Mining and Web Usage Analysis (in conjunction with KDD'06)*, 2006.
8. W. M. Rand. Objective Criteria for the Evalluation of Clustering Methods. In *Journal of the American Statistical Association*, 66, pages 846–850, 1971.
9. C. Ruiz, M. Spiliopoulou, and E. Menasalvas. User Constraints Over Data Streams. In J. Gama, J. Aguilar-Ruiz, and R. Klinkenberg, editors, *Workshop Notes of IWKDDS Workshop "Knowledge Discovery from Data Streams" at ECML/PKDD'06*, Berlin, Germany, Sept. 2006.
10. C. Ruiz, M. Spiliopoulou, and E. Menasalvas. C-DBSCAN: Density-Based Clustering with Constraints. In *RSFDGrC'07: Proc. of the Int. Conf. on Rough Sets, Fuzzy Sets, Data Mining and Granular Computing held by JRS'07*, 2007.
11. P.-N. Tan, M. Steinbach, and V. Kumar. *Introduction to Data Mining*. Pearson Education, 2004.
12. K. Wagstaff and C. Cardie. Clustering with Instance-level Constraints. In *ICML'00: Proc. of 17th Int. Conf. on Machine Learning*, pages 1103–1110, 2000.
13. K. Wagstaff, C. Cardie, S. Rogers, and S. Schroedl. Constrained K-means Clustering with Background Knowledge. In *ICML'01: Proc. of 18th Int. Conf. on Machine Learning*, pages 577–584, 2001.
14. S. Wedig and O. Madani. A Large-scale Analysis of Query Logs for Assessing Personalization Opportunities. In *KDD'06: Proc. of the 12th ACM SIGKDD Int. Conf. on Knowledge Discovery and Data Mining*, pages 742–747, 2006.
15. J.-R. Wen, J.-Y. Nie, and H.-J. Zhang. Clustering user queries of a search engine. In *WWW'01: Proc. of the 10th Int. Conf. on WWW*, pages 162–168, 2001.
16. J.-R. Wen, J.-Y. Nie, and H.-J. Zhang. Query clustering using user logs. *ACM Transactions on Information Systems*, 20(1):59–81, 2002.
17. Q. Zhao, T.-Y. Liu, S. S. Bhowmick, and W.-Y. Ma. Event detection from evolution of click-through data. In *KDD'06: Proc. of the 12th ACM SIGKDD Int. Conf. on Knowledge Discovery and Data Mining*, pages 484–493, 2006.

A Hybrid Framework for a Personalized Document Filtering System

AliReza Sadjadieh

Computer Engineering Department, KhomeiniShahr Branch, Islamic Azad University, Isfahan, Iran
sadjadieh@iaukhsh.ac.ir

Summary. As more information becomes available electronically, tools for finding information of interest to users become increasingly important. In this research, a personalized document filtering system with learning ability is designed and implemented. Each user puts one query in the system and five intelligent software agents rank each new document from their points of view. Three of these agents rank the document according to user's query, user's documents of interest in the profile, and user's group. The forth agent merges these scores and provides another score. Finally, the last agent corrects the obtained score by using previous estimated ranks and the judged ranks by the user. Using this agent that learns the users' behavior and adjusts the combination of the outputs of the other agents to estimate a rank for a document, results in a more accurate score than using each one (query, profile, or group) alone. We used LSI (Latent Semantic Indexing) method for documents and queries.

Keywords: LSI, Latent Semantic Indexing, Intelligent Agents, Information Filtering, Web Personalization, Recommendation System.

1 Introduction

As more information becomes available on the Internet, the need for effective personalized information filters becomes more critical. In particular, there is a need for tools to capture user's information needs, and to find documents relevant to these needs. The ultimate goal of any user-adaptive system is to provide users with what they need without them asking for it explicitly [1]. In this research, the representation used for queries and documents is based on the vector space representation, commonly used in the information retrieval literature. Cosine similarity is used to find similarity between vectors. A standard method of indexing text consists of recognizing individual words, eliminating the commonly used words included on a stop-word list and stemming the remaining words for content identification of the texts. The weight of the term depends on its frequency of occurrence in the text and the number of documents it appears in (tfidf method) [2]. Latent Semantic Indexing [3] is used to infer the structure of relationships between documents and words. In the next section, we describe the

K.M. Węgrzyn-Wolska and P.S. Szczepaniak (Eds.): Adv. in Intel. Web, ASC 43, pp. 310–315, 2007.
springerlink.com

standard data used for evaluating our system. Then, a new criteria for ranking is proposed and later the information filtering system is described which uses five agents to score a document for each user. At last, a conclusion will summarize the whole process.

2 Implementation of Information Filtering System

2.1 Test Data

The standard document source dealt with in this research is the Cranfield corpus[4], which contains a collection of 1400 documents with 225 judged queries. Each query represents one user. These queries have on average 8 relevant documents, and each document is relevant to one or more queries. In order to have a reasonably sized set of documents for each query, we used a subset of 26 queries for this experiment: queries 1, 2, 23, 46, 47, 57, 65, 67, 72, 73, 90, 125, 132, 156, 157, 186, 201, 202, 203, 204, 212, 217, 218, 219, 220, and 221. These test queries have between 15 and 40 relevant documents each, with an average of 19. To test group filtering, we added 14 queries from 27 to 40 that were made by combining other queries. The relevant documents for each are a collection of relevant documents in ingredients queries that produce this new query. The new made queries are presented in the form of numbers standing before the parenthesis and the numbers within, show the ingredients of the new queries. The representation is as follows: 27(25,26), 28(7,8,9), 29(22,23,24), 30(1,2,6), 31(23,24), 32(22,23), 33(4,5), 34(15,17), 35(18,19), 36(2,6) 37(10,15), 38(5,9), 39(7,9), 40(19,20). These test queries use 327 relevant documents from 1400 documents for 40 users. These standard documents in the Cranfield [4] collection are technical scientific abstracts, and are all quite short, between 100 and 4200 bytes in length. They are supposed to use 30000 terms. Using LSI, we get 100 highest singular values (i.e. k=100). Therefore we represent our documents and queries in lower dimensions.

2.2 Criteria

Usually the precision-recall curve, the area under it [5], or its average is used to evaluate retrieval systems. As our system is a filtering system with machine learning capabilities, we suggest a special new formula to evaluate it. The formula which is pseudo average in the precision-recall curve, computes error for each method. We made its rules as follows:

- If the document is retrieved and it is relevant and the difference between the estimated similarity and the real rank is less than 0.2 then the error is zero.
- If the document is retrieved and it is relevant and the difference between the estimated similarity and the real rank is not less than 0.2 then the error is that difference.
- If the document is not retrieved and it is relevant then the error is the real rank multiplied by 3.
- If the document is not retrieved and it is not relevant then the error is zero.

2.3   Implementation of Agents

Here, we implement each agent and evaluate our system based on our criteria. Step by step we add more agents to the system and they collaborate to rank the documents and in each step we measure the system's error which is the area under the corresponding curves.

QAgent

At first, the system uses the user's query to estimate the ranks. We named this suggested agent as QAgent (Query's Agent). Each new document that enters the system is routed to users' QAgent. QAgent computes cosine similarity of this document with user's query and returns the similarity value as a rank. QAgent in figure 1 shows the average error for each user. Vertical axis represents the average error for each user and horizontal axis represents users sorted according to their errors.

PAgent

Now, the system uses another agent that uses user's profile to estimate the ranks. We named this suggested agent as PAgent (Profile's Agent). To build user's profile there are some methods like in [6][7] but we used a simple approach. Each new document that enters the system is routed to users' PAgents. PAgent computes cosine similarity of this document with the user's profile and returns the similarity value as a rank. A profile is a set of documents with their ranks that specify user's interests in them. To have a vector for a profile to compare with a new document, the system multiplies the terms in each document in the profile by the rank of that document and then averages the results. It is a vector in LSI space with 1*100 dimensions, which specifies user's terms of interest. Then, the system computes cosine similarity of the new document with this vector and returns the similarity value as a rank. PAgent curve in figure 1 shows the average error for each user. The produced error by PAgent is close to the error produced by QAgent.

ArbiterAgent

We can use an agent that combines QAgent's and PAgent's values. We name this agent as ArbiterAgent. It computes coefficients for QAgent and PAgent and then ranks the documents with weighted summation of those agents. In [1][8][9] there are some methods to calculate coefficients but our method is simpler and is as follows: if the number of documents in the user's profile is very low, the weight of PAgent will be very low and that of QAgent will be very high. Gradually, as more documents enter the system, user's profile increases, and system increments the weight of PAgent and decrements the weight of QAgent. By using ArbiterAgent to rank the new documents, average error for each user is computed and shown in figure 1. The produced error by the ArbiterAgent is near the error of PAgent and that of QAgent.

GAgent

Now, we suggest another agent which ranks the documents based on the user's group and name it GAgent (Group's Agent). Each new document that enters the system is routed to users' GAgents. GAgent computes the cosine similarity of this document with the user by group's approach. To compute GAgent's values, the similarity between this document and other users is computed and is multiplied by the similarity between this user and other users. Obtained values specify similarity of this document with the user by group's view. By averaging these similarities, the GAgent ranks the document for the user. GAgent must compute the similarity of the new document to other users and the similarity of user to other users. To compute new document's similarity to other users, it will use the user's rank for this document if the user ranks this document, otherwise the rank of QAgent, PAgent, or ArbiterAgent will be used. To compute the user's similarity to other users, GAgent can use one of these three ways:

1. Similarity based on terms in profiles: GAgent multiplies the rank of each document in the user's profile by terms' rank in that document and averages these values. Then, each user has a vector with 1*100 dimensions that describes the user's terms of interest. By computing the similarity of this vector between users, GAgent selects the most similar ones as a group that user belongs to. Gradually, by getting more documents, user's group may be changed.
2. Similarity based on queries in LSI space: GAgent multiplies the query vector of user by query vector of other users to compute cosine similarities. Then it selects the most similar ones as a group for this user.
3. Similarity based on documents' ranks in profiles: GAgent computes the similarity among the profiles, which contains ranks for each document to the profiles of other users. Gradually, by getting more documents, user's group will be changed.

In [10][11][12]there are some more methods for group ranking. By using GAgent, which is built by first method, to rank the new documents, average error for each user is shown in figure 1. As the figure shows, the error is increased a little. Now we add GAgent to ArbiterAgent. The new ArbiterAgent's coefficient for GAgent is very low for a few number of documents in user's profile and it increases gradually as the number is increased. Using new ArbiterAgent to rank the documents, the produced error is near to the errors produced by other agents.

EstimatorAgent

Much of the error in the system may be from the high difference between agent's suggested score for the document and the real rank from the user's view. To estimate user's rank for the new document, we suggest a new agent named EstimatorAgent that estimates rank by previous behaviors of the user. When the new document enters the system, QAgent, PAgent, and GAgent will rank it.

EstimatorAgent computes the similarity between these ranks and the ranks of previous documents. It selects the most similar ones and multiplies the similarity values by the difference between the user's rank and the ArbiterAgent's suggested rank for those documents and then averages the results. The result is the estimated error for the new document's suggested rank and is added to the rank that produced by the ArbiterAgent. This operation makes the estimated rank closer to the user's rank. The average error for users is shown in figure 1 as EstimatorAgent. Here we can see that the error is reduced considerably.

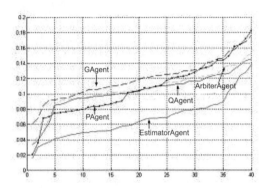

Fig. 1. Errors of QAgent, PAgent, GAgent, ArbiterAgent, and EstimatorAgent

3   Conclusion

Combining user's query, user's profile, or user's group to estimate a rank for a document will lead to almost same results in comparison of using each of them alone. The method proposed for suggested EstimatorAgent uses user's behavior in the past to learn and compute a bias value to adjust itself and the error is reduced considerably.

References

1. Mobasher, B., Dai, H., Luo, T., Sun, Y., Zhu, J.: (2000) Integrating web usage and content mining for more effective personalization. In: E-Commerce and Web Technologies: Proceedings of the EC-WEB 2000 Conference. Lecture Notes in Computer Science (LNCS) 1875, Springer (September 2000) 165Ű176
2. Micarelli, A., Sciarrone, F., Marinilli, M.: (2006) Web document modeling. In Brusilovsky, P., Kobsa, A., Nejdl, W., eds.: The Adaptive Web: Methods and Strategies of Web Personalization. Volume 4321 of Lecture Notes in Computer Science. Springer-Verlag, Berlin Heidelberg New York
3. Deerwester, S., et. al. (1988). Improving information retrieval using Latent Semantic Indexing. Proceedings of the 1988 annual meeting of the American Society for Information Science.
4. The Cranfield Collection Test Data, http://www.dcs.gla.ac.uk/idom/

5. Mobasher, B., et al.: (2002) Discovery and evaluation of aggregate usage profiles for web personalization. Data Mining and Knowledge Discovery 6(1)61Ű82
6. Gauch, S., Speretta, M., Chandramouli, A., Micarelli, A.: (2006) User profiles for personalized information access. In Brusilovsky, P., Kobsa, A., Nejdl, W., eds.: The Adaptive Web: Methods and Strategies of Web Personalization. Volume 4321 of Lecture Notes in Computer Science. Springer-Verlag, Berlin Heidelberg New York
7. Middleton, S.E., et al.: (2004) Ontological user profiling in recommender systems. ACM Transactions on Information Systems 22(1) 54Ű88
8. Burke, R.: (2006) Hybrid web recommender systems. In Brusilovsky, P., Kobsa, A., Nejdl, W., eds.: The Adaptive Web: Methods and Strategies of Web Personalization. Volume 4321 of Lecture Notes in Computer Science. Springer-Verlag, Berlin Heidelberg New York
9. Burke, R.: (2005) Hybrid systems for personalized recommendations. In Mobasher, B., Anand, S.S., eds.: Intelligent Techniques in Web Personalization. LNAI 3169. Springer-Verlag 133Ű152
10. Pazzani, M.J., Billsus, D.: (2006) Content-based recommendation systems. In Brusilovsky, P., Kobsa, A., Nejdl, W., eds.: The Adaptive Web: Methods and Strategies of Web Personalization. Volume 4321 of Lecture Notes in Computer Science. Springer-Verlag, Berlin Heidelberg New York
11. Sinha, R., Swearingen, K.: (2002) The role of transparency in recommender systems. In: CHI Š02 extended abstracts on Human factors in computing systems. 830Ű831
12. Hofmann, T.: (2004) Latent semantic models for collaborative filtering. ACM Transactions on Information Systems 22(1) 89Ű115

A Proposal for a Web Information Extraction and Question-Answer System

José Saias[1] and Paulo Quaresma[2]

Departamento de Informática
Universidade de Évora, Portugal
{jsaias,pq}@di.uevora.pt

Summary. The Web is part of today's life and offers all kind of content. We present a system that can help the user to extract information from web documents and to find the answer for simple questions in natural language. This work is focused on newspaper articles and it is based on an ontology knowledge representation, natural language processment and a logic-programming framework.

1 Introduction

In the last decade the volume of available information on the web has grown exponentially. As an effect of globalization, the news we hear from a remote point of the globe have now gained importance and may influence some aspects of our life. In the other hand, most of the information taken in media resources may not be relevant to the end citizen. Nowadays, the main newspapers have an online RSS[1] service where they publish the latest news to all Internet users. Computer based systems can help people, allowing a quick and broader analysis on the available sources. This paper proposes an ontology based methodology for news article processing[2] in order to cover a large amount of documents, try to automatically understand some information in those documents and get automatic answers to some simple questions.

2 Common Sense Knowledge Base

When we have an isolated sentence it's usually difficult to automatically capture the semantics in it. Ontologies allow the definition of class hierarchies, object properties and relation rules, such as, transitivity or functionality. Our approach uses an ontology as the starting knowledge base with semantic information that helps to perform the sentence analysis and the subsequent inferences and interrogations. The ontology is expressed in OWL[3]. This language has the intended

---

[1] Really Simple Syndication (sometimes also used for Rich Site Summary), is a popular XML format for Web content publication.

[2] This paper is an extension of previous work described in [5].

[3] OWL is the short name for Web Ontology Language. It's a language proposed by the W3C Consortium for the *Semantic Web* and ontology representation.

K.M. Węgrzyn-Wolska and P.S. Szczepaniak (Eds.): Adv. in Intel. Web, ASC 43, pp. 316–321, 2007.
springerlink.com   © Springer-Verlag Berlin Heidelberg 2007

semantic features and it is suitable for web publications, allowing us to share parts of our knowledge base in a direct and appropriate manner. Besides the formal concept definitions and "IsA" relations, there are a few simple facts about everyday life that might be very useful for document analysis. Some of those are also expressed by ontology relations. Our current ontology contains about 3500 concepts and has several relations connecting them: *isA*, *usedFor*, *locatedAt*, *capableOf* and *madeOf*. These concepts and relations represent a small common sense knowledge base about places, entities and events. Some of the top-level concepts are: *AbstractConcept* (the root concept), *Event*, *Time* and *Entity*. The next section explains the document analysis performed by the system.

3   Fetching and Processing the News

Some popular newspapers like *Público* or *Correio da Manhã* have a "last hour" news section in their web site[4], including an RSS channel. This is suitable for an automatic search for any recently added news article.

We used a program to periodically collect the recent news from *Público*'s RSS channel. As we can see in figure 1, each news item has some metadata fields: title, description, author, category, publication date and hour, and of course, the link to the web document containing the information. The category gives us a first simple classification for the document, placing it in Economy, Politics, International or Sports (in Portuguese Desporto - like the item listed in figure 1). The publication date gives the temporal context to the semantic content we find in the document, as we will see later. Each document imported to the system has a text body. That text is processed, following a methodology based on natural language processing techniques, namely, a syntactical parser and a semantic analyzer able to obtain a partial interpretation of the document. The tool used for the syntactical analysis is PALAVRAS [1]. It's a syntactical parser based in the Constraint Grammars formalism and it is able to cover a large percentage of the Portuguese language.

Let us consider a sentence in the above sports news item:

"*Marcus Grönholm venceu neste domingo o Rali da Grécia.*" (in English: "Marcus Grönholm won the Greece Rally, this Sunday.")

The parser identifies the subject, the predicate and direct object with extra details that are stored on a Prolog structure and passed to the semantic analysis module. The technique used for this module is based on Discourse Representation Structures (DRS) [2]. The partial semantic representation of a sentence is a DRS built with two lists, one with the rewritten sentence and the other with the sentence discourse referents. We are only dealing with a restricted semantic analysis and we are not able to handle every aspect of the semantics: our focus is on the representation of concepts (nouns and verbs) and the correct extraction of its properties (modifiers, agents, objects). The previous news item is stored in the system with the details on figure 2.

---

[4] http://www.publico.pt/ and http://www.correiodamanha.pt

```
<?xml version="1.0" encoding="ISO-8859-1"?>
<rss version="2.0" xmlns:msxsl="urn:schemas-microsoft-com:xslt"
xmlns:t="http://www.publico.pt">
<channel>
...
<title>Publico.pt Desporto</title>
<link>http://www.publico.clix.pt</link>
...
<item>
<title>Marcus Grönholm vence Rali da Grécia</title>
<link>http://www.publico.clix.pt/shownews.asp?id=1259478</link>
<description><![CDATA[<h3>Sébastien Loeb, líder do Mundial, foi segundo< ...
h3><br/>
Marcus Grönholm venceu neste domingo o Rali da Grécia.
Ao volante de um Ford Focus, o piloto finlandês reduziu para 29 pontos a dist
que o separa, na classificação geral do Mundial, para o bicampeão e actual lí
francês Sébasten Loeb (Citroën Xsara), que terminou em segundo na prova g
></description>
<author>AFP</author>
<category>Desporto</category>|
<pubDate>Sun, 04 Jun 2006 16:09:00 GMT</pubDate>
</item>
...
  more items
...
</channel>
</rss>
```

```
item(publico1259478,
    'Desporto',
    'Sun, 04 Jun 2006 16:09:00 GMT').

sentence(publico1259478,
    [ name(A, 'Marcus_Grönholm' ,
            ['M/F', 'S', 'Marcus_Grönholm' ],
        [ ] ),
    name(B, 'Rali_da_Grécia' ,
            ['M', 'S', 'Rali_da_Grécia' ],
        [ ] ),
    'vencer'(A,B,
        [ modif(verb,'vencer',
                    ['PS','3S','IND'] ) ] ),
    [ modif(temp,'domingo', ['M','S'],
        [ modif(pronDet,'este',
                ['M','S'], []) ] ) ]     ],
    [ ref(A), ref(B) ] ).

... (other sentences)
```

Fig. 1. RSS document from *Público*   **Fig. 2.** An item captured semantics

4 Using the System

Once the news documents are obtained and analyzed they become part of the second knowledge base: the facts knowledge base. The Question-Answer module receives a natural language written query, in Portuguese. The query is processed using the same natural language tools used for the news texts. The search for an answer is done by a logic-programming based module that performs a pragmatic interpretation of the query DRS over the full system knowledge base (the ontology and the news facts).

The inference process is done with the Prolog resolution algorithm, which tries to unify the referent from the query with facts extracted from the documents and expressed in DRS structures.

4.1 Who/What Questions

As an example, we could enter a query like:

"Quem ganhou o Rali da Grécia?" (in English: "Who won the Greece Rally?")

The DRS for such query is presented in figure 3. This logic structure is checked against each sentence DRS. The result displayed by the system web interface is given in figure 4. It may include zero or more values considered valid as response to the query. For each possible response value there is also a document link list, pointing to the news item(s) where the system found the answer, and a numeric value with an estimated weight for that answer. Each sentence DRS component (subject, verb and object) match is given a weight (100 for direct match or less for dictionary and ontology driven cases). The weight for the document answer is calculated as the average weights of their matched sentence components. Finally, the weight assigned to an answer is the maximum value from their documents weights plus $\#docs - 1$. The previous question was answered because the concept

```
query(q281,
    [ q(X, 'quem' , ['M/F', 'S', 'quem' ],
        [ ] ),
    name(Y, 'Rali_da_Grécia' ,
        ['M', 'S', 'Rali_da_Grécia' ], [ ]),
    'ganhar'(X, Y,
        [ modif(verb,'ganhar',
            ['PS','3S','IND'] ) ] ),
    [ ] ],
    [ ref(X), ref(Y) ] ).
```

Resposta(s)

Valor	pontos	Documentos que suportaram a resposta
Marcus Grönholm	101	ut1151971470255 publico1259478 (2)
Marta	100	ut1160424366497 (1)

(2 respostas em 3 documentos relacionados)
Nota: o motor de inferência Senso pode necessitar de uma actualização para

Fig. 3. A *Who-Question* DRS   **Fig. 4.** Question-Answer result

vencer is defined as a synonym of *ganhar*. Another note is that there are two answers in the result. In this case, the reason is that we have a document with a sentence identifying last year winner. We could now follow the links and check the best solution by reading the text. The precise query for this year winner would be:

"Quem venceu o Rali da Grécia no ano de 2006?" (in English: "Who won the Greece Rally in the year 2006?")

That would introduce a temporal modifier on the query DRS expression to be checked against the date of publication of the document, such as:

```
... [ modif(temp,'ano', ['M','S'],
        [modif(num,'2006',
        [modif(prp,'de') ] )] ) ]    ], ...
```

The system answer is now only *Marcus*, as seen in the newspaper article.

4.2 When Questions

Another example of query about time is:

"Quando é que Marcus Grönholm venceu o Rali da Grécia?" (in English: "When did Marcus Grönholm won the Greece Rally?")

Once again, the interrogative term *quando*'s referent is matched against the temporal modifier on the sentence DRS: *"este domingo"* (in English: this Sunday). This information is then related with the news item publication date, the ontology and the sentence verb time (future, present or past), by the question-solver logic module. This allows the system to infer the desired date answer for the question, 2006-06-04. Similar treatment is given to temporal expressions like *today*, *this month*, *last year* and other. The next sentence list has several cases for temporal expressions:

- A Feira da Luz é em Setembro.
- A Feira da Luz é em Setembro de 1958.
- A Feira da Luz decorreu no último ano.

- A Feira da Luz foi no mês passado.
- A Feira da Luz é amanhã.
- A Feira da Luz é a 14 do próximo mês.

Each of these document sentences will produce an answer to the query:

"Quando é a Feira da Luz?" (in English: "When takes place the Feira da Luz?")

The question-solver logic module infers the offset relative to the document publication date. Then presentation module gives a formatted date value. As an example, for the year 2006, if the document date is *Monday, October 2* and the sentence has *"... is on Thursday."* then the answer is next Thursday on that week: *2006-10-05*. If the document date is *Sunday, October 8* and the sentence has *"... will be in January."* then the answer is next January: *2007-01*.

4.3 Where Questions

For this kind of interrogations the sentence information near a preposition is taken into account and it is related with the ontology concepts below *"lugar"* (*place*, such as a *city* or *country*). If the term found in the selected sentence is a possible place, then it can be used as an answer. Lets consider the three documents in figure 5 and their assertions. Asking where is *Feira da Luz*, with a query:

"Onde é a Feira da Luz?"

will give us the expected answers. Figure 6 has the result, one answer value per document and having equal weights. Those were direct answer cases. The system

```
ut1160492122437 A Feira da Luz é no Alentejo.
ut1160492097626 A Feira da Luz é em Montemor.
ut1160492138407 A Feira da Luz é em Évora.
```

Resposta(s)

Valor	pontos	Documentos que suportaram a resposta
Évora	100	ut1160492138407 *(1)*
Montemor	100	ut1160492097626 *(1)*
Alentejo	100	ut1160492122437 *(1)*

(3 respostas em 3 documentos relacionados)

Fig. 5. Natural language assertions   **Fig. 6.** QA result: Where case

can also infer the answer for some nontrivial cases. Having the previous three assertions, we can ask if *Feira da Luz* is in a certain city:

"A Feira da Luz é em Montemor-o-Novo?"

and the answer is *yes*, because there is an indication, given by the ontology, stating that *Montemor* is an alias to *Montemor-o-Novo*. In the case where we ask if the event takes place in Portugal:

"A Feira da Luz é em Portugal?"

the answer is *yes*. This time it was not so immediate. The question-solver had to look for a place where *Feira da Luz* is happening and then check on the ontology or fact knowledge base if that place is located in Portugal.

5   Related Work

There are other initiatives related to the semantic content search. Ontologies are used in [3] for the specific domain of International Affairs. It has a natural language interface also, but works with RDQL[5] instead of the Prolog logic resolution environment we adopted. The semantic archive features provided by [4] include means to annotate news materials and semantic search and browsing capabilities. This system runs inside the newspaper environment and uses a

---

[5] RDQL is a query language for RDF based on SquishQL. For more detail visit http://jena.sourceforge.net/tutorial/RDQL/

newspaper library specialized ontology, while the system we present works alone and outside the newspaper, allowing the use of many independent news sources, and our ontology is about common sense knowledge and not about a specific domain.

6 Conclusions and Future Work

We presented a web fact learning system focused on news texts from media sites. The system captures natural language texts, in Portuguese language, and performs information extraction for the question-answer feature. This feature is supported by the logic inference module, whose accuracy is affected by the quality of the ontology and the precision of the semantic information taken from the text sentences.

The ontology should be manually revised and extended. The semantic analysis can be improved if we add a tool to identify the inter-sentence anaphoric references. Along with this, some disambiguation tool is needed for better precision when a sentence concept is being related with an ontology existent term or when the system is trying to match a sentence with a query structure. Finally, the automatic question-answer system needs to be fully evaluated.

References

1. Eckhard Bick. *The Parsing System "Palavras". Automatic Grammatical Analysis of Portuguese in a Constraint Grammar Framework.* Aarhus University Press, 2000.
2. Kamp, H. and Reyle, U. *From Discourse to Logic.* Kluwer: Dordrecht. 1993
3. J. Contreras, V. Richard Benjamins, M. Blázquez, S. Losada, R. Salla, J. Sevilla, D. Navarro, J. Casillas, A. Mompó, D. Patón, Óscar Corcho, P. Tena, I. Martos. *A Semantic Portal for the International Affairs Sector.* In Proceedings of the EKAW 2004. pages 203-215. Springer, 2004
4. Pablo Castells, F. Perdrix, E. Pulido, M. Rico, V. Richard Benjamins, J. Contreras, J. Lorés. *Neptuno: Semantic Web Technologies for a Digital Newspaper Archive.* In 1st European Semantic Web Symposium, Greece, pages 445-458, 2004.
5. José Saias and Paulo Quaresma. *A proposal for an ontology supported news reader and question-answer system.* Solange Oliveira Rezende et al. (Eds): 2nd Workshop on Ontologies and their Applications (WONTO'06) in the Proceedings of International Joint Conference, 10th IBERAMIA, ICMC-USP, Ribeirão Preto, Brazil, 2006. ISBN: 85-87837-11-7.

Improved Trust Metrics and Variance Based Authorization Model in e-Commerce

Jayamsakthi Shanmugam[1] and M. Ponnavaikko[2]

[1] Research Student - BITS
[2] Director of Research and Virtual Education - SRM Institute of Science and Technology

Summary. The problem of Authentication and Authorization is studied with an aim to trust the customer's transactions and to authorize the payment. Considering the limitation of the available methods and procedures, an improved trust metrics and variance based authorization model in e-commerce is proposed. The solutions proposed assess the deviation of the customers' transactions to calculate the Standard Deviation and employs normal distribution to assess the transaction to authorize. The model was applied on the customers' transactions and the results were studied that are promising to employ in e-commerce systems.

1 Introduction

The rapid proliferation of the Internet and the cost effective growth of its key enabling technologies are revolutionizing online electronic transactions and creating unpredicted opportunities for developing large scale distributed applications like e-commerce with multiple technologies[1]. But these transactions are not with out problems. When an e-commerce transaction is initiated by a customer, there are no ways by which the financial institution can decide whether this transaction is originated from a genuine card holder or by a hacker.

1.1 Payment Acceptance and Processing

Payment card transactions in the e-commerce systems go through the following steps of action once the merchant receives a consumer's payment card information through SSL protected page. The merchant/e-commerce systems must authenticate the payment card to ensure that it is both valid and not stolen. Within a few days following the consumer's request for purchase, settlement occurs, this means that funds travel through the e-commerce system into the merchant's account after the purchase has been shipped. As millions of customers participate in e-commerce, a very large number of transactions take place with varied quantity and value, and hence quantifying the risk becomes more tedious[2].

Researches in the past have addressed this issue and proposed a few models for solving these problems. The Authorization based on Evidence and trust model suggested by Bharat Bhargava and Yuhui Zhong, proposes a framework to characterize the probability that a user will not carry out harmful actions.

K.M. Węgrzyn-Wolska and P.S. Szczepaniak (Eds.): Adv. in Intel. Web, ASC 43, pp. 322–327, 2007.
springerlink.com © Springer-Verlag Berlin Heidelberg 2007

However a hacker who steals the credentials of a credit card holder can enter into the system using the opinion parameters as a genuine user. Thus, this model doesn't cover the application level hacking and doesn't prevent the hacking by using opinion parameters [3].

The Authenticate if trust violated (ATV) model [4] proposed by Daniel W.Manchals used the randomization techniques and trust metrics to verify the transactions. Randomization techniques would fail, if many of the credit cards are hacked at one instance. Thus this model does not prevent the occurrence of harm to the system.

The authors of this paper have developed a new approach with improved trust metrics for recognition and authorization process in e-commerce which provides solution to the problems unaddressed in the earlier works.

2   Improved Trust Metrics

Trust Metrics are represented by a 2-tuple with two elementary names id and attributes where id is the identifier of the customer and attributes are the trust parameters. i.e. (Id, Attrs). Possible attributes for a trusted model is represented as a1, a2, a3 ... etc. For each attribute three possible linguistic values are assigned as a[1,2,3....n] = {Min, Max, Mod}.

The following trust parameters are defined as attributes for considerations in the process of authorization.

Cost: Cost is considered as one of the main trust parameters in the proposed solution procedure. The amount transacted in each transaction, the mean and the standard deviation of the transaction over a selected period of time forms the basis for the authentication procedure suggested.

Location: This parameter is not used to track the intermediaries as defined in the Daniel Manchal's work. Instead this parameter represents the transaction from where it is requested. The possible locations of a customer can either be collected from the customer during the registration process or by tracking the transactions over a period time to restrict the bogus transactions. When the customer makes an online transaction, the IP can be tracked to verify the location of the transaction. If the location is too far away from the location of the immediate last transaction not justifying the time interval of the travel then the transactions are considered as initiated by a hacker. The distances of each transaction from the base station of the customer, its mean, variance and standard deviation forms the basis for the proposed method.

Frequency of Transactions: Frequency of Transactions is another important trust metric in the process suggested. The frequency of transactions per day over a selected period of time, the daily mean and the standard deviation of the frequency of transactions forms the basis for assessing the risk factor in the proposed solution procedure.

Password reset history: Normally if customers are prompted to set a new password after a certain period of time then the customers would reset the password. When the threat is more, the customer is advised to reset the password in a defined interval. Thus the behavioral pattern of the password reset history when considered as a separate metric would add value to assess the risk factor. The interval between the dates on which passwords were reset are obtained for a customer. The mean, variance, standard deviation and average of the time intervals between password reset forms the basis for the proposed method.

For effective and efficient implementation of the trust metric model, a term called Risk Factor or Control limit is defined. The Risk factor describes the degree to which the transaction can be trusted. It also defines the maximum tolerance limit determined by the standard deviation of the trust parameters.

2.1   Proposed Application Procedure

By making use of the trust parameters defined above, authentication of a particular transaction can be processed using the method proposed below:

It is assumed that the values of the trust metrics are approximately normally distributed. The random variable X of a trust metric is defined as a function of real numbers. The probability of the variable X having a value'a', is denoted by $P(X=a)$

When X assumes any value in an interval $a<X<b$, the probability is denoted by $P(a<X<b)$. In a normal distribution, the following equations show how X has the following distribution pattern.

(a) About 2/3 values lie between μ-SD and μ+SD
(b) About 95% of the values lie between μ-2SD and μ+2SD
(c) About 99% of the values lie between μ-3SD and μ+3SD

Where μ is the mean and SD is the Standard Deviation. This is known as 68-95-97 rule. Based on this rule, 68% of the transactions will lie within a standard deviation of 1.0 and hence the transaction can be permitted without verification. The remaining transactions which deviate from the mean by more than one standard deviation need to be verified when the transaction deviates beyond μ+SD for authorization.

Under this assumption a state of a customer at any instance of time is represented as

a1:v1, a2:v2, ... an:vn where a1, a2, a3... an are the trust metrics and v1, v2, v3.... are the corresponding values.

As defined earlier, the risk factor is a function of the trust metrics. The value of the Risk Factor is determined using the formula defined below:

f[a1(NoRisk, Min, Mod, Max)+ a2(NoRisk, Min, Mod, Max)+.... + an (NoRisk, Min, Mod, Max)]$\rightarrow$ [0,1].

Where Min, Mod, Max represents the Minimum, Moderate and Maximum value of each of n trust metrics. In this case n=4. The Risk Factor 'f' results in either 0 or 1. The factor 'f' will be assigned a value of 0 when all the trust metrics are assigned "No Risk" state. (i.e. when the trust metric value is lesser than μ+SD). The value one is assigned when any one of the trust metric is assigned a value Min, Mod or Max. The risk factor value '0' means that the transaction is in the trusted state and hence the transaction will automatically be permitted.

2.2   Determination of the Parametric Values of the Trust Metrics

Based on the estimated standard deviation the Min, Mod or Max values are assigned to each trust metrics as discussed below:

Let the current value of a trust metrics be denoted by Xi. The level of risk of the transaction Xi is assessed by checking how much it varies beyond the standard deviation SD as defined below:

In the case of a trust metrics a risk factor of Min is assigned to the transaction if $(\mu+2SD)>Xi>(\mu+SD)$ and Mod is assigned to the transaction if $(\mu+3SD)>Xi>(\mu+2SD)$. Any transaction Xi is assigned a Max risk factor when $Xi>(\mu+3SD)$. We take till 3 SD because the confidence interval for 3SD is 99.7.

3   Implementation Strategy

The authorization level is determined based on the risk factor and depending up on Min, Mod, Max values assigned to the trust metrics. The transaction will be further verified and decision on levels of authorization is taken manually when f=1. When the trust metrics have a combination of values of Min, Mod and Max, then the transaction is subjected to authorization at different levels of authorities as proposed in Table 1. The levels are explained in Authorization flow section.

Table 1. Part of payment verification matrix

Combinations of Trust Metric		Authorization Level
	1 Min	Primary Level
3 Min	1 Mod	Primary Level
1 Min	3 Mod	Intermediate Level
1 Min	3 Max	Final Level
3 Mod	1 Min	Intermediate Level
1 Mod	3 Min	Primary Level
3 Max	1 Mod	Final Level

3.1   Operational Access Matrix Construction

Possible authorization matrix for the different trust metrics can be constructed for authorization as depicted in Table 1. Authorizations with different levels of authorization are defined under the following three categories.

Primary Level: Primary level of authorization is assigned to a transaction when the trust metrics have values with a combination of Min and/or Mod, at least 3 trust metrics having assigned with Min values.

Intermediate Level: Intermediate level of authorization is assigned to a transaction when the trust metric of the transaction have values with a combination of Mod or Max, at least 2 or 3 metrics having assigned with a value Mod. Final or Terminal Level: Final level of authorization is assigned to a transaction when 3 or more trust metrics have a value of Max.

Table 2. Part of Operational Access Matrix constructed for implementation

States	Cost			Frequency of transaction			Location			Password reset History			Authorization level
	Mn	Md	Mx	Mn	Md	Mx	Mn	Md	Mx	Mn	Md	Mx	
1	√			√			√			√			P
2	√				√		√				√		I
3	√					√	√					√	F

Table 3. Mean and Standard deviation of a customer

Trust Metrics	Mean	Standard Deviation
Cost	674.07	819.9
Frequency of Transactions	3.36	2.07
Location	29.16	80
Password Reset History	46.25	20.8

3.2   Implementation of the Proposed Approach

The authors have implemented the proposed approach using the macros programming in excel.

Note: Mn indicates Min value, Md indicates Mod value, Mx indicates Max value, P indicates Primary level, I indicates intermediate level and F indicates Final level. The mean, standard deviation of the trust metrics Cost, Frequency of transactions and Location were estimated. The estimated parameters of a customer are given in table 3.

Consider the following two transactions of the above customer:

The standard deviation was calculated based on all the previous months' transactions and was revised for every transaction for authorization of payments. Here we present 2 transactions that required authorization from the sample.

Table 4. Calculated Risk Factors for the transactions that needed authorization for the customer

Transaction	Date of Transaction	Cost	Frequency of Transactions for the day	Location (Kms)	Password reset history
1	14 Jan 2005	3,199.0	2	38	45
2	24 Jul 2005	816.27	4	25	45

Using the authorization level matrix given in Table 1, the risk factors are derived and the authorization level is determined for the transactions as could be seen below in Table 5.

The metrics frequency, location, and password reset history for the two transactions are assigned 'No risk'. Since Cost parameter value satisfies the equation $Xi > (\mu + 3SD)$ it is assigned 'Max' value and hence the level of authorization is

Table 5. Transactions and the derived authorization levels out of payment verification matrix

Transaction	Trust Metrics				Authorization Level
	Cost	Frequency of transactions	Location	Password re-set history	
1	Max	No Risk	No Risk	No Risk	Intermediate
2	No Risk	No Risk	No Risk	Min	Primary

set as intermediate. The actual deviation from the mean transaction in this case is 2525. The intermediate authorizer can verify the authenticity of the user by obtaining his social security code and also by checking the maximum amount of his earlier transaction.

Like wise the rest of the matrix is constructed out of the payment verification matrix as described in Table 1.

4 Conclusion

A new improved trust metrics based on authorization model for e-commerce developed by the authors takes care of the all possible risks of hacking. In the pro-posed model mean and standard deviation of the four trust metrics of a customer is periodically updated including last transaction in a secured internal environment of e-commerce network, thus the model is not prone to contour analysis as en-countered by Daniel W. Manchala. In the authors opinion the proposed authorization model will save the e-commerce transactions from the hands of hackers. Redefining the location trust metric and including Password reset history in trust metric over comes the risks encountered by the earlier trust metrics based authorization techniques.

References

1. Vijay Ahuja "Building Trust in Electronic Commerce", IT Professional, Volume 2, Issue 3, May-June 2000, pp 61 - 63.
2. Barry W. Boehm, Defense Advanced Research Projects Agency, "Software Risk Management: Principles and Practices" IEEE, January/February 1991, Volume. 8, No. 1 pp. 32-41.
3. Bharat Bhargava and Yuhui Zhong, "Authorization based evidence and Trust", Proceedings of the 4th International Conference on Data Warehousing and Knowledge Discovery, 2002, pp: 94 - 103.
4. Daniel W. Manchala, "e-commerce Trust metrics and Models" , IEEE Internet Computing, Volume 4 , Issue 2, March 2000, pp: 36-44

Semantic Analysis of Web Pages Using Cluster Analysis and Nonnegative Matrix Factorization

Václav Snášel[1], Hana Řezanková[2], Dušan Húsek[3], Miloš Kudělka[1], and Ondřej Lehečka[1]

[1] Department of Computer Science, VSB-Technical University of Ostrava, 17. listopadu 15, Ostrava, Czech Republic
vaclav.snasel@vsb.cz
[2] Department of Statistics and Probability, University of Economics, Prague, W. Churchill sq. 4, 130 67 Prague, Czech Republic
rezanka@vse.cz
[3] Institute of Computer Science, Academy of Sciences of the Czech Republic, Prague, Czech Republic
dusan@cs.cas.cz

Summary. In this paper, the web pages concerning products sale are analyzed with the aim to create clusters of similar web pages and characterize these by GUI patterns. We applied GD-CLS (gradient descent - constrained least squares) method which combines some of the best features of other methods. Both traditional methods for searching clusters and nonnegative matrix factorization are used.

1 Introduction

One of the key tasks in a problem of orientation in large space of weak-structured data like web is to find out what information it is possible to find on the web page and which methods can be used. There are many approaches which lead more or less to the same destination. It is to offer effective orientation in big amount of weak-structured data. A deeper analysis must be performed to obtain detail page information [8]. This detail information is something what can be said about the page. Simple example can be occurrence of particular word (with its frequency and position). Next example can be fact that the page belongs to particular domain (for example, selling products). The field for evolving page analysis methods is very wide. One feature of all the approaches is independency on form which is the information written in. If we focus on methods based on the full-text approach then the key problem seems to be the loss of information about structure. On the other hand there are methods which prefer page structure and hence they use HTML code elements structure as one of the information source. Trouble with both approaches is changing web, markup language standards and evolution of requirements on web page design [5]. In this paper we introduce our perspective on mentioned problems which comes out from an interesting connection of two fields - information retrieval methods and semantic web on one side and on the other side web design practices and web patterns [12, 14].

K.M. Węgrzyn-Wolska and P.S. Szczepaniak (Eds.): Adv. in Intel. Web, ASC 43, pp. 328–336, 2007.
springerlink.com

This perspective allows us to efficiently combine both mentioned approaches. The form in which the user can find information on web page is changing (but some features remains). The content is more or less still the same. This implies that the most effective approaches must handle on key aspects of the form and the content which is held in the form. In our perspective we are strictly focused on the way the user perceive web page. This sensation then motivates web page designers to create pages conforming user requirements. This never-ending and invisible interaction between web designers and users is projected to patterns [1]. The evidence of patterns on web pages with the same content but created by different designers is demonstrated by more or less the same look. This fact shows key feature of our approach and is essential for us. If we realize this fact then we can find a lot of methods in information retrieval and semantic web fields which can be used very efficiently for finding ways to fulfill user requirements. On the other hand this perspective opens new motivation for domain and GUI pattern experts. They formulate requirements on user interface with a view to web designers. We are binging new motivation to them because patterns can be found and described with regard to their usage in description of web page semantics.

Web pages concerning products sale can contain individual parts with different kinds of information (GUI patterns [4, 12, 13, 14]). There are short description and price of the product, possibility to buy it, detail description (technical data), review, discussion etc. We can distinguish different types of these web pages according to degrees of detection of GUI patterns [5, 6, 7]. The aim of this paper is to classify such web pages and search typical patterns which are contained in them.

The paper is organized in the following way. In the chapter 2, we will show how the web pages were described and which data were used as an input for the further analyses. In the chapter 3, we will describe the analyses by methods of cluster analysis. We will mention both clustering GUI patterns and clustering web pages and characterization of obtained clusters by their centroids. Chapter 4 concerns the analysis by nonnegative matrix factorization. The GD–CLS method was applied for finding clusters of web pages.

2 Web Page Description

We analyzed more then 20 thousands Czech web pages. We searched for 10 GUI patterns [2] which could be contained in them. There were price, sale, discount, credit, description, opinion, discussion, login, bazaar and questionnaire. Therefore, we described each web page by a 10–dimensional vector of values which express degrees of detection of individual GUI patterns contained in it. By this way we obtained 23 422 vectors $S = (v_1, \ldots, v_{10})$ characterizing individual web pages where $0 \leq v_j \leq 1$. Average values for individual patterns based on all web pages are the following:

	price	sale	discount	credit	description	opinion	discussion	login	bazaar	questionnaire
Average	0.482	0.342	0.235	0.077	0.113	0.135	0.135	0.215	0.081	0.049

For further analyses, the data were transformed so that $\sum_j^{10} v_j = 1$. In this case, average values for individual patterns based on all web pages are the following:

	price	sale	discount	credit	description	opinion	discussion	login	bazaar	questionnaire
Average	0.254	0.164	0.095	0.027	0.053	0.098	0.113	0.114	0.053	0.030

3   Analysis by Traditional Methods

The base image on web pages structures can be obtained by investigation of associations between GUI patterns (i.e. attributes). Because of common occurrence of higher values is more important than common occurrence of lower (especially zero) values, the cosine measure is better than correlation coefficient for this purpose [10]. Cosine measure matrix for the set of patterns is shown in Table 1. Cosine similarity measure is expresses by the formula

$$s_C(X_k, X_l) = \frac{\sum\limits_{i=1}^{n} v_{ik} \cdot v_{il}}{\sqrt{\sum\limits_{i=1}^{n} v_{ik}^2} \cdot \sqrt{\sum\limits_{i=1}^{n} v_{il}^2}}$$

where X_k and X_l are k^{th} and l^{th} patterns, n is the number of web pages and v_{ik} is a value for the i_{th} web page and the k_{th} pattern.

The highest value is 0.416 for the pair price, discount; the second high value is 0.410 for the pair price, sale.

Table 1. Cosine measure matrix for the set of 10 patterns (obtained by the SPSS system)

	price	sale	discount	credit	description	opinion	discussion	login	bazaar	questionnaire
price	1.000	0.410	0.416	0.184	0.142	0.039	0.037	0.151	0.140	0.040
sale	0.410	1.000	0.316	0.160	0.142	0.031	0.027	0.166	0.048	0.029
discount	0.416	0.316	1.000	0.169	0.100	0.023	0.019	0.115	0.077	0.026
credit	0.184	0.160	0.169	1.000	0.097	0.030	0.013	0.062	0.026	0.019
description	0.142	0.142	0.100	0.097	1.000	0.029	0.026	0.067	0.026	0.016
opinion	0.039	0.031	0.023	0.030	0.029	1.000	0.206	0.125	0.033	0.055
discussion	0.037	0.027	0.019	0.013	0.026	0.206	1.000	0.069	0.011	0.065
login	0.151	0.166	0.115	0.062	0.067	0.125	0.069	1.000	0.041	0.059
bazaar	0.140	0.048	0.077	0.026	0.026	0.033	0.011	0.041	1.000	0.010
questionnaire	0.040	0.029	0.026	0.019	0.016	0.055	0.065	0.059	0.010	1.000

By complete linkage method, the distance between two different clusters is the greatest distance between two objects in the clusters. On the base of this similarity matrix, we can analyze the data structure by hierarchical cluster analysis. We applied different linkage methods and we obtained two main clusters by complete linkage (dissimilarity was computed as $1 - cosine\ measure$). The dendrogram is shown on Figure 1.

We compared results mentioned above with the results obtained by means described in [10]. We applied factor analysis and we used factor loadings for two components as an input for fuzzy cluster analysis in the S-PLUS system. This technique was designed for binary data but it can be used also in other cases. We

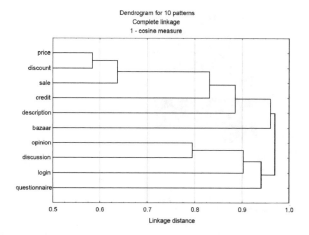

Fig. 1. The dendrogram for the set of 10 patterns (obtained by the STATISTICA system)

Table 2. Membership coefficients for 2 clusters (obtained by the S-PLUS system)

	[Cluster 1]	[Cluster 2]	Closest hard clustering
price	0.736	0.264	1
sale	0.738	0.262	1
discount	0.805	0.195	1
credit	0.748	0.252	1
description	0.603	0.397	1
opinion	0.164	0.836	2
discussion	0.187	0.813	2
login	0.470	0.530	2
bazaar	0.444	0.556	2
questionnaire	0.288	0.712	2

Table 3. The size of clusters for 2 - 5 clusters (obtained by the SPSS system)

Cluster	2 clusters	3 clusters	4 clusters	5 clusters
1	17354	10909	10310	5262
2	6068	6695	4669	4533
3		5818	2733	2806
4			5710	5679
5				5142
Total	23422	23422	23422	23422

used it from two reasons. Firstly, interpretation of factor loadings is disputable and secondly, it is difficult to apply fuzzy cluster analysis to clustering attributes (in S-PLUS only objects can be clustered). For two clusters, the patterns were assignment in the way shown in Table 2.

Graphical output from fuzzy cluster analysis is the silhouette plot. It is shown in Figure 2 for 2 clusters. Patterns bazaar and login can be assigned both to cluster 1 and to cluster 2; therefore are displayed by different way.

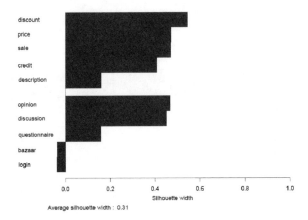

Fig. 2. The silhouette plot for the set of 10 patterns (obtained by the S-PLUS system)

Table 4. Clustering by two-step cluster analysis for 2 - 5 clusters (centroids of clusters)

	Price	Sale	Discount	Credit	Description	Opinion	Discussion	Login	Bazaar	Questionnaire
1 / 2	0.333	0.216	0.128	0.036	0.070	0.019	0.018	0.074	0.068	0.039
2 / 2	0.029	0.014	0.001	0.000	0.004	0.324	0.384	0.231	0.009	0.004
1 / 3	0.411	0.281	0.161	0.000	0.028	0.009	0.011	0.089	0.009	0.002
2 / 3	0.200	0.107	0.071	0.093	0.137	0.038	0.034	0.058	0.163	0.100
3 / 3	0.022	0.011	0.001	0.000	0.002	0.333	0.394	0.227	0.008	0.002
1 / 4	0.418	0.283	0.164	0.000	0.023	0.008	0.012	0.089	0.001	0.002
2 / 4	0.241	0.125	0.106	0.133	0.046	0.030	0.010	0.055	0.250	0.005
3 / 4	0.143	0.100	0.015	0.000	0.287	0.058	0.074	0.076	0.008	0.239
4 / 4	0.022	0.011	0.001	0.000	0.001	0.334	0.399	0.227	0.005	0.000
1 / 5	0.313	0.209	0.330	0.000	0.038	0.010	0.012	0.076	0.007	0.005
2 / 5	0.239	0.123	0.103	0.137	0.046	0.031	0.010	0.055	0.251	0.006
3 / 5	0.148	0.107	0.011	0.000	0.287	0.057	0.071	0.081	0.008	0.230
4 / 5	0.021	0.010	0.001	0.000	0.001	0.335	0.400	0.226	0.005	0.000
5 / 5	0.522	0.354	0.000	0.000	0.003	0.006	0.013	0.100	0.002	0.000

However, our main aim was the identification of clusters of web pages and characterization of these clusters by mean values of attributes. First, we used two-step cluster analysis (in the SPSS system) with log-likelihood dissimilarity measure. This method is suitable for large data files. The results correspond from 1 to 5 clusters are shown in Table 3 and 4. In Table 4, the values greater than 0.09 are considered as significant. The procedure in SPSS determined 3 clusters as an optimal number (both by Schwarz's Bayesian and Akaike's information criterions). From the variants from 1 to 15 clusters, the variant of 10 clusters (with the sizes from 1322 to 5152 web pages) was determined as an optimal.

4   Analysis by Nonnegative Matrix Factorization

The nonnegative matrix factorization (NMF) method for text mining is a technique for clustering that identifies semantic features in a document collection and groups the documents into clusters on the basis of shared semantic features [3]. A collection of documents can be represented as a term-by-document matrix. Since each vector component is given a positive value if the corresponding term is present in the document and a zero value otherwise, the resulting term-by-document matrix is always nonnegative. This data non-negativity is preserved by the NMF method as a result of constraints that produce nonnegative lower rank factors that can be interpreted as semantic features or patterns in the text collection.

4.1   NMF Method

With the standard vector space model a set of documents S can be expressed as an $m \times n$ matrix V, where m is the number of terms and n is the number of documents in S. Each column V_j of V is an encoding of a document in S and each entry v_{ij} of vector V_j is the value of i-th term with regard to the semantics of V_j, where i ranges across the terms in the dictionary. The NMF problem is defined as finding an approximation of V in terms of some metric (e.g., the norm) by factoring V into the product WH of two reduced-dimensional matrices W and H [11]. Each column of W is a basis vector. It contains an encoding of a semantic space or concept from V and each column of H contains an encoding of the linear combination of the basis vectors that approximates the corresponding column of V. Dimensions of W and H are $m \times k$ and $k \times n$, where k is the reduced rank. Usually k is chosen to be much smaller than n. Finding the appropriate value of k depends on the application and is also influenced by the nature of the collection itself.

Common approaches to NMF obtain an approximation of V by computing a (W, H) pair to minimize the Frobenius norm of the difference $V - WH$. The matrices W and H are not unique. Usually H is initialized to zero and W to a randomly generated matrix where each $W_{ij} > 0$ and these initial values are improved with iterations of the algorithm.

4.2   GD-CLS Method

GD-CLS is a hybrid method that combines some of the better features of other methods. The multiplicative method, which is basically a version of the gradient descent optimization scheme, is used at each iterative step to approximate the basis vector matrix W. H is calculated using a constrained least squares (constrained least squares - CLS) model as the metric.

Algorithm

1. Initialize W and H with nonnegative values, and scale the columns of W to unit norm.
2. Iterate until convergence or after l iterations:
   - $W_{ic} = W_{ic} \frac{(VH^T)_{ic}}{(WHH^T)_{ic} + \epsilon}$, for c and i $[\epsilon = 10^{-9}]$
   - Rescale the columns of W to unit norm

- Solve the constrained least squares problem where $min_{H_j}\{||V_j-WH_j||_2^2+\lambda||H_j||2^2$ the subscript j denotes the j-th column, for $j = 1,\ldots,m$. Any negative values in H_j are set to zero. The parameter k is a regularization value that is used to balance the reduction of the metric $||V_j - WH_j||_2^2$ with the enforcement of smoothness and sparsity in H.

For any given matrix V, matrix W has k columns or basis vectors that represent k clusters, matrix H has n columns that represent n documents. A column vector in H has k components, each of which denotes the contribution of the corresponding basis vector to that column or document. The clustering of documents is then performed based on the index of the highest value of k for each document. For document i $(i = 1,\ldots,n)$, if the maximum value is the j-th entry $(j = 1,\ldots,k)$, document i is assigned to cluster j. We used the GD-CLS method for searching $k = 2,3,4,5$ clusters. Results are in Table 5. We can see that in the table are very readable results. For example in the first row is a vector which represents sale - Price, Sale, Discount, Credit, Description, and Login. In the second row is a vector which describes information cluster - Opinion, Discussion, Login, and Questionnaire. Limit for a successful pattern we set up to 0.05. The method was unstable for 6 and more clusters. In the table there are clusters-vectors with only one higher value (for example row 5). These clusters do not have a good information value because we expect at least two patterns on each page.

Table 5. Clustering by NMF

	Price	Sale	Discount	Credit	Description	Opinion	Discussion	Login	Bazaar	Questionnaire
1 / 2	0	0.01	0	0.01	0.03	0.29	0.27	0.337	0	0.06
2 / 2	0.33	0.25	0.19	0.06	0.07	0	0	0.05	0.03	0.01
1 / 3	0.35	0.26	0.21	0.07	0.07	0	0	0	0.04	0.01
2 / 3	0.02	0	0	0.01	0.01	0.41	0.46	0	0.02	0.06
3 / 3	0	0.08	0	0	0.08	0.04	0	0.76	0	0.04
1 / 4	0.42	0	0.41	0.04	0	0	0	0	0.13	0.01
2 / 4	0.01	0	0	0.02	0.02	0.42	0.47	0	0.01	0.06
3 / 4	0	0	0.01	0	0.05	0.04	0	0.85	0	0.05
4 / 4	0.26	0.49	0	0.09	0.15	0.01	0	0	0	0
1 / 5	0	0	0	0.02	0.02	0.42	0.47	0	0	0.06
2/ 5	0.25	0.50	0	0.09	0.16	0	0	0	0	0
3 / 5	0.31	0.01	0.57	0.06	0	0	0	0	0.04	0.01
4 / 5	0.62	0	0	0	0	0	0	0	0.36	0.02
5 / 5	0.01	0	0	0	0.04	0.04	0	0.85	0	0.05

5 Conclusion

From cluster analysis of GUI patterns, we found the most similar (from the point of view of degree of detection) price, discount and sale, and further opinion and discussion. These patterns appear important together in the characterizations of web page clusters obtained both by two-step cluster analysis and NMF. The NMF method reflects better the fact that the similarity of price and discount is a litter higher than the similarity of price and sale. By two-step cluster analysis, we did not obtain any combination of price and discount without sale. By NMF, we found such a combination in the cases of 4 and 5 clusters. Further, by NMF we found a combination opinion, discussion, login and questionnaire as important (for 2 clusters). Contained patterns correspond with members of the second cluster obtained by hierarchical cluster analysis of attributes. On the other hand, we found all clusters obtained by two-step cluster analysis with average membership degree with the value 0.1 and higher for 3 and more patterns. Experimental results on Web pages suggest the effectiveness of our approach. In future, we will also provide a unified view on binary clustering [10, 9] by establishing the connections among various clustering approaches.

Acknowledgement. This work was partially supported by grant 201/05/0079 awarded by the Grant Agency of the Czech Republic.

References

1. Alexander, Ch.: A Pattern Language: Towns, Buildings, Construction, Oxford University Press, New York 1977.
2. Dearden, A., Finlay, J.: Pattern Languages in HCI: A critical review, Human Computer Interaction, Vol. 21, No. 1, Pages 49–102. January 2006.
3. Ding, C., Li, T., Peng, W., and Park, H. Orthogonal nonnegative matrix t-factorizations for clustering. In Proceedings of the 12th ACM SIGKDD international Conference on Knowledge Discovery and Data Mining (Philadelphia, PA, USA, August 20 - 23, 2006). KDD '06. ACM Press, New York, NY, Pages 126–135.
4. Van Duyne D. K., Landay J. A., Hong J. I.: The Design of Sites: Patterns, Principles, and Processes for Crafting a Customer-Centered Web Experience. Addison-Wesley Professional, 2002.
5. Ivory, M. Y., Megraw, R.: Evolution of Web Site Design Patterns. ACM Transactions on Information Systems, Vol. 23, No. 4 (2005) Pages 463–497.
6. Kudělka, M. Snášel, V., Lehečka, O., El-Qawasmeh, E.: Semantic Annotation of Web Pages Using Web Patterns. IEEE/WIC/ACM conference WI-2006, Hong Kong 2006. Pages 329–333.
7. Kudělka, M. Snášel, V., Lehečka, O., El-Qawasmeh, E., Pokorný, J.: Semantic Annotation of Web Pages Using Web Patterns SITIS, IEEE/ACM Springer Verlag, Tunisia, 2006, in print
8. Labský, M., Svátek, V., Šváb, O., Praks, P., Krátký, M., Snášel, V.: Information extraction from HTML product catalogues: from source code and images to RDF, Web Intelligence, 2005. Proceedings. The 2005 IEEE/WIC/ACM International Conference on 19-22 Sept. 2005 Pages 401–404.
9. Li, T.: A general model for clustering binary data. In Proceeding of the Eleventh ACM SIGKDD international Conference on Knowledge Discovery in Data Mining (Chicago, Illinois, USA, August 21 - 24, 2005). KDD '05. ACM Press, New York, NY, Pages 188–197.
10. Řezanková, H., Húsek, D., Frolov, A. A.: Overlapping Clustering of Binary Variables. Anacapri. In: Knowledge Extraction and Modelling. Italy: TILAPIA Edizion, 2006, Pages 1–7.
11. Shahnaz, F., Berry, M., W., Pauca, P.V., Plemmons, R.J.: Document clustering using nonnegative matrix factorization. Information Processing and Management, Volume 42, Issue 2 (March 2006). Pages 373–386.

12. Tidwell, J.: Designing Interfaces: Patterns for Effective Interaction Design. O'Reilly Media, Inc. 2006.
13. Van Welie M., van der Veer G. C.: Pattern Languages in Interaction Design: Structure and Organization. Proceedings of Interact '03, Zürich, Switzerland. IOS Press, Amsterdam 2003.
14. Wellhausen, T.: User Interface Design for Searching. A Pattern Language. http://www.tim-wellhausen.de/papers/UIForSearching/UI   ForSearching.html (May 29, 2005)

CAEP: A Method Based on Expert Profiles for Recommend System*

Jingyu Sun, Xianhua Li, Xueli Yu, Jianlin Li, and Rui Wang

College of computer and software, Taiyuan Univ. of Tech., China, 030024
whitesunpersun@163.com

Summary. In this paper, we present a method based on Expert Profiles (EP) for recommend system in ontological knowledge community in order to improve the efficiency of obtaining accurate knowledge in current Web. At first we improve Computable Context-Awareness Approach (CCAA) to fit a recommend system. Secondly, we design an indexing method to store links between keywords, which a domain expert often uses, and URLs in Search Profile (SP). Through computing semantic similarity between keywords, which a regular user inputs, and ones stored in SP, Search Engine (SE) can recommend pages, which are experts' searching history. Finally, we give an instance running on Nutch Search Engine.

Keywords: Recommend System, Expert Profile, CCAA, Context-Awareness, CAEP.

1 Introduction

In current web, there are a lot of pages, which can be understood by human, but "Most of these sites do not yet make their data available in a machine understandable form" [1]. When one types keywords in search engine, it often returns millions of links he or she can choose. So he or she must spend much time clicking links and browsing many pages to find some pages he or she wanted. But there is an obviously different situation for domain experts searching in the same search engine. They can quickly find pages they wanted with their domain knowledge. So if a regular user has similar searching goal with one expert's, we can let search engine recommend some links based on this domain expert's searching history.

In order to recommend pages preferably, the critical issue is enabling users and search engine to understand the precision semantic meaning of the words or phrases chosen by users as well as to locate the user's requirements exactly. In fact, gaining the meaning of the word in essence is not easy, because it needs not only to analyze the syntax and morphology but also to grasp the semantic content.

* Sponsored by the Natural Science Foundation of China (No. 60472093), Scientific Research Foundation for the Returned Overseas Scholars in Shanxi Province (No. 2006-30) and Natural Science Foundation of Shanxi Province (No. 20041043).

K.M. Węgrzyn-Wolska and P.S. Szczepaniak (Eds.): Adv. in Intel. Web, ASC 43, pp. 337–342, 2007.
springerlink.com   © Springer-Verlag Berlin Heidelberg 2007

In this paper, we improve the Computable Context-Awareness Approach (CCAA) [2] to fit a recommend system and design an indexing method to store links between keywords, which a domain expert often uses, and URLs in Search Profile (SP). Through computing semantic similarity between keywords, which a regular user inputs, and ones stored in SP, Search Engine (SE) can recommend pages, which are expert's searching history.

For implementing and verifying this method, we design a demo (the url is http://www.whitesun.cn:8080/rindex.htm) to search our campus resources based on Nutch [3] During we put forward the method and design the system (named Whitesun Semantic Search Engine), we have referenced some application of semantic web technologies to improve search engines and knowledge management systems [4].

2 Design Philosophy

2.1 Context and Context-Awareness

As defined in paper[5],"Context is any information that can be used to characterize the situation of an entity. An entity is a person, place, or object that is considered relevant to the interaction between a user and an application,including the user and applications themselves".

Context-Awareness is used to design new user interfaces, and is often a part of ubiquitous and wearable computing. It is also beginning to be felt in the Internet with the advent of hybrid search engines. Ted Selker of MIT and Dey of Carnegie-Mellon are leading experts in this bleeding edge field[6].

2.2 CCAA and Expert

In Computable Context-Awareness Approach (CCAA) model[2], user profile includes Basic Profile (BP), represents as a set BP $\{b_1, b_2, b_3, ..., b_n\}$, and Character Profile (CP), represents as CP $\{p_1, p_2, p_3, ..., p_m\}$.The model combines BP and CP with the user's inputs which are his or her best interesting domain knowledge, represents as IKD $\{d_1, d_2, d_3, ..., d_h\}$ to compute the Role List of Community(RLC).We also can use CCAA to help a regular user to find similar domain experts.

In current web, Search Engine (SE) provides the same chance for everyone. But we all know that someone can find information they want due to they have searching skills and specific knowledge while someone can not. We define the former as Expert. If SE had stored the Experts' searching history in their profiles according to semantic relationship between keywords and URLs, a regular user could use them and the experts' searching history when the user's searching actions and experience could match the experts' profile by some methods.

In order to implement this good idea, we improve CCAA and give an extended Context-Awareness model based on Expert Profile, named CAEP. In a recommend system, we depend on CAEP model aiding regular users to gain knowledge in the corresponding ontology communities.

3 Context-Awareness Model Based on Expert Profile

3.1 Hypothesis of Context-Awareness Model

The extended context-awareness model is based on a hypothesis of searching information between regular users and experts: experts could find information they want due to they have searching skills and specific knowledge while regular users can not. When a regular user inputs his or her keywords, firstly SE analyzes his or her action and experience combined with expert profile DB (EPDB), secondly, SE finds all experts who have similitude keywords and experience through the model. Thirdly, SE recommends pages from EPDB and WebDB. Fig.1 is the process of query information without context-awareness; Fig.2 is that with context-awareness.

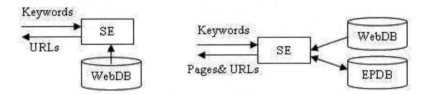

Fig. 1. Process of query information without context-awereness

Fig. 2. Process of query information with CAEP

3.2 Context-Awareness Model

This model includes two parts: (1) Computing which domain a regular user should choose in order to find domain experts exactly, named improved CCAA; (2) Computing which experts' keywords is similar to a regular user's inputs, named CAEP.

According to CCAA model [2], while a regular user logs in, he or she can input or select the most interest knowledge domain (IKD), then the Role List of Community (RLC) can be formed though the process of context-awareness on the regular users' profiles and the IKD inputting newly. Then, he or she can choose one community, which usually is the first one in the RLC list to log in [2]. So domain experts can be fixed who had searched in one community, which he or she is interesting.

In order to compute semantic similarity between keywords of a regular user and one of domain experts', we have enhanced users' profile.We give some definitions about CAEP as follow:

Definition 1. *Expert Profile (EP):* includes Basic Profile (BP), represents as a set BP$\{b_1, b_2, ..., b_n\}$, Character Profile (CP), represents as CP$\{p_1, p_2, ..., p_m\}$, and Searching Profile(SP), represents as SP $\{s_1, s_2, ..., s_t\}$. BP and CP are same to user profile, but SP is a special profile which can store and index experts' searching history in one domain. We design a new data structure to represent SP shown as Fig.3.

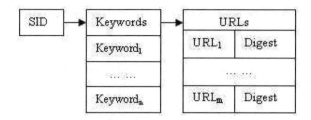

Fig. 3. An element of SP's Data Structure

Definition 2. *Searching Frequency (SF)*: depending on the situation of choosing searching history of an expert, system can designate a value for every SID, called Searching Frequency (SF), and provide the value, which is 0 in default. Administrators of SE or experts can also specify a proper value according to semantic relationship between experts' keywords and URLs. So we can represent $s_r \in SP$ as a 2-Tuple(SID, SF).

Based on definitions above, we give two algorithms to compute RLC and aid a regular user to get recommend pages from experts according to SP.

Algorithm 1: the target of CCAA is to choose domain experts and narrow down the searching scopes of a regular user greatly and the whole executing course of CCAA includes 3 steps. There is more information in Paper [2] in detail. So we need only find domain experts in which community a regular user chooses.

Algorithm 2: When a regular user inputs keywords in order to get his or her wanted knowledge, SE can recommend pages, which domain experts often browse. The target of this algorithm is to compute semantic similarity between a regular user's keywords and one domain expert's. The steps are as follows:

1. Making keywords' set:

We take $UKW(u_1, ..., u_i, ..., u_n)(i = 1, ..., n)$ to represent keywords set which a regular user inputs. When there are no more than $N(N \leq 5)$ experts in one community, we must search SID with maximal semantic similarity for keywords as step 2.

2. Computing Semantic Similarity for Keywords (abbr. SSK):

(1) Choose one SID to Compute SSK:

We take $EKW(e_1, ..., e_j, ..., e_m)(j = 1, ..., m)$ to represent a SID of one domain expert's keywords set, choose the SID with maximal SF and compute some expression as follows:

  a. $n_1 = max(n, m), n_2 = min(n, m)$;

  b. $f(u_t, e_t), t = 1, ..., n_2$;

  c. $s = sum(f(u_t, e_t))/n_1$.

Function $f(u_t, e_t)$ is a fuzzy function and the domain of value is $[0, 1]$. For simple computing, $f(u_t, e_t) = 1$ when $u_t = e_t$, otherwise, $f(u_t, e_t) = 0$; Function $sum(f(u_t, e_t))$ is used to compute sum of all $f(u_t, e_t), t = 1, ..., n_2$; the value of s is SSK of choose SID and if $s < S_{recommended}$(threshold for recommended pages and usually 0.6), we must choose another SID as follow step.

(2) Choose SID which $s \geq S_{recommended}$:

We search one SID in choose expert's SP in order and compute SSK as above step until there is one SID with $s \geq S_{recommended}$.

(3) When system couldn't find one SID with $s \geq S_{recommended}$, we must choose next domain expert and compute SSK with above step.

(4) In order to get satisfied SID, we can compute maximal SSK or s using all domain experts' SP and store links between SSK and keywords which users often use with Hash Function in SP.

3. Recommend Pages:

Finding the SID with maximal SSK, SE recommends URLs at Browser in right and shows other pages (at Browser in left) as usually.

4   A Recommend System Based on Expert Profile

We design a recommend system(http://www.whitesun.cn:8080/rindex.htm) to search our campus resources based on Nutch [3] Search Engine with above method. In the system, we use OWL to represent communities and take Jena as an inference tools to manage them.

For simplifying situation we choose community amount of our system to 14. These communities can be organized with hierarchical tree and made by Protege. In order to example easily, we represent communities a set :(Education, Music, Movie, Computer, Software, Database, Oracle, SQL2000, OS, DOS, Windows, Unix, Internet, Management).

For example, Mr. Li is an expert at OS (He is an OS teacher in fact.), and he often searches pages about OS in this system. So we take him as a domain expert at OS, and system allocates him an expert account and stores his searching history in his profile as above CAEP model.

Mr. White has completed his registration as a regular user successfully, and his profile has been produced. When he logs in the system and inputs "OS" and "Windows" as his interesting knowledge domain, the system will analyze and produce a RCL automatically depended on BP, CP and IKD sets as follows:

Windows, Education, Music, Movie, Computer, Software, Database, Oracle, SQL2000, OS, DOS, Unix, Internet, Management

Now, White can choose "Windows" community role to log in the system, and system chooses domain experts (for example, Mr. Li) for recommending pages, who are in community Windows or father community (such as OS), which is an inference of communities represented in OWL by Jena. Therefore, White can query pages about "Windows" under the backgrounds "Computer" and "OS".

Let's suppose that the user "White" wants to search some pages about "Apple Brand Computer". So he can take "Apple" as a keyword to submit, and the system can return some links at left and some recommend pages at right in browser.

The process which White logs in and queries is shown as http://www.white sun.cn:8080/rindex.htm.

5 Conclusion

In this paper, the improved CCAA and CAEP are the core method for a recommend system. So firstly we improve CCAA to fit a recommend system and get domain experts. Secondly, we design CAEP model to store links between keywords, which a domain expert often browses, and URLs in Search Profile (SP); and through computing SSK between a regular user and domain experts, Search Engine (SE) recommends pages, which are expert's searching history. Finally we have a demo running on Nutch Search Engine to search our campus resources.

However, we need find good method to find domain experts and speed up computing SSK in order to recommend better pages for a regular user. So there are a lot of works and this is our future works.

References

1. R.Guha, R. McCool, etc,Semantic search. In WWW2003, Proc.of the 12th international conference on World Wide Web, ACM Press, 2003, pp 700-709.
2. Xueli Yu, Jingyu Sun,etc, Studying on the Model of Context-Awareness and Content-Awareness In Ontological Knowledge Community, Lecture Notes in Computer Science, Volume 3528 / 2005,p.468-p.474 ,2005.
3. Nutch (version 0.8.1), http://lucene.apache.org/nutch/index.html, 2006.9.12.
4. Li Ding,etc. Swoogle: A search and metadata engine for the semantic web. In Proceedings of the Thirteenth ACM Conference on Information and Knowledge Management. ACM Press, 2004.
5. Anind K.Dey, Gregory D.,Abowd,Towards a Better Understanding of Context and Context-Awareness, Presented at the CHI 2000 Workshop on the What, Who, Where, When, Why and How of Context-Awareness, April 1-6, 2000.
6. http://www.answers.com/topic/context-awareness, 2006.12

Web Co-citation: Discovering Relatedness Between Scientific Papers

Thanh-Trung Van and Michel Beigbeder

Centre G2I/Département RIM
Ecole Nationale Supérieure des Mines de Saint Etienne
158 Cours Fauriel, 42023 Saint Etienne, France
{van,mbeig}@emse.fr

Summary. In this paper we review two well-known citation methods to find relatedness between scientific papers: co-citation and bibliographic coupling. We propose a practical method to estimate the co-citation relatedness using the Google search engine. We call this method Web co-citation. We conducted experiments on a collection of scientific papers to compare the performances of different methods. The experimental results show that our approach, despite its simplicity, is efficient in discovering the relatedness between scientific papers.

1 Introduction

For a long time, citation-based methods have been used to find relatedness between scientific papers beside content-based methods. In 1963 Kessler [1] proposed the *bibliographic coupling* method. In this method, the similarity between two papers is based on the number of their *co-references*. He supposed that if two papers have common references in their bibliographies, they may focus (entirely or partially) on the same topic. In 1973 Marshakova [2] and Small [3] independently proposed another method called "co-citation". In this method, the relatedness between two papers is based on their *co-citation frequency*. The co-citation frequency is the frequency that two papers are *co-cited*. Two papers are said to be *co-cited* if they appear together in the bibliography section of a third paper.

The two methods bibliographic coupling and co-citation have been used widely since about 40 years for different purposes. The digital library CiteSeer[1] uses these methods to find related papers. In [4] the co-citation method is used to create a patent classification system for conducting patent analysis and management. Recently, these methods are used in hyperlinked environments to find the relatedness between Web pages [5, 6] because of the similarity between the notion of "citations between scientific papers" and "links between Web pages". However, both of these methods have their limits. In the bibliographic coupling method, the relatedness between two papers is fixed since their publication date because they are based on the number of their co-references which is unchanged.

---

[1] http://citeseer.ist.psu.edu/

K.M. Węgrzyn-Wolska and P.S. Szczepaniak (Eds.): Adv. in Intel. Web, ASC 43, pp. 343–348, 2007.
springerlink.com

In the co-citation methods, with the time two related papers may receive more and more citations and their co-citation frequency can increase. However if we want to know this citation information, we have to extract from the *citation graph* of the actual library or read from a *citation database*[2] which are usually limited; i.e. we can only know citing papers of a given paper if these citing papers exist within the same digital library or citation database. That is why in this paper we propose an approach to compute co-citation relatedness between scientific papers which can overcome this limit.

The rest of this paper is organized as follows. In Sec. 2 we describe two approaches to compute co-citation relatedness between scientific papers: traditional approach using the Web of Science citation database and our new approach using the Google search engine. Sec. 3 presents our experiments: simulation of personalized searching using different citation methods. The paper concludes in Sec. 4.

2   Methodology

2.1   Using Web of Science as Citation Database

Actually, there are many citation databases like Web of Science[3], Scopus[4] and digital libraries like CiteSeer, ACM Digital Library which provide citation information about scientific papers. After regarding in detail these sources, we decided to choose Web of Science (WoS) as a citation database in our experiments. The Web of Science of Thomson ISI is an important citation database which is used widely for citation studies [7]. Besides, it also provides an API which facilitates the access to its database without using an Web browser. Another important reason for using WoS is that it contains most of journals used in our experiments (see Sec. 3.)

In WoS, an article is represented by a primary key called **UT**. Its API supports many operations on its database. Thanks to the search service of ISI, if we know some information about a paper (like title, year of publication, journal etc.) we can use these information to find the UT primary key of this paper in WoS database by calling the *searchRetrieve* function. Then using this UT primary key we can find all papers that cite this paper with the *citingArticles* function. From these information we can know the number of times that a paper is cited or the frequency that two papers are co-cited in WoS database. More documentation about ISI search service could be found in its support site[5].

2.2   Web Co-citation Method

With the explosion of the World Wide Web, Web search engines have to be more and more complete in order to satisfy information needs of users and their

---

[2] A citation database is a system that can provide bibliographic/citation information of papers.

[3] http://portal.isiknowledge.com

[4] http://www.scopus.com/scopus/home.url

[5] http://scientific.thomson.com/support/faq/webservices/

databases become bigger with the time. With their huge databases, Web search engines could be a good source for many data mining tasks.

Recently, a new method for citation analysis called Web citation analysis begins attracting the research community. Web citation analysis finds citations to a scientific paper on the Web by sending the query containing the title of this paper (as phrase search using quotation marks) to a Web search engine and analyze returned pages [8]. Because a Web search engine can index many kinds of documents in many different formats, the notion of "citation" used here is a "relaxation" in comparison with traditional definition

In our Web co-citation method, we compute the co-citation similarity of two scientific papers by the frequency that they are "co-cited" on the Web; i.e. the frequency that they are mentioned by a Web page. The notion of "co-citation" used here is also a "relaxation" in comparison with the traditional definition. If the Web document that mentions two scientific papers is another scientific paper then these two papers are normally co-cited. However, if this is a table of content of a conference proceeding, we could also say that these two papers are co-cited and have a relation because a conference normally has a common general theme. If these two papers appear in the same conference, they may have the same general theme. Similarly, if two papers are in the reading list for a course, they may focus on the same topic of this course. In summary, if two papers appear in the same Web document, we can assume that they have a (strong or weak) relation. The search engine used in our experiment is the Google search engine. To find the number of times that two papers are "co-cited", we send the titles of these two papers (as phrase search and in the same query) to Google and note the number of hits returned. In our experiments, we use a script to automatically query Google instead of manually using a Web browser.

3 Experiments

As stated above, in this work we conduct experiments for evaluating performance of two methods: bibliographic coupling and co-citation with Web of Science and Google. The experiments described here are simulations of *personalized searching* in a digital library using *user profiles*. Users of information retrieval systems generally use short queries to describe their information need. Because of the polysemy and synonym problems of natural language, these short queries become ambiguous and lead to wrong answers. However if the system knows about user, it can use these information to improve searching performance. The information about each user is called *user profile*. Generally, a user profile is a set of information that represent interests and preferences of a user.

3.1 Test Collection and Evaluation Procedure

The test collection that we use in our experiments is the collection used in INEX 2005 (version 1.9[6]). In the first step we remove all elements that are not scientific papers. After this process, the collection contains 14237 documents. Then

---

[6] http://inex.is.informatik.uni-duisburg.de/2005/

we extract all necessary information for our experiments from these documents (title, journal, publication year, bibliography etc.). There are also many topics with relevance assessments distributed with the collection. Each topic represents an information need and the relevance assessments were done by INEX participants. In our experiments we use only CO topics which do not contain structure of documents to create user queries. INEX uses a two-dimensional, multi-valued scale for relevance assessments of each topic. However in our experiments we use precision/recall metrics with binary scale relevant/non-relevant). Therefore we did a transformation on the relevance assessments of INEX: if a document has at least one element which is judged relevant (entirely or partially), this document will be considered as relevant; otherwise it will be considered as non-relevant. There are 29 original CO topics but only 20 topics that have more than 30 relevant documents will be used for experiments.

As mentioned above, our experiments are simulations of personalized searching using user profiles. In this case, 20 topics represent different information needs of 20 different people. For each topic, we choose some relevant papers as "pseudo user profile" of this person (5 in average in our experiments). (Please note that our goal is not to learn user profiles but to evaluate citation-based methods). The selected papers are chosen among the highly relevant papers to the correspondence topic and those that receive many citations from other documents. The papers which are included in these profiles are removed from the collection to avoid effect on the experimental results.

After the preparation step, we use the **zettair**[7] search engine to index the INEX collection (the default model is *Dirichlet-smoothed*), then we send 20 queries (which are formed from above topics) to **zettair**; with each query we take the first 300 documents for re-ranking using "user profiles" of correspondence topic. The similarity between a document d and a user profile p is computed as:

$$similarity(p, d) = \sum_{d' \in p} similarity(d', d) \tag{1}$$

In Eq. 1, $similarity(d', d)$ is the similarity (bibliographic coupling and co-citation) between a document d' in profile p and document d. The co-citation similarity between two papers is defined as:

$$cocitation\_similarity(d', d) = ln(\frac{cocitation(d', d)^2}{citation(d') \cdot citation(d)}) \tag{2}$$

In Eq. 2, $cocitation(d', d')$ is the number of times that these two papers are co-cited, $citation(d')$ and $citation(d)$ are respectively the citation frequency that papers d' and d received. The bibliographic coupling similarity is computed by a similar formula. The final score of a document is obtained by combining its original score computed by **zettair** and the similarity document-profile. In our experiments we tried two combining functions: a linear function and a product

---

[7] http://www.seg.rmit.edu.au/zettair

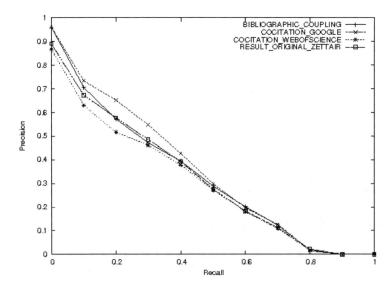

Fig. 1. Experimental results: re-ranking search results of zettair with different citation-based methods

Table 1. Precision at 5, 10, 20, 30 documents

	Original Result	Bibliographic coupling	Co-citation using WoS	Co-citation using Google
At 5 docs	0.6600	0.7300	0.6300	0.7100
At 10 docs	0.6150	0.6050	0.5900	0.6800
At 20 docs	0.5375	0.5600	0.5150	0.6025
At 30 docs	0.4867	0.4883	0.4567	0.5600

function. However, in our experiments the product combination seems to be better than linear combination, thus it is used in final results which are presented in the next part.

3.2   Results and Discussion

The experimental results are presented in Fig. 1 (precision/recall) and Tab. 1 (precision at 5, 10, 20, 30 documents). The **trec_eval**[8] program is used for evaluation.

From the experimental results, we can see that the co-citation method using the WoS database does not bring any improvement, it even causes a slight performance decrease. The bibliographic coupling method performs better but not very clearly. The co-citation method using Google is the best, it brings 15.06% improvement for the precision at top 30 documents.

Now we will analyze the experimental data to explain these results. To compute the similarity between documents and "profiles" for re-ranking, we have to

---
[8] http://trec.nist.gov/trec_eval/

compute the co-citation (or co-reference) frequency of 25497 pairs of documents (each pair consists of a document to be re-ranked and a document in a "user profile"). In the co-citation methods using Web of Science database, only 213 pairs are co-cited with the average co-citation frequency of each pair is 1.94. This small number of co-cited pairs is the reason why it could not bring any improvement and even becomes a noisy source which causes bad effect on the final result. In the bibliographic coupling method, there are 1126 pairs of documents which have co-references with the average number of co-references of each pair is 1.69. This is a little better than the first case and it is able to make some improvement. In the co-citation method using Google, there are 4845 pairs of documents which are "co-cited" with the average co-citation frequency of each pair is 4.84. This is much better than the first two cases. That is why it gains the best performance.

4   Conclusions and Future Work

In this paper we consider two famous citation-based methods: bibliographic coupling and co-citation. We propose new approach to compute co-citation relatedness between scientific papers using the Google search engine. Experimental results show that such approach could be more efficient than the traditional approach. We believe that this new approach could be successfully applied to other applications like classification, clustering of scientific papers, finding related papers etc. Another approach we are considering is to combine multiple different citation databases that could lead to better performance of co-citation method.

References

1. Kessler, M.M.: Bibliographic coupling between scientific papers. American Documentation (1963) 10–25
2. Marshakova, I.: System of document connections based on references. Nauchno-Tekhnicheskaya Informatsiya Seriya 2 – Informatsionnye Protsessy i Sistemy (1973) 3–8
3. Small, H.G.: Co-citation in the scientific literature: A new measure of the relationship between two documents. Journal of American Society for Information Science **24**(4) (1973) 265–269
4. Lai, K.K., Wu, S.J.: Using the patent co-citation approach to establish a new patent classification system. Information Processing and Management **41**(2) (2005) 313–330
5. Pitkow, J., Pirolli, P.: Life, death, and lawfulness on the electronic frontier. In: CHI '97: Proceedings of the SIGCHI conference on Human factors in computing systems, New York, NY, USA, ACM Press (1997) 383–390
6. Dean, J., Henzinger, M.R.: Finding related pages in the world wide web. In: WWW '99: Proceeding of the eighth international conference on World Wide Web, New York, NY, USA, Elsevier North-Holland, Inc. (1999) 1467–1479
7. Jacso, P.: As we may search : Comparison of major features of the web of science, scopus, and google scholar citation-based and citation-enhanced databases. Current Science **89**(9) (2005) 1537–1547
8. Vaughan, L., Shaw, D.: Bibliographic and web citations: what is the difference? J. Am. Soc. Inf. Sci. Technol. **54**(14) (2003) 1313–1322

An Integrated System of Semantic Web Reasoning and Argument-Based Reasoning

Toshiko Wakaki[1], Hajime Sawamura[2], and Katsumi Nitta[3]

[1] Shibaura Institute of Technology, Faculty of Systems Engineering,
307 Fukasaku, Minuma-ku, Saitama-City, Saitama, 337–8570 Japan
`twakaki@sic.shibaura-it.ac.jp`
[2] Niigata University, Faculty of Engineering,
8050, 2-cho, Ikarashi, Niigata, 950-2181 Japan
`sawamura@ie.niigata-u.ac.jp`
[3] Tokyo Institute of Technology, Department of Computational Intelligence
and Systems Science, 4259 Nagatsuta, Midori-ku, Yokohama 226–8502, Japan
`nitta@dis.titech.ac.jp`

Summary. So far only rule-based knowledge has been taken into account in multi-agent systems based on argumentation. Recent progress of the Semantic Web technology provides expressive ontology languages. In this paper, we present an integrated system of Semantic Web reasoning and argument-based reasoning, where the Logic of Multiple-valued Argumentation-based agent system (specialized to two values $\{f, t\}$) can inquire of the description logic reasoning system for ontologies that are lacking in agent's knowledge bases. Our approach contributes to advanced argumentative reasoning system with capabilities such as semantic web-based reasoning, which is shown by illustrating an interesting example of argumentation handling both rules and and ontologies.

1 Introduction

Argumentation is a powerful tool in our daily life as well as in the agent's world. A Logic of Multiple-valued Argumentation (LMA, for short) built on rule-based knowledge bases expressed by extended annotated logic programs (EALPs) was recently proposed to formalize multiple-valued argument models to express different kinds of uncertainty such as vagueness and paraconsistency in agent's argumentation [8]. Though it allows to specify various types of truth values depending on application domains, it cannot handle ontological knowledge.

In our daily life, however, there are a lot of human argumentation where both ontological and rule knowledges are used. For example, in e-commerce, a seller and a buyer usually need to use ontologies about products along with their respective strategic rules for buying and selling in their argumentation.

In this study, in order to meet such requirements, we challenge a novel attempt to integrate monotonic Semantic Web reasoning and non-monotonic argument-based reasoning, taking into account of recent progress of the Semantic Web technologies providing expressive ontology languages such as OWL DL. In our

K.M. Węgrzyn-Wolska and P.S. Szczepaniak (Eds.): Adv. in Intel. Web, ASC 43, pp. 349–356, 2007.
springerlink.com

hybrid approach, Semantic Web reasoning is established as the description logic reasoning system which makes use of answer set programming [1, 4, 5], whereas the Logic of Multiple-valued Argumentation-based agent system (specialized to two values $\{f, t\}$) can inquire of the description logic reasoning system for ontologies that are lacking in agent's knowledge bases.

In this paper, after introducing the overview of our *WebArg* system, we briefly address theoretical aspects w.r.t. integration of Semantic Web reasoning and LMA-based agent argumentation. Section 3 concludes with our future works.

2   Integration of Semantic Web Reasoning and Argument-Based Reasoning

We assume that a huge amount of ontologies expressed by OWL DL on the Semantic Web are given to every agent as the common knowledge, whereas each agent has its own rule-based knowledge expressed by EALP, and communicates with each other in LMA-based multi-agent argumentation with being allowed to consult the description logic reasoning system for ontologies if required. Figure 1 shows our *WebArg* system architecture, which consists of two components as follows:

1. Semantic Web reasoning system including ontology translation from OWL DL to DL $\mathcal{SHOIN}(D)$ (or vice versa) as well as the DL reasoning system that is, the theorem prover for description logics.
2. LMA-based multi-agent argumentation system which has the communication interface with the DL reasoning system.

2.1   Semantic Web Reasoning System

Based on the semantic equivalence between OWL DL and the DL $\mathcal{SHOIN}(D)$ [7], OWL DL ontologies on the semantic web are translated into DL $\mathcal{SHOIN}(D)$

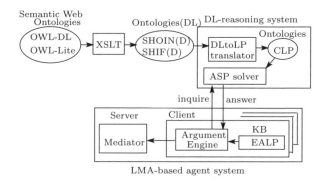

Fig. 1. WebArg system architecture

ontologies by means of XSLT [9] whose translation rules are described as template rules in the style sheet. The inverse translation of ontologies from the DL $\mathcal{SHOIN}$(D) to OWL DL is also available in our *WebArg* system.

(1) Syntax of $\mathcal{SHOIN}$(D)

We first describe the syntax of $\mathcal{SHOIN}$(D)[7].

Definition 1. *(Concepts and Roles)*
We assume a set **D** of *elementary datatypes*. Let **A**, $\mathbf{R}_A$, $\mathbf{R}_D$,and **I** be nonempty finite sets of *atomic concepts*, *abstract roles*, *datatype roles*, and *individuals*, respectively. Let $\mathbf{R}_A^-$ be the set of all inverses R^- of abstract roles $R \in \mathbf{R}_A$. A *role* is an element of $\mathbf{R}_A \cup \mathbf{R}_A^- \cup \mathbf{R}_D$. *Concepts* are inductively defined as follows:

$$A \mid \neg\, C \mid C \sqcap C' \mid C \sqcup C' \mid \{o_1, \ldots\} \mid \forall R.C \mid \exists R.C \mid\, \geq nR \mid\, \leq nR \mid$$
$$\forall U.D \mid \exists U.D \mid\, \geq nU \mid\, \leq nU$$

where C, C' are *concepts*, and $A \in \mathbf{A}$, $D \in \mathbf{D}$, $R \in \mathbf{R}_A \cup \mathbf{R}_A^-$, $U \in \mathbf{R}_D$, $o \in \mathbf{I}$.

Note: In our system, a *datatype role* $U \in \mathbf{R}_D$ is not supported.

We define *a DL knowledge base* expressing ontological knowledge as follows.

Definition 2. *(A DL knowledge base)*
An *axiom* is an expression of one of the following forms: $(1)C \sqsubseteq D$ (*concept inclusion*), $(2)R \sqsubseteq S$ (*role inclusion*), (3)Trans(R), where $R \in \mathbf{R}_A$ (*role transitivity*). $(4)C(a)$, where C is a concept and $a \in \mathbf{I}$ (*concept membership*), $(5)R(a,b)$, where $R \in \mathbf{R}_A$ and $a,b \in \mathbf{I}$ (*role membership*), $(6)a = b$ (resp., $a \neq b$), where $a,b \in \mathbf{I}$. Tbox (resp., Abox) is a finite set of axioms whose form is $(1) \sim (3)$ (resp., $(4) \sim (6)$). A DL knowledge base is defined by Tbox $\cup$ Abox.

Example 1. The curriculum ontologies of the university are expressed as the DL knowledge base $\mathcal{K}_1$ in Figure 2, where $C \equiv D$ denotes both $C \sqsubseteq D$ and $D \sqsubseteq C$.

\<TBox\>
$uni\_curriculum \equiv course$
$course \equiv faculty\_e \sqcup faculty\_s \sqcup faculty\_h$
$cs\_dept \sqsubseteq faculty\_e$
$ee\_dept \sqsubseteq faculty\_e$
$math\_dept \sqsubseteq faculty\_s$
$philo\_dept \sqsubseteq faculty\_h$
$we\_logic \equiv western\_logic \sqcup eastern\_logic$
$we\_logic \sqsubseteq logic$
$mp\_logic \equiv math\_logic \sqcup philo\_logic$
$mp\_logic \sqsubseteq logic$
$easy\_get\_credit \equiv \exists reg.(\neg pass)$
$hard\_get\_credit \equiv \forall reg.pass$

\<ABox\>
$hard\_get\_credit(philosophy)$
$reg(ai,\ st\_0)$
$cs\_dept(l1)$
$math\_dept(l2)$
$phili\_dept(l3)$
$cs\_dept(ai)$
$cs\_dept(c\_programming)$
$cs\_dept(prolog\_programming)$
$cs\_dept(technical\_english)$
$math\_logic(l2)$
$philo\_logic(l1)$
$\neg pass(st\_0)$

Fig. 2. The curriculum ontologies: $\mathcal{K}_1$

(2) Semantics of $\mathcal{SHOIN}$(D)

Let $\mathcal{K}$ be a DL $\mathcal{SHOIN}$(D) knowledge base. We say $\mathcal{K}$ is *satisfiable* (resp., *unsatisfiable*) iff $\mathcal{K}$ has a (resp., no) model [7]. An axiom F is a *logical consequence* of the DL knowledge base $\mathcal{K}$, denoted $\mathcal{K} \models F$, iff every model of $\mathcal{K}$ satisfies F.

(3) DL Reasoning System

To reason with a DL knowledge base $\mathcal{K}$, we adopted Heymans et al's method [5]. They proposed (extended) *conceptual logic programs* (CLPs) which not only enable to simulate monotonic reasoning in the description logic $\mathcal{ALCHOQ}$ ($\sqcup, \sqcap$) with DL-safe rules but also allow nonmonotonic reasoning in locally closed subareas of the Semantic Web. The DL $\mathcal{ALCHOQ}(\sqcup, \sqcap)$ differs from the DL $\mathcal{SHOIN}$(D) that adds transitive roles, inverse roles, and data types to $\mathcal{ALCHOQ}(\sqcup, \sqcap)$, while removing support for role constructors and qualified number restrictions from it, and allowing only unqualified number restrictions. Reasoning with the CLP transformed from a DL knowledge base, is reduced to finite, normal Answer Set Programming (ASP)[4, 1] thanks to its restricted syntax as well as the newly introduced anonymous constants, which are based on their theorem as follows.

Theorem 1. *[5] For an $\mathcal{ALCHOQ}(\sqcup, \sqcap)$ knowledge base Σ, a DL-safe program P, and a ground atom α, we have*

$$(\Sigma, P) \models \alpha \quad \text{iff} \quad \Phi_1(\Sigma, \ P) \cup \Phi_2(\Sigma, \ P) \models \alpha, \tag{1}$$

Roughly speaking, Σ is a set of terminological axioms (i.e. Tbox) in terms of $\mathcal{ALCHOQ}$ ($\sqcup, \sqcap$), and P is a set of DL-safe rules including axioms in Abox. Then, by interpreting Σ as a first order theory $\pi(\Sigma)$, $(\Sigma, P) \models \alpha$ means that that α is a *logical consequence* of $\pi(\Sigma) \cup P$, whereas $\Phi_1(\Sigma, \ P) \cup \Phi_2(\Sigma, \ P)$ in the right part of (1) denotes the CLP transformed from given Σ and P based on Heymans et al.'s method. In our system, since a DL knowledge base $\mathcal{K}$ is expressed by DL $\mathcal{SHOIN}$(D), not only $\geq nR$ and $\leq nR$ are evaluated as $\geq nR.\top$ and $\leq nR.\top$ allowed in $\mathcal{ALCHOQ}(\sqcup, \sqcap)$ respectively, but also ASP rules expressing transitive roles, inverse roles are added to such a transformed CLP. Using the ASP solver DLV [1], our WebArg system evaluates the right part inference of the formula (1) by deciding if the following ASP program:

$$\Phi_1(\Sigma, \ P) \cup \Phi_2(\Sigma, \ P) \cup \{cte(x_i) \leftarrow \mid 1 \leq i \leq k\} \cup \{not \ \alpha \leftarrow\}$$

is inconsistent, where cte is a newly introduced predicate symbol to introduce k anonymous constants, x_i $(1 \leq i \leq k)$ in the program [5].

Example 2. According to Heymans et al.'s method, the axiom,
$$easy\_get\_credit \equiv \exists reg.(\neg pass)$$
from Tbox in $\mathcal{K}_1$ is translated into CLP rules in $\Phi_1(\Sigma, \ P)$ as follows:

$$\leftarrow easy\_get\_credit(X), not \ \{\exists reg.(\neg pass)\}(X),$$
$$\leftarrow \{\exists reg.(\neg pass)\}(X), not \ easy\_get\_credit(X),$$
$$\{\exists reg.(\neg pass)\}(X) \leftarrow reg(X,Y), \{\neg pass\}(X),$$

$$\{\neg pass\}(X) \leftarrow not\ pass(X).$$
$$easy\_get\_credit(X) \vee not\ easy\_get\_credit(X) \leftarrow,$$
$$pass(X) \vee not\ pass(X) \leftarrow,$$
$$reg(X, Y) \vee not\ reg(X, Y) \leftarrow,$$

and $cs\_dept(ai)$ from Abox in $\mathcal{K}_1$ is translated into the following:
$$cs\_dept(ai) \leftarrow, \qquad cs\_dept(ai) \vee not\ cs\_dept(ai) \leftarrow .$$
which are CLP rules in $\Phi_2(\Sigma,\ P)$.

2.2 LMA-Based Multi-agent Argumentation System

A LMA-based agent system is a multi-agent system which simulates the process of human argumentation based on the dialectical proof theory [6, 8]. It consists of a server program as a mediator which controls the multi-agent argumentation, and client programs as agents which communicate with each other through the exchange of arguments. In our approach, it has been slightly augmented with the reasoning interface of the DL reasoning system as shown in Figure 1, through which it is allowed to consult the DL reasoning system for unknown or unresolved literals in putative arguments in the process of argumentation. Consulting in this manner amounts to the theoretical setting as follows.

(1) Knowledge Representation and Integration

Our argument-based agent system is formalized as follows. A LMA-based agent A_i ($1 \leq i \leq n$) has its own knowledge base KB_i expressed by an EALP, whose rule has the following form:

$$H \leftarrow L_1\ \&\ \ldots\ \&\ L_k\ \&\ not\ L_{k+1}\ \&\ \ldots\ \&\ not\ L_m, \qquad (2)$$

where H, L_j ($1 \leq j \leq m$) are annotated object literals (or annotated literals, for short). An EALP has an ideals-based semantics [3], especially on the complete lattice $\mathcal{TWO} = (\{\mathbf{t}, \mathbf{f}\}, \leq)$ in our case (see the right part of Figure 3 where $\mathbf{f} \leq \mathbf{t}$, and the ideal of ν is defined as $\|\nu\| = \{\rho \in \mathcal{TWO} \mid \rho \leq \nu\}$).

In our system, however, given the DL knowledge base $\mathcal{K}$, each agent A_i is regarded as virtually possessing the following knowledge base P_i,

$$P_i = KB_i \cup \{l : \mathbf{t} \leftarrow\ |\ \mathcal{K} \models l\ \} \qquad (1 \leq i \leq n) \qquad (3)$$

where $\mathbf{t}$ is an annotation, and l is an axiom whose form is $C(a)$ or $R(a, b)$ such that C (resp. R) is a concept (resp. a role), and a, b are individuals. Hereafter, we call $l : \mathbf{t}$ an annotated DL atom. It should be noted that, though no DL atoms are allowed to occur in the head (i.e. H) of a rule (2), it is possible for annotated DL atoms to occur as literals L_j in the body.

The left part of Figure 4 shows an example of the rule-based knowledge bases for a Lecturer agent and a Student agent where an annotated DL atom $\alpha : \mathbf{t}$ is described by $DL[\alpha]$ called a *DL expression*, due to implementability issues.

(2) Argumentation Framework of Mutli-agent Systems

Though our argumentation framework of mutli-agent systems is formalized as usual [8], it is slightly extended to handle ontologies. An *argument* is a finite

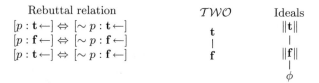

Fig. 3. The rebuttal relation based on the complete lattice for ideals of $\mathcal{TWO}$

sequence of rules from an EALP. In our approach, each agent has virtually a knowledge base: $P_i = KB_i \cup \{l : \mathbf{t} \leftarrow | \mathcal{K} \models l\}$, in which the ontological knowledge base $\mathcal{K}$ is shared among agents concerned. So, let $Args_{P_i}$ be the set of arguments associated with the knowledge base P_i for each agent A_i $(1 \leq i \leq n)$. Then, for multi-agents $MAS = \{A_1, \ldots, A_n\}$ associated with the set of EALPs $\{P_1, \ldots, P_n\}$, we define the set $Args_{MAS}$ of arguments as follows:

$$Args_{MAS} = Args_{P_1} \cup \ldots \cup Args_{P_n}.$$

(3) Justified Arguments and the Dialectical Proof Theory

Given $Args_{MAS}$ for multi-agents where DL atoms may occur in its argument, an *annotated attack relation* x on $Args_{MAS}$ is defined as a binary relation on $Args_{MAS}$, i.e. $x \subseteq Args_{MAS} \times Args_{MAS}$. There are various annotated attack relations such as *rebut, undercut, attack, defeat* and so on [8].

For example, for $Args_{MAS} = \{[p : \mathbf{t} \leftarrow], [p : \mathbf{f} \leftarrow], [\sim p : \mathbf{t} \leftarrow], [\sim p : \mathbf{f} \leftarrow]\}$[1], there exist three pairs in the rebuttal relation based on the complete lattice for ideals of $\mathcal{TWO}$ as shown in Figure 3.

Now, let x and y be variables to denote annotated attack relations, A be an argument, and S be a set of arguments. Then A is x/y-acceptable wrt. S if for every argument B such that $(B, A) \in x$ there exists an argument $C \in S$ such that $(C, B) \in y$. The monotonic function $F_{Args_{MAS}, x/y}$ mapping from $\mathcal{P}(Args_{MAS})$ to $\mathcal{P}(Args_{MAS})$ is defined by $F_{Args_{MAS}, x/y}(S) = \{Arg \in Args_{MAS} | Arg \text{ is } x/y\text{-}acceptable \text{ wrt. } S\}$. It has a *least fixed point*: lfp($F_{Args_{MAS}, x/y}$), which is a set of justified arguments denoting winning arguments. In our system, given $Args_{MAS}$, such a justified argument is computed based on the dialectical proof theory [6, 8].

As an example of agent-argumentation referring to ontologies, we briefly explain how our integrated system constructs the argument Arg including the first rule of the knowledge base $KB_{Lecturer}$ shown in Figure 4 where two DL expressions, $DL[course](ai)$ and $DL[logic](l1)$ occur in the body, with accessing to the curriculum ontologies $\mathcal{K}_1$ as follows.

For queries, $course(ai)$ and $logic(l1)$ issued by the LMA-based agent system, DL-reasoning system draws an inference with respect to $\mathcal{K}_1$ such that $\mathcal{K}_1 \models course(ai)$, $\mathcal{K}_1 \models logic(l1)$, which are immediately returned to the agent system. As a result, two rules, $course(ai) : \mathbf{t} \leftarrow$ and $logic(l1) : \mathbf{t} \leftarrow$ are

---

[1] $\sim$ is an ontological negation [8], i.e. the classical negation in answer set semantics[4].

$KB_{Lecturer} = \{$
$\quad \sim take\_credit\_in(ai,\ st\_1,\ second):\mathbf{t} \leftarrow$
$\quad DL[course](ai)\ \&\ DL[logic](l1)$
$\quad \&\ \mathbf{not}\ study(st\_1,\ l1,\ first):\mathbf{t}. \quad \}$

$KB_{Student} = \{$
$\quad take\_credit\_in(ai, st\_1,\ second):\mathbf{t} \leftarrow$
$\quad DL[course](ai)\ \&\ DL[easy\_get\_credit](ai).$
$\quad study(st\_1,\ l1,\ first):\mathbf{t}.$
$\quad \leftarrow study(st\_1,\ l2,\ first):\mathbf{t}\ \&\ DL[course](l2).$
$\quad study(st\_1,\ l2,\ first):\mathbf{t}.$
$\quad study(st\_1,\ c\_programming,\ first):\mathbf{t}. \quad \}$

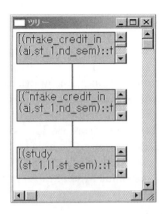

Fig. 4. The dialogue tree for "$take\_credit\_in(ai,\ st\_1,\ second) :: t$" using agent's knowledge bases $\{KB_{Lecture}, KB_{Student}\}$ along with the curriculum ontologies $\mathcal{K}_1$

dynamically generated based on the formula (3) in the process of argumentation, which eventually leads to generation of the following argument:

$Arg = [\sim take\_credit\_in(ai, st\_1, second):\mathbf{t}$

$\quad \leftarrow course(ai):\mathbf{t}\ \&\ logic(l1):\mathbf{t}\ \&\ \mathbf{not}\ study(st\_1, l1, first):\mathbf{t},$

$\quad course(ai):\mathbf{t} \leftarrow, \qquad logic(l1):\mathbf{t} \leftarrow \quad]$

The right part of Figure 4 is the dialogue tree depicted in the execution screen of our *WebArg* system. It shows that "$take\_credit\_in(ai, st\_1,\ second)$" denoting "the student st_1 can take a credit for Artificial Intelligence in the second semester" is justified in the the dialogue tree.

3 Conclusion

We presented an integrated system of the Semantic Web reasoning and LMA-based argument reasoning, where ontologies expressed by OWL DL on the Semantic Web are given to every agent as the common knowledge, whereas each agent has its own rule-based knowledge expressed by EALP, and communicates with each other. Our hybrid system has been implemented using C++, ASP solver DLV, SICStus Prolog and XSLT to do the feasibility study of our approach.

So far, several approaches which combines description logics with nonmonotonic logic programs have been proposed (e.g. [2], [5]). To our knowledge, however, our system is the first which combines nonmonotonic, multiple-valued agent argumentation with ontological reasoning, i.e. description logic reasoning.

As mentioned in the introduction, the main advantage of our hybrid approach is that ours can meet the requirement such that both rule and ontological knowledges can be handled in agent's argumentation as needed for agents such as a seller and a buyer in e-commerce, whereas it is hardly realized if both kinds of reasoning technique are used in a separate way.

Though various expressive knowledge representations such as EALP, LMA, OWL DL and the description logic $\mathcal{SHOIN}$(D) are available for multiagent argumentation in our hybrid system, its applicable domains may be rather small since the complexity of reasoning with OWL DL, or its variant, $\mathcal{SHOIN}$(D) is NEXPTIME-complete as recognized to be intractable.

Thus our future work is not only to enhance our multi-agent system based on argumentation to be able to handle static, dynamic and hierarchical preferences on rules and ontologies as needed in the legal reasoning, but also to make the DL reasoning system more practical reasoner by using many kinds of heuristics.

References

1. T. Eiter, N. Leone, C. Mateis, G. Pfeifer and F. Scarcello: A deductive system for nonmonotonic reasoning, Proc. of LPNMR 1997, LNCS 1265, pp. 364-375, Springer, 1997, URL http://www.dbai.tuwien.ac.at/proj/dlv/
2. T. Eiter et al.: Combining Answer Set Programming with Description Logics for the Semantic Web, Proc. of KR 2004, pp. 141-151, 2004.
3. M. Kifer and V. S. Subrahmanian: Theory of generalized annotated logic programming and its applications, Journal of Logic Programming, Vol.12 (4), pp. 335-367, 1992.
4. M. Gelfond, and V. Lifschitz: Classical Negation in Logic Programs and Disjunctive Databases. *New Generation Computing 9*, pages 365-385, 1991.
5. S. Heymans, D. V. Nieuwenborgh, and D. Vermeir: Nonmonotonic Ontological and Rule-Based Reasoning with Extended Conceptual Logic Programs, ESWC 2005, pp.392-407, 2005.
6. H. Prakken and G. Sartor: Argument-based extended logic programming with defeasible priorities, J. of Applied Non-Classical Logics, 7(1):25-75, 1997.
7. I. Horrocks and P. F. Patel-Schneider: Reducing OWL entailment to description logic satisfiability, Proc. of ISWC-2003, LNCS 2870, pp. 17-29, Springer, 2003.
8. T. Takahashi, H. Sawamura: A Logic of Multiple-Valued Argumentation, Proc. of AAMAS 2004, pp.789-805, 2004
9. URL http://www.w3.org/TR/xslt

Behavior Analysis Based Automatic Composition of Semantic Web Services

Rui Wang, Xueli Yu, Yingjie Li, Jianlin Li, and Jingyu Sun

College of Computer and Software, Taiyuan University of Technology
Taiyuan, Shanxi, P.R. China
boystanford@yahoo.com.cn

Summary. With more Web services emerging on Internet, it becomes a challenge for Web Service community to integrate existing Web services to meet the complex requirement of users with little intervention of human. Although semantics added into Web service promises us more powerful measures to automate the service discovering, matching, and integrating, how to realize the process of automatic Service Composition with semantics is still an urgent problem. In virtue of the fruits of the study of behaviorism, this paper analyzes and models the complex services. And through the behavior analysis and behavior partition, this paper reaches a better combination of services, reasoning algorithms with architecture-Service Agent Community(SAC), and makes Service Composition more automatically and intelligently.

Keywords: Web service, Service Composition, behavior analysis, behavior partition, SAC.

1 Introduction

Recently, many researches contribute to service composition with different approaches, which can be classified into three categories: dataflow-oriented, control flow-oriented, and AI planning based.

The first two approaches focus on the workflow pattern. Lassila Dixi's approach[4] is based on a workflow pattern with one input and one output. The last approach relies on AI planning. An approach for pro-active Web services selection and composition with AI planning suggested by [3]. In addition, there are some interesting approaches that model the service with Description Logic[8].

However, the first two approaches rarely consider the composite service as a complex behavior, thus the system lack of the ability of self-awareness. And behavior analysis can better help the third approach to instruct the goal-direction design for actions in reasoning. Consider the deficiency of these three approaches, this paper provides a new approach—behavior analysis—to instruct the partition of a complex service, and the function design of agents in system, and combines the approaches mentioned above together. At the same time it utilizes richer semantic information from OWL-S[6] during reasoning process for a service food chain.

K.M. Węgrzyn-Wolska and P.S. Szczepaniak (Eds.): Adv. in Intel. Web, ASC 43, pp. 357–362, 2007.

The contents of this paper are organized as follows. Section 2 introduces how to use behavior-based analysis to partition the complex service. Section 3 details the mechanism of Service Composition. Section 4 introduces an application.

2 Behavior-Based Analysis for Service Composition

2.1 Behavior Analysis and Behavior Partition

Behavior Analysis, a scientific study of animal and human behavior and learning, provides a strong conceptual framework for intelligent computer systems known as adaptive autonomous agents [1]. With the behavior analysis applying in this paper, we can model the behavior taking place in complex task, such as composite behavior. In fact, composite service as a behavior isn't easily measurable and realized. So behavior partition is needed to fractionalize it into atomic ones. And Edward[2] used to give definition of Behavior-partition(BP) that BP in a set of act-types that are mutually exclusive and, in general, jointly exhaustive of all instances of behavior.

2.2 The Grounding for Behavior Partition

Skinner's analysis[5] said that learning in a network of input-output relationships (operant learning) could count for behavior of any complexity. Therefore, a composite behavior performed by the agent can be divided into three parts: 1), sense(input, self-awareness), 2) process(the behavior of agent), 3) response(output, effect).

Hence, the composite behavior in Web service can be defined by three sub-behaviors at the beginning of the partition: 1). Discovering behavior, just like sensor, senses and matches the dataflow and checks state change. 2). Processing behavior, assumes the work of reasoning and organizing for processing. 3). Recording behavior, responds, records the middle results of reasoning. In the process of behavior partition, I follow three rules[2] for dividing behavior in the purpose of this paper:

Rule 1. The low level behavior can be combined to realize the function of the high level behavior. **Rule 2.** The sub-behaviors of the same behavior are sequentialčňbut the sub-behaviors of different partition can be parallel or sequential. **Rule 3.** Atomic behavior can be implemented by several functional actions.

3 The Service Composition Mechanism

According to the functional information in service profile of semantic Web service, a service can be formalized as[6]: u=Servicename < Precondition, Dataflow, Effect>=< $\{fact_1, ..., fact_n\}$, $< input, output >$, $\{change_1, ..., change_n\}$ > "*Precondition*" is the necessary condition for service execution."*Dataflow*" relates to the semantics of service input data and output data. *"Effect"* is a set that is composed of several changes to this service world by executing this service. A query is formalized as: q=Queryname<precondition, input, output, constraints>, in which constraints are rules for checking the data of input and output.

3.1 Actions for Reasoning Behavior

By fractionalizing the function of reasoning behavior, this paper can design actions to realize its whole function. The definition of action takes Situation calculus as reference. The worldview of agent is represented as "state".

Definition 1 (Relate action). The action binds the dataflow of a service u with a set of domain class in Domain Ontology Depository(DOD).

1) The predicate definition:
class(u.dataflow): the class relates with service u's output in DOD; in(uclass, Class_tree):"uclass" can be matched in "Class_tree"(Class_tree represents the relationship of ontology in DOD); match_set(uclass, Class_tree, Set): the classes in "Class_tree", which match the "uclass" will be added into "Set"; bind(u.dataflow, Class_tree): binding the class set matching with u.dataflow.

2) Description of Relate action and knowledge base updating

Possibility Axiom
$$in(class(u.dataflow), class\_tree) \Rightarrow Poss(bind(u.dataflow, service), state)$$

Effect Axiom
$$Poss(bind(u.dataflow, service), state) \Rightarrow$$
$$match\_set(class(u.dataflow), Class\_tree, Set)$$
$$KB \xleftarrow{add} K(relate(u.dataflow, set)).$$

Definition 2 (match action). The action matches the class of u.dataflow with another service dataflow. $Match\_Degree(u_1.dataflow, u_2.dataflow)^{[7]}$ represents the matching degree of the dataflow of service u_1 and u_2.
$(u_1.dataflow, u_2.dataflow) = \{(u_1.output, u_2.input), (u_1.input, u_2.output)\}$

1) Prediction definition
range(constraints): represents the range of value according to pre-constraints; track(u): if u.input can be matched in the action, then u will be add to ACT_set. table(item): adding the item into the Onto_relation that records the dataflow chain among services;$item =< u_1, out[i], u_2, in[j], Match\_Degree(u_1.out[i], u_2.in[j]) >$
2) Description of match action and knowledge base updating.

Possibility Axiom

$(u_2.precondition \subseteq poststate(u_1)) \wedge value(u_2.output) \in range(constraints)$
$\Rightarrow Poss(match(u_1.output, u_2.input), state)$

Effect Axiom

$Poss(match(u_1.output, u_2.input), state) \Rightarrow table(u_1) \wedge track(u_2)$
$KB \xleftarrow{add} K(matched(u_1, u_2)) \wedge K(Onto\_relation) \wedge K(ACT\_set).$

Definition 3 (check_state and update_state action). The actions check and update the state for other services' execution and the worldview of agent.

1) Prediction description
check(u,state): with changes of current state, the tag State_changed will be set.

2) Axiom for action and knowledge base updating:
$check(u, state) \xrightarrow{State\_changed} update(state),$
$updata(state) = True \; iff \; State\_changed = True.$

Effect Axiom

$Poss([check(u, state_1), update(state_1)], state_1) \Rightarrow state_2$
$KB \xleftarrow{change} K(state_2)$

3.2 Bi-directional Chain Algorithm for Service Composition

Through the Bi-directional chain algorithm, we can conclude a service food chain stored in the matrix Onto_relation.

The Process of One-Time Reasoning

The defined actions above compose the complete process of one-time reasoning by a modified ADL(Activity Description Language) that is applied in AI planning.

```
ADL Reasoning ( u₁.dataflow = u₁.output, u₂.dataflow = u₂.input ) returns True or False
    Goal(find(u₂) ∧ matched( u₁.dataflow, u₂.dataflow))
    Plan: Init(u₁.dataflow ∧ state₁)
        Action( bind(u₁.dataflow, service),
                Precond: in(class( u₁.dataflow),Class_tree),
                Effect: set)
        Foreach u₂ in set do:
        Action( match(u₁.dataflow, u₂.dataflow),
            Precond:
                (u₂.precondition ⊆ poststate(u₁)) ∧ (value(u₂.dataflow ) ∈ range(constraints))
            Effect: Onto_relation ∧ ACT_set) )
        Action( check(u₂, state₁),
            Precond:
            Effect: When <State_changed=True>:
                Action( update( state₁ ),
                    Precond: State_changed,
                    Effect: state₂)).
```

```
Function Bi-di-chain(start, end) returns Onto_relation
    Def set F_mem={start}, F_set={}, F_temp={}, B_mem={end}, B_set={}, B_temp={};
    Def matrix Onto_relation=();
    Repeat until (F_temp={} and B_temp={} or Forward:Reasoning=false and Backward:Reasoning=false)
        Forward Reasoning:
            Foreach service in F_mem do
                Reasoning (u₁.dataflow = service.input,u₂.dataflow = u₂.output);
                F_temp =F_temp∪ACT_set;
                F_set =F_set∪ACT_set;
                table(service);
            F_mem=F_temp;
        Backward Reasoning:
            Foreach service in B_mem do
                Reasoning (u₁.dataflow = service.output,u₂.dataflow = u₂.input);
                B_temp=B_temp∪ACT_set;
                B_set=B_set∪ACT_set;
                table(service);
            B_mem=B_temp;
        F_mem=F_mem/(F_mem∩(B_mem∪B_set));
        B_mem=B_mem/(B_mem∩(F_mem∪F_set));
    return Onto_relation.
```

Bi-directional Chain Algorithm

In this algorithm, there are two directional reasoning to generate service food chain.

4 Application

In this section, there is a simple example of buying a book on line. S_1 registers service. S_2 is book-searching service. S_3 searches user's information. S_4 verifies user's Credit Card. S_5 charges the customs. S_6 sends the book to the custom.

$S_1 = UserRegister < \{\}$
   $\{in_1 = UserName, in_2 = Password\}$
   $\{out_1 = UserID\}$
   $\{change_1 = UserVerified\} >$
$S_2 = SearchBook < \{\}$
   $\{in_1 = BookName\}$
   $\{out_1 = BookPrice, out_2 = BookSummary\}$
   $\{\} >$
$S_3 = SearchUserprofile < \{fact_1 = UserLogin\}$
   $\{in_1 = UserID\}$
   $\{out_1 = UserInformation\}$
   $\{\} >$

$S_4 = AccountRegister < \{fact_1 = UserLogin\}$
   $\{in_1 = CreditID, in_2 = Password\}$
   $\{\}$
   $\{change_1 = CreditVerified\} >$
$S_5 = Charge < \{fact_1 = CreditVerified, fact_2 = UserLogin\}$
   $\{in_1 = BookName, in_2 = BookPrice\}$
   $\{out_1 = PayID\}$
   $\{change_1 = BookPayed\} >$
$S_6 = SendBook < \{fact_1 = UserLogin, fact_2 = BookPayed\}$
   $\{in_1 = UserName, in_2 = Address, in_3 = SendWay\}$
   $\{out_1 = RecieveDate, out_2 = PosterName\}$
   $\{effect_1 = SendVerified\} >$

The service query is represented as:

$BuyBook <$
$\{in_1 = BookName, in_2 = RegisterInformation, in_3 = CreditInformation\}$
$\{out_1 = BookPrice, out_2 = RecieveData\}$
$\{Constraints : RecieveData \leq 2007.1.19\} >$

The course of service composition with SAC is represented as Figure 1.

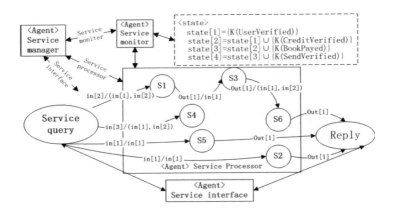

Fig. 1. The course of service composition in SAC(The final reply to the service requester is $\{S_5.out[1], S_2.out[1], S_6.out[1]\}$)

5 Conclusion

In this paper, we rely on behavior-based analysis to partition a complex service into several realizable parts and to integrate them with reasoning for a service food chain ,which is acted in SAC. And architecture of system and reasoning algorithm can be amalgamated together to help us to find the better answer for complex problems.

Acknowledgement

Sponsored by the Natural Science Foundation of China (No. 60472093).

References

1. William R. Hutchison (2002) A Behavior Analytic Paradigm for Adaptive Autonomous Agents.The Cambridge Center For Behavioral Studies
2. Edward G. Rozycki (1974) Human Behavior: Measurement and Cause. Can there be a science of education? Ed.D. Dissertation. Temple U.
3. Mihhail Matskin (2005) From Proceeding Internet and Multimedia Systems, and Application.
4. O.Lassila (2004) Interleaving discovery and composition for simple workflows. In semantic Web Service, 2004 AAAI Spring Symposium Service
5. Chomsky, N (1959) Verbal Behavior by B.F. Skinner. Language, New York
6. http://www.daml.org (2003) OWL-S: Semantic Markup for Web Services. The OWL Services Coalition
7. Li Wang, Yingjie Li (2005) The Semantic Matching of Semantic Web Services. KGGI04
8. Yingjie Li (2006) Research on Reasoning of the Dynamic Semantic Web Services Composition. WIC conference

Discovering Web Services Based on Match History

Yang Xu, Shengqun Tang, Youwei Xu, Ruliang Xiao, and Lina Fang

State Key Laboratory of Software Engineering
Wuhan University, Wuhan 430072, China
xuyang_cn@hotmail.com

Summary. This paper presents an approach for distributed discovery of web services based on the match history. A service routing table (SRT) is introduced into each registry node to record the history of matches for selecting the repositories against incoming requests. The approach does not depend on the publishing mechanism of web service and the classification system for managing registries, and has not the constraint that all registries must conform to one shared ontology of concepts.

1 Introduction

The increasing number of web services demands for distributed discovery infrastructures. It is a challenge to select the most appropriate services from numerous services from various providers existing in various registries, which all declare themselves fulfilling user's request. To solve this problem, a system has to provide means for describing web services with expressive semantics and locating distributed registries with web services satisfying user's requests.

In this paper, we present an approach to address the issue how to search distributed registries that potentially fulfill the user's requirements. In our solution, Web service discovery within a registry node consists of two aspects: Goal-to-Goal discovery that is responsible for searching distributed registries and Goal-to-Web service discovery for local matchmaking. Goal-to-Goal discovery is based on the history of successful matches of users' requests with web services descriptions provided by registries. We introduce a service routing table (SRT) into each registry node to record the history.

In this paper, we assume that web service are being described semantically including user request, for example, using OWL-S[1] or WSMO[2].

The rest of this paper is organized as follows: section 2 briefly discusses related work. Section 3 introduces our model of distributed discovery. And Goal-to-goal discovery is depicted in section 4. We conclude in section 5.

---

[1] http://www.w3.org/Submission/OWL-S
[2] http://www.w3.org/Submission/WSMO/

K.M. Węgrzyn-Wolska and P.S. Szczepaniak (Eds.): Adv. in Intel. Web, ASC 43, pp. 363–368, 2007.
springerlink.com   © Springer-Verlag Berlin Heidelberg 2007

2 Related Work

A number of distributed discovery approaches, such as [1], based on P2P technologies have been proposed. In these approaches, the selecting of an applicable registry to store and search for a web service description depends on classification of the service. However, it is difficult for users to search for a specific service without knowing details of the classification. Other proposals, such as [2] and [3] use index terms for web service descriptions to publish web services to registries and quickly identify the registries against user request. The choosing of registries in these methods lies on the publishing methods of web services and all registries need agree on one ontology of concepts.

Our solution of searching specific registries is based on the history of successful matches. The problem of selecting registries becomes the question of matching user's requests stored in the SRTs against incoming user requests. Searching of registries does not depend on a publishing mechanism of web service, as well as a classification system for managing the registries in the network. Also our approach has not the constraint that all registries must conform to one shared ontology of concepts.

3 Model of Discovery Based on Match History

In our approach, the Web service discovery within a registry consists of two aspects: Goal-to-Goal discovery (G2G discovery) and Goal-to-Web service discovery (G2S discovery). G2G discovery is responsible for selecting the registries that can potentially fulfill the user's request (goal). We call these registries as candidate registries. G2S discovery is responsible for matching the user's goal against the capabilities of the Web services within the registry, i.e. local matchmaking. The approaches of local matchmaking will not be discussed in this paper. More information about them can be found in [4], [5] and etc.

For G2G discovery, we introduce a service routing table (SRT) in each register node to record the history of successful matches. A record in SRT contains the requester's goal and corresponding registries which ever succeeded in fulfilling the goal. Fig. 1 shows the conceptual model of our distributed service discovery.

A registry implements G2S discovery first. If G2S discovery fails, G2G discovery is executed. Within G2G discovery, the user's goal is matched with the goals already kept in the SRT (2). If matched, the corresponding registries will be put into a list of candidate registries. The current registry forwards users request message to all candidates in the list (3). In case that all the candidates failed to match against the user's goal, the originated registry will broadcast the request message to all other known registries in the network (7).

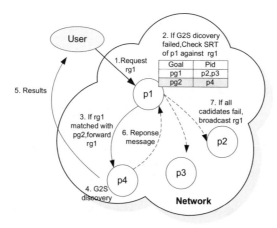

Fig. 1. Conceptual model of distributed discovery based on match history

4   Goal-to-Goal Discovery

4.1   Description of Discovery History

A Service Routing Table is defined as a set of routes used to record the discovery history. A route R records a user goal and its matching registries. It is a tuple $R_i = (\ G_i,\ P_i\ )$. Where, R_i represents the route of user goal i, G_i is the vector for user's goal i and P_i is the set of identifiers of registries which ever fulfilled the goal i.

Users' goal G_i is encoded as a multi-key vector called the characteristic vector as described in [3]. Eventually, the question of G2G discovery becomes the problem of matching the vector of goals kept in SRT with the vector of the incoming user's goal. Ontological concepts representing outputs (or post-condition) of a service will be categorized into different Concept Groups based on their semantic similarity [6] and can be mapped into numerical key values in order to support semantic reasoning efficiently . A Concept Group can be associated with a Bloom key built by applying k hash functions h_1, h_2, ..., h_k to the key of each concept member. Using Bloom filters, the step of checking the membership of a concept in certain concept groups can be done fast and with very high accuracy level. In Fig. 2, for example, a goal g_1 with concepts C_2, C_3 which belong to concept groups CG_1, CG_2, is then represented by the characteristic vector $V_{g1} = \{k_1, k_2\}$, where k_i is CG_i's Bloom key.

For each incoming user goal, we first find the concept group CG_is that it belongs to. The characteristic vector is obtained and then sent to match with those of the goals in the SRT. When matched, we get the identifiers of candidate registries the matched vectors refer to.

The strategy used for partitioning the ontological graph will not affect the correctness but mainly the efficiency of the discovery algorithm. Each registry node can have it own strategy of ontological graph partitioning. Therefore, all

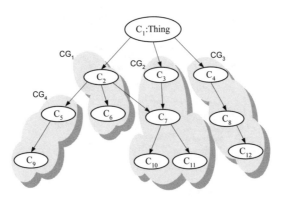

Fig. 2. Ontological Concept Groups

registries need not comply with only one ontology in our approaches, which is a restriction in some other approaches, such as [3].

4.2   Types of G2G Matchmaking

The outcome of G2G matchmaking between incoming user's goal RG and the goal PG in a SRT could be one of the types below:

- **Match** - If the characteristic vector of RG and that of PG are equivalent, i.e. $V_{RG} = V_{PG}$, PG is match with RG.
- **Fail** - Failure occurs when the characteristic vector of PG is not equivalent that of RG, marked as $V_{RG} \neq V_{PG}$.

In Fig. 2, Let us suppose that the SRT of a registry node has three routes (records). $R_1=(pg_1, \{p_1, p_2\})$,$R_2=(pg_2, \{p_3\})$, and $R_3=(pg_3, \{p_1, p_4\})$. pg_1 operates on two concepts C_2, C_7, pg_2 on C_6, C_3, C_8, and pg_3 on C_9, C_8, then V_{pg1} = $\{k_1, k_2\}$, $V_{pg2} = \{k_1, k_2, k_3\}$, $V_{pg3} = \{k_3, k_4\}$. When a new user request rg1 with concept C_6 and C_{11} is incoming, the characteristic vector of rg_1 is $\{k_1, k_2\}$. The characteristic vector of rg_1 is equivalent to that of pg_1, i.e. $V_{rg1} = V_{pg1}$, so rg_1 is match with pg_1. Similarly, pg_1 fails to match with pg_3. As a result, p_1, p_2 are candidate registries for rg_1.

4.3   Goal Graph

We organize the vectors of the goals in SRT into a graph according to containership relationship, such as $\{k_1, k_2\} \subset \{k_1, k_2, k_3\}$, to improve the performance of match between characteristic vectors. The containership relation is the partial order relation and can be represented by a directed acyclic graph (DAG). We model the vectors of goals in SRT and their containership relations as a DAG called as Goal Graph.

A sample Goal Graph is shown in Fig. 3. In Fig. 3, node n_i (i =1, 2, ..., 10) represents the characteristic vector of goal i, and the edge from n_6 to n_2 refer to the relation $V(n_6) \subset V(n_2)$, i.e. $\{k_1, k_4, k_5\}$ contains $\{k_4, k_5\}$. n_7, n_8, n_9 and n_{10}

are leaves with no incoming edges. No edge between n_5 and n_7 means no direct relation between them. Once the strategy used for partitioning the ontological graph is decided, the Goal Graph can be established before G2G matchmaking happens. So we can devote more time on the establishment of the Goal Graph to reduce later discovery time.

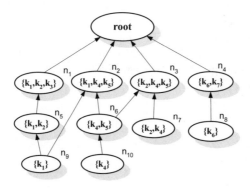

Fig. 3. A sample Goal Graph

4.4   G2G Matchmaking

G2G Matchmaking is based on Goal Graph. A G2G matchmaking starts from the matching the vector of incoming user's goal RG with those of leaves of the Goal Graph. The algorithm for G2G matchmaking is listed below, where set_{leaves} is the set of all leaves in a Goal Graph, and $set_{path(i,root)}$ is the set of nodes in all the path from i to root except for node i and root.

```
initialize the candidate list L to NULL
calculate the characteristic vector of RG, V_rg
DAG ← GoalGraph
repeat
    get all leaves in DAG, set_leaves
    for each leaf i ∈ set_leaves
        if V_rg = V(i) then
            get the registry List P_i
            L ← L ∪ P_i
            break;
        else
            if V_rg ⊂ V(i) then
                calculate the set of nodes set_path(i,root)
                DAG ← DAG - {u | u ∈ set_path(i,root)}
                      ∪ {(u,v) | u ∈ set_path(i,root) ,v ∈ N }
                      ∪ {(v,u) | u ∈ set_path(i,root) ,v ∈ N }
            DAG ← DAG - {i} ∪ {(i,v) | v ∈ N} ∪ {(v,i) | v ∈ N }
until (only root in DAG)
return L
```

5   Conclusion

In this paper we propose a distributed Web service discovery approach based on match history. We introduce a SRT as a resort to record the history. Basing on SRT, Web service discovery is consisted of G2G discovery and G2S discovery. We present how to search applicable registries through SRT. In our solution, the searching of registries does not depend on a publishing mechanism of web service, as well as a classification system for managing the registries in the network. Also our approach has not the constraint that all registries must conform to one shared ontology of concepts. In future, we will investigate to enhance SRT to take into account QoS characteristics in G2G discovery, such as reliability, response time, etc.

References

1. Verma K., Sivashanmugam K., Sheth A., Patil A., Oundhakar S., and Miller J. (2005) METEOR-S WSDI: A scalable P2P infrastructure of registries for semantic publication and discovery of web services. Inf. Tech. and Management, 6(1): 17-39
2. Kashan F.B.i, Chen C.-C., and Shahabi C.(2004) WSPDS: Web services peer-to-peer discovery service. In Proceedings of International Conference on Internet Computing, USA.
3. Vu L.-H., Hauswirth M. and Aberer K. (2005) Towards P2P-based Semantic Web Service Discovery with QoS Support. In proceeding of Workshop on Business Processes and Services (BPS), Nancy, France
4. Paolucci M., et al. (2002) Semantic matchmaking of web services capabilities. In proceeding of 6th International Symposium on Wearable Computers, USA
5. Srinivasan, N., Paolucci, M., Sycara, K. (2004) An Efficient Algorithm for OWL-S Based Semantic Search in UDDI. In proceedings of 1st International Workshop on Semantic Web Services and Web Process Composition (SWSWPC 2004), San Diego, USA
6. Castano S., Ferrara A., Montanelli S., and Racca G. (2004) Matching techniques for resource discovery in distributed systems using heterogeneous ontology descriptions. In proceedings of International Conference on Information Technology: Coding and Computing, USA

Using Segmentation Method for Better Ontology Inference

Youwei Xu, Shengqun Tang, Yang Xu, Ruliang Xiao, lina Fang, and Ling Li

State Key Laboratory of Software Engineering, Wuhan University
xuyouweiz@163.com,
tangshengqun@whu.edu.cn,
xuyang_cn@hotmail.com,
xiaoruliang@163.com,
dpetfln@gmail.com,
liling8126@163.com

Summary. Ontologies become more and more big, which is an important factor of low-efficient reasoning. In this paper, we propose an ontology segmentation method for alleviating the inference burden. The method is based on ontologies described by OWL DL. Unlike the traditional method, the optimized method will "chip off" a part of knowledge from the big ontology according to user's query and only uses this knowledge for reasoning. We implemented the algorithms by ontology editor and explain it is very useful in practice.

1 Introduction

Although many semantic web applications have been developed, it seems there is still a long way to substitute them for the traditional systems. One of the reasons is the expensive cost of inference. Commonly, inference system has to read the back-ground knowledge into memory to construct the inference model at first. When user proposes a query, system uses certain kind of reason algorithm to create new implicated logical assertions based on the model. At the end, the system will search the relative answer in result assertions set according to the query. Unfortunately, this method always failed because of the large scale of background ontologies. So the efficiency of inference algorithm becomes a bottleneck of the development of semantic-based applications.

To alleviate the burden of inference, direct method is optimizing the inference algorithm. Much endeavor had been done through this approach. Seidenberg J and Rector A[5] recently present an ontology segmentation method. The method is based on travesal algorithm. A certain concept is considered as the target node for ontology extract, upwards & downwards travesal will then start from this concept node for gain related ontology, at last a sub-set of this ontology was acquired to be an abstract of this knowledge ontology. Besides Seidenberg J

K.M. Węgrzyn-Wolska and P.S. Szczepaniak (Eds.): Adv. in Intel. Web, ASC 43, pp. 369–374, 2007.
springerlink.com

and Rector A's method, Stuckenschmidt H and Klein M[6] present the method of structure-based partitioning. Noy N and Musen M[3] present their extraction methodology in PROMPT suite. At last, Harmelen F[2] propose an incremental selection function. his thinking largely impacts our method.

We propose a newly ontology segmentation method based on OWL DL[4]. The main idea of our method is acquiring a segmented knowledge from the whole big ontology according the user's query. Nine segmentation algorithms are used to guarantee the 'proper' segment. We consider that a complex query can be decomposed into simple binary queries and simple query can furthermore classified into 9 kinds which will be discussed below. When a query is asked, our method can determine its type and then use corresponding algorithms to segment the whole ontology. Result is the cutdown ontology ready for inference.

The paper is structured as follows. The next section we give the basic definitions for the algorithms. In Section 3, nine kind of segmentation algorithms are described. The following Section 4 introduce the implementation of algorithms. At last, we address the future work.

2   Basic Definitions

This section describes the basic definitions which involve the segmentation algorithm.

Definition 1 (Logical Constructor). *We call a class, a property or an individual as the logical constructor.*

Definition 2 (Query). *A query is a question being asked by outside relative to the logic knowledge base. It can be represented as a tuple $<a>$ or $<a,b>$, which a and b are sets of logical constructors.*

Definition 3 (Class Tree). *A model of ontology only contains classes and their hierarchy can be seen as a class tree.*

Definition 4 (Property Tree). *A model of ontology only contains properties and their hierarchy can be seen as a property tree.*

Definition 5 (Basic Problem). *We call 9 problems as basic problems.*

1. C Problem. *Query about the description of a Class.*
2. P Problem. *Query about the description of a Property.*
3. I Problem. *Query about description of an Individual.*
4. C-C Problem. *Query about relation between two Classes.*
5. C-P Problem. *Query about relation between a Class and a Property.*
6. C-I Problem. *Query about relation between a Class and an Individual.*
7. P-P Problem. *Query about relation between two Properties.*
8. P-I Problem. *Query about relation between a Property and an Individual.*
9. I-I Problem. *Query about relation between two Individuals.*

3   Algorithms

Here we propose 9 segmentation algorithms.

3.1   C-C, C-P, C-I, P-I and I-I Problem Algorithm

Firstly, we propose the C-C problem algorithm. The C-C problem segmentation is the most important algorithm because there are 4 algorithm(C-P, C-I, P-I and I-I problem) need use it and experience shows that query about C-C problem maybe the most frequently being asked questions.

Procedure C-C Inference $(A, B, O_S=(O_T, O_A))$

/*1. Initialization: M_c is the class-tree of O_S in memory. A, B are the formal parameters which represent two classes. O_S is the source ontology, O_R is the result target ontology. Function readin() is used for reading an ontology file into memory. Function classTree() is used for getting the class tree model.*/

> $M_C:=classTree(readin(O_S))$

/*2. Get the sub-tree(descendant) of A, B from M_C, delete the sub-tree(root A) sub-tree(root B) from M_C (except A and B roots), so we get an new ontology model called M', which is a mediator model for segmentation.*/

> $M'=delete(M_C,subtree(root\ A))$
> $M'=delete(M',subtree(root\ B))$

/*3. Upwards traverse (depth-first) the M' from A and B nodes respectively, until reach the root of the M' (the top concept). Then label the classes which have been traversed and delete all the unlabelled classes in M'.*/

> LabelListA:=upwardsTraverse(A,M',root)
> LabelListB:=upwardsTraverse(B,M',root)
> UnlabelList:=sub(sub(class(M'),LabelListA))
> **for all** class X in UnlabelList **do**
>   M'=delete(M',X)
> **end for**

/*4. Consider the properties. Add the properties in O_S which are directly relate to the classes in M' into a list PL. For properties in PL, if there are properties which domain and range are both in the scope of M', then add all these properties' assertions into M'. the remainder properties are sent to list PNL. */

> $M_P:=property(readin(O_S))$
> $PL:=listProperty(M_P)$
> **for all** property p in PL **do**
>   **if** (inModel(domain(p),M')&&inModel(range(p),M')) **then**
>   M':=union(M',assertion(p));
>   **end if**
> **end for**

//5. obtain the result ontology O_R from the M' for inference.

> $O_R:=readout(M')$

From the above description, it can be found that the C-C algorithm cares only about TBOX and takes no account of ABOX. The reasoning is obvious, query of C-C problem expects for the answer of the relations between two classes. Sometimes individual here is inessential.

C-P problem algorithm answers for the query of what's the relationship between a class and a property. The basic process is determining the domain and range of property p, then the problem transforms as two C-C problems. C-I problem is similar with C-C problem. The algorithm firstly gets the closest class of individual i, then invokes C-C problem algorithm to obtain model M'. The last result is the union of M' and all the assertions about i. P-I problem will use the C-P problem algorithm. Individual i's closest class is checked out and the problem becomes a C-P problem. Similarly, I-I problem can transform to C-C problem by checking out the closest class of two individuals.

3.2 C, I, P and P-P Problem Algorithm

Queries come from users usually want to know what a concept is. C problem segmentation algorithm aims at getting a proper model about the concept. Proper means the model contain all the necessary assertions about the concept while controlling the scale of cutdown model in possible.

Procedure C Inference (A, $O_S=(O_T,O_A)$)

/*1. Initialization: Construct the model M_C, which is the class-tree of O_S in memory.*/
    M:=model(reading(O_S))
    M_C:=classTree(reading(O_S))
/*2. Upwards&Downloads traverse the M_C from class A. Add all labeled classes and their class-trees into $M_{ąr̀}$*/
    LabelListA:=upwardsTraverse(A,M_C,root)
    LabelListB:=downwardsTraverse(A,M_C,⊥)
    LabelList:=union(LabelListA,LabelListB)
    $M'=M_C$
    for all class X in LabelList **do**
        M'=add(M',classTree(X))
    end for
/*3. If p's domain or range is in the $M_{ąr̀}$, then include the range or domain class and p's assertions into M. p is any property in model $M_{ąr̀}$.*/
    for all property p in M **do**
        if (inModel(domain(p),M')) **then**
            M':=union(union(M',range(p)), propertyassertion(p))
        end if
        if (inModel(range(p),M')) **then**
            M':=union(union(M',domain(p)), propertyassertion(p))
        end if
    end for

/*4. If individual i is the instance of class A or Aq's childClasses, then add the assertion, declaring i isInstanceOf certain class into Mq'. */
  for all *instance i* **do**
      if *(isInstanceOf(i) is A or child of A)* **then**
          M'=union(M',assertion(i,subclassOf))
      end if
  end for
  O_R:=readout(M')

I problem usually desires the result about the information of individual i's closest class and i itself assertions. Therefore, the algorithm will use the C algorithm for the closest class of i. P problem likes the C problem. All the assertions about the target property p and its domain and range class hierarchy should be considered. Property p's hierarchy contains its ancestor properties and descendent properties. For the former, their domain and range classes should be added in result. For the later, only assertions about themselves need to included. P-P problem algorithm's target is acquiring the relationship between two properties. Firstly, two properties' domain and range classes and their subtrees will be added into M'. Secondly, two properties' descendent and ancestor properties will be considered.

4   Implementation

The algorithms have been implemented by VO-Editor[8], which is a visual ontology editor using Jena 2.0[7]. Figure 1 shows the example of using segmentation algorithms for tourist ontology (which is developed by our VO-Editor). To the left part, we can see the complex structure of tourist ontology. When user wants to know the relation of class Inn and Hotel, he or she can fill the query form of C-C type in editor's menu, then the editor can located the relevant classes in red bolt rectangle (see left part of Figure 1). After the C-C algorithm is executed, we will get the result of right part of Figure 1. It is asserted that class Inn and Hotel are subclasses of TourOrganization. Although the ontology is so complex

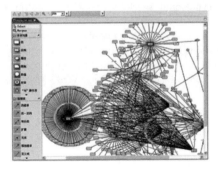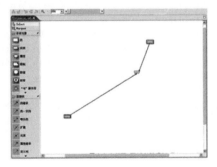

Fig. 1. Using Segmentation Algorithm for Tourist Ontology in VO-Editor

(48 classes,78 properties and 482 individuals), using segmentation algorithms, we can easy chip off the mass into a little ontology we interested in. Obviously, inference such little result ontology is a very easy task for reason engine. Moreover, we can find that the algorithm is also very benefit for good understanding of ontology structure.

5   Conclusion and Feature Work

In this paper, we have introduced a newly ontology segmentation method. The method can be seen as the optimization technique for reasoning and searching.We argue that it is not wise to take the whole ontology into reason engine for inference, because that will be too much for solving a certain inference problem. It is more advisable to analyse only a piece of ontology just concerning about the problem. We uses 9 algorithms for segmenting the ontology correspond to query's requirement and implemented them.

Our work on this issue is just at beginning. So far the method we proposed hasn't been proved by strict formal thoery. In spite of obvious efficiency increment, we have to work hard to proof the method in the feature.Another important work is improving the algorithms fitting for more domains requirements.At last, being the emergence of OWL1.1[1], there's lot of modifications to made for according to new standard.

References

1. Grau B, Horrocks I, Parsia B et al(2006) Next Steps for OWL.In: Proceedings of the OWLED06 Workshop on OWL: Experiences and Directions,CEUR-WS.
2. Harmele F(2006) Where does it break?or:Why Semantic Web research is not just "Computer Science as usual". In: keynotes of European Semantic Web Conference.
3. Noy N, Musen M(2004) Specifying ontology views by traversal. In: International Semantic Web Conference. Springer,Berlin Heidelberg
4. McGuinness D, Harmelen F(2004) OWL Web Ontology Language Overview. W3C Recommendation.
5. Seidenberg J, Rector A(2006) Web Ontology Segmentation: Analysis, Classification and Use. In: Proceedings of the 15th International World Wide Web Conference. Springer
6. Stuckenschmidt M, Klein M(2004) Structure-Based Partitioning of Large Class Hierarchies. In: Proceedings of the 3rd International Semantic Web Conference. Springer
7. Carroll J, Dickinson I, Dollin C et al(2004) Jena: Implementing the Semantic Web Recommendations. In: Proceeding of the 13th International World Wide Web conference. ACM Press
8. Xu Y, Tang S, Yang Y, et al(2006) Towards a Selective Inference Platform Based on OWL.In: Proceedings of Web Intelligence 2006. IEEE Press

Semantic Search on Cross-Media Cultural Archives

Zhixian Yan, François Scharffe, and Ying Ding

Digital Enterprise Research Institute (DERI) Innsbruck,
University of Innsbruck, Austria
firstname.secondname@deri.org

Summary. With the emergence and development of semantic web, traditional archive service meets a new challenge to provide more intelligent and interactive services for web users. To take good advantage of semantic web technologies, in particular ontology and semantic inference, this paper proposes a semantic search portal for cross-media cultural archives involving document, image, audio and video. With such semantically-enhanced search portal, implicit multimedia archives can be retrieved under the support of ontology modeling and semantic reasoning. Furthermore, the retrieved cross-media archives can be semantically navigated and repacked in a more meaningful and integrated way.

1 Introduction

Semantic web, invented by Tim Berners-Lee a half decade ago, aims to provide a new generation of web with machine understandable meanings [1]. With the starting point of semantics rather than syntactics, information can be organized and described with a fully-fledged schema like ontology. Based on the comprehensive information description, semantic web supports reasoning with logic foundation, in particular description logic. Furthermore, due to the agent technology, more intelligent and interactive services can be realized.

Besides theoretical studies on ontology, logic and agent, semantic web come forth with many domain scenarios to further demonstrate its advantages. Several domain-oriented ontolgoies have been constructed to achieve semantic resources organization and retrieval, such as DOAP[1] (Description Of A Project) for open source projects, SIOC[2] (Semantically-Interlinked Online Communities) and FOAF[3] (Friend Of A Friend) for social networks. Besides those domain-specific ontolgoies, there are some generalized or so-called upper-level ontolgoies crossing different domains such as Dublin Core, WordNet, SHOE (Simple HTML Ontology Extensions) and SKOS (Simple Knowledge Organisation

---

[1] DOAP, http://usefulinc.com/doap/
[2] SIOC, http://sioc-project.org/
[3] FOAF, http://www.foaf-project.org/

K.M. Węgrzyn-Wolska and P.S. Szczepaniak (Eds.): Adv. in Intel. Web, ASC 43, pp. 375–380, 2007.

System). However, though with so many existing ontology construction works, there is still less study on real-world semantic applications, especially for semantically-enhanced multimedia services. Related to multimedia, we have to mention here is some highlight EU semantic projects referring to the audio domain, such as SIMAC[4], EASAIER[5], SMaRT[6]. Different from those projects focusing on media ontology construction, our work mainly emphasize the cross-media archives searching service with the semantic cornerstone.

Based on the real world and valuable cross-media cultural archives, this paper proposes semantic search to validate the semantic advantages on boosting archive search services. The paper proceeds as follows: Section 2 provides an integrated semantic model; Section 3 presents the ontology-enhanced culture archives modeling; Section 4 shows how semantics can achieve the intelligent search services for culture achieves; and finally the conclusion is discussed.

2   Semantic Model for Cultural Archives

Hereby, we briefly analyze the semantic description requirements for cultural archives. Furthermore, we propose an integrated semantic model with the support of providing more intelligent and interactive archive search service.

2.1   Semantic Description for Cross-Media Cultural Archives

Ontology plays a crucial role in semantic web[5][7], and in archive modeling either. Cross-media culture archives should be semantically described to support the subsequent intelligent archive search service. We take traditional Dunhuang culture, with two thousand years of history, in west China as our use case [2]. For simplicity but efficiency, there are only five core concepts in our ontology model, i.e. MEDIA (the cross-media characteristics for cultural archives, involving document, image, audio, video), DYNASTY (embodying the individual history of each archive, such as the thriving Tang dynasty in ancient China), CAVE (there are many cultural caves, which store abundant culture legacies comprising frescos and sculptures etc.; of course, the cave itself is also a kind of culture legacy), CONTENT (the significant concept embodies the cultural and artistic characteristics, such as the frescos and sculptures mentioned before); finally, the most important concept for archives is DATA (which models all concrete culture archives in a comprehensive concept).

We have developed an ontology tool for cultural archives modeling, shown in Fig.1. In the figure, the five red rectangles express the five core concepts discussed before; whilst, the green small ones are some example instances belonging to those concepts respectively.

---

[4] SIMAC, http://www.semanticaudio.org

[5] EASAIER, http://www.easaier.org/

[6] SMaRT, http://www.k-space.eu/

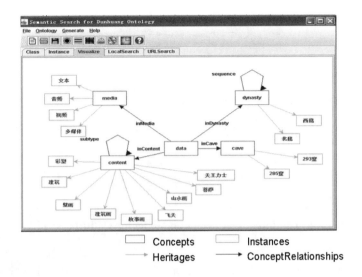

Fig. 1. The Snapshot for Cultural Archives Ontology Visualization

2.2 Semantically-Enhanced Service Model

From previous semantic schema of culture archives, we can see plentiful information among archives. How to provide those information with fully-fledged semantic description? Furthermore, how to achieve more interactive and intelligent service? To answer them, we propose an integrated semantic model, including archives modeling (both concept and instance level), storing, and semantic searching. The original model shown in Fig.2 is domain-independent, and can be adapted to other domains like research community [3][4].

There are four layers in the model, including "semantic storage", "semantic annotation", "semantic search" and "user interface". In addition, "semantic inference" is a vertical component to further infuse semantic technologies. More details about this domain-oriented but domain-independent semantic model can be found in [4]. In this paper, the emphasis is the adaptation of such model for culture archives to attain more intelligent search services.

3 Semantically-Enhanced Cultural Archives Modeling

From the model in Fig.2, there are three core issues have to be considered referring to cultural archives modeling, i.e. archives ontology model, supporting semantic annotation and storage strategy.

Conceptualized Model. To achieve comprehensive schema for cultural archives, we apply previous semantic scenario. As shown in Fig. 1, there are five core concepts should be modeled. In consist with the ontology structure, all the relevant data should be modeled, including data-type properties

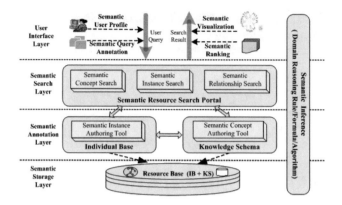

Fig. 2. The Integrated Semantic Model for Cultural Archives

and object properties [7]. In the meanwhile, the schema of their instances, even the storage mechanism, ought to be considered seriously. The detailed storage strategies will be discussed in the following section.

Semantic Annotation. Semantic annotation for multimedia content has been identified as an important step towards more efficient retrieval of multimedia data [9] [8]. Basically, we apply authoring tools for semantic annotation, involving two kinds of archive annotation interfaces (shown in Fig. 3 a, the top-left is semantic annotation about basic archives information, whilst the right-down is for annotating semantic relationships with other instance resources).

Semantic Storage. There are two kinds of information about cultural archives need to be stored, namely concept information and concrete instances. For concept, we mainly take OWL files as the serialization layer as only five core concepts and rich semantics; whilst, the relational database is more advisable for the abundant real world cultural archives in instance level.

4   Intelligent Search for Cross-Media Cultural Archives

With the essential semantic model and annotation preconditions, more intelligent services as semantic search can be attained on cultural archives. Besides basic semantic search, we further highlight semantic inference to enhance search service with more intelligence, which is our distinguishing feature from some existing semantic search studies on multimedia data [10][11].

4.1   Semantic Search for Archives

As shown in the semantic model and different from traditional search engine, the query-keys input to search portal can be denoted with certain semantic information, for example a concrete concept. With this semantic denotation, the

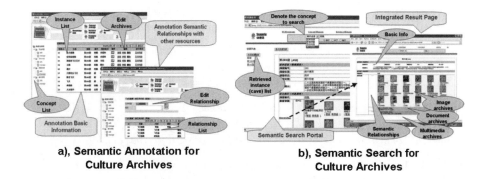

| a), **Semantic Annotation for Culture Archives** | b), **Semantic Search for Culture Archives** |

Fig. 3. Snapshots for Semantic Search on Archives

query results can be profoundly-repacked with the inward semantic schema of the concept, including basic information and semantic relationships with other concepts.

In stead of listing search algorithm, we address the search portal more clearly by a concrete search example. When users query Cave285, the portal feedbacks all the relevant information in the CAVE schema asserted before. Firstly, the basic information of the 285th cave includes the cave number, name etc. and some beyond description information like protection-level; secondly, semantic associated information must be extracted. There are two main associated information for CAVE, i.e. DYNASTY and DATA. For Cave285, there is only one dynasty referred, namely the building time in the XIWEI dynasty. However, many other caves are concerned with more than one dynasty, including the first-building time and some restoring milestones in other dynasties. About DATA instances in Cave285, they are the core elements as cultural archives. To embody the cross-media features of cultural archives, the archives are demonstrated in three main categories, namely document, image and multimedia (including audio and video). Furthermore, every concrete archive (DATA instance) should contain relevant semantic information about its belonging CONTENT instance (such as *Murals* or *Sculptures* etc, even their inherited sub-contents, such as *Feitian* and *Pushan*). The previous Fig.3 b) is the snapshot for the 285th cave example. The left is about semantic search portal, whilst the right (navigated from the left) is the integrated result page for a certain retrieved cave.

4.2 Inference-Enhanced Search Service

Besides semantically-enhanced (or rather ontology-supported) resource description, another main advantage of semantics is semantic inference with logic foundation. With inferences, more implicit information can be extracted, and the search results can be more comprehensive and rational. As shown in Fig. 2, the vertical inference mechanism is attained gradually by reasoning rules, formulas and detailed algorithms. According to the space limitation, we do not discuss much details about inference, but mention two representative reasoning rules and one formula having been concerned in our implementation.

Rule 1. *the transitivity of subContent property for the CONTENT instances*
$< c_1, subContent, c_2 >, < c_2, subContent, c_3 > \Rightarrow < c_1, subContent, c_3 >$
$< data, isContent, c_2 >, < c_1, subContent, c_2 > \Rightarrow < data, isContent, c_1 >$

Rule 2. *the transitivity of DYNASTY sequence*
$< dyn_1, after, dyn_2 >, < dyn_2, after, dyn_3 > \Rightarrow < dyn_1, after, dyn_3 >$
$< dyn_1, before, dyn_2 >, < dyn_2, before, dyn_3 > \Rightarrow < dyn_1, before, dyn_3 >$

Formula 1. *calculate the archives belonging to CONTENT instance c*
$getArchives(c) = annotatedArchives(c) + subContentArchives(c)$
$annotatedArchives(c) = \{arh_k\}, \forall k < arh_k, isContent, c >$
$$subContentArchives(c) = \begin{cases} \cup getArchives(c_k), & \forall k < c, subContent, c_k > \\ \emptyset, & \text{with no tuple above} \end{cases}$$

5 Conclusion

In the paper, we propose a semantic model providing semantic search on cross-media cultural archives. With the analysis and implementation of this model, especially the support of semantic inference, we can validate the advantages of semantic web technologies for traditional multimedia services, which can further provide more intelligent and interactive archive search services.

References

1. T. Berners-Lee, J. Hendler, O. Lassila, Semantic Web, Scientific America, 2001
2. China Knowledge Grid Group, Dunhuang Cultural Grid, http://www.culturegrid.net/
3. J. Zhang, Z. Yan, Semantic-driven Management and Search for Resource on Research Community, Intl. Conf. Semantics, Knowledge and Grid, China, 2006
4. Z. Yan, Semantic-based Resource Management, Search and Application, Master Thesis, Institute of Computing Technology, Chinese Academy of Sciences, 2006
5. D. Fensel., Ontologies: Silver Bullet for Knowledge Management and Electronic Commerce. Springer. 2001
6. F. Baader et cl., Description Logic Handbook: Theory, Implementation and Application, Cambridge University Press, 2002
7. S. Staab, R. Studer, Handbook on Ontologies, Springer, January 2004.
8. S. Bloehdorn et cl, Semantic Annotation of Images and Videos for Multimedia Analysis, ESWC2005, Heraklion, Greece, 2005
9. L. Hollink et cl. Semantic annotation of image collections, the K-CAP 2003 Workshop on Knowledge Markup and Semantic Annotation, Florida, 2003.
10. I. Celino et cl., Squiggle: a Semantic Search Engine for indexing and retrieval of multimedia content, SAMT 2006, Athens, Greece
11. M. Worring et cl. The MediaMill semantic video search engine. ICASSP 2007, Hawaii, USA, April 2007.

Service-Oriented Geographic Information Sharing and Service Architecture for Digital City

Xin Zhang[1,2], Huabin Chen[1], Huasheng Hong[2], and Tianhe Chi[1]

[1] Institute of Remote Sensing Applications, Chinese Academy of Sciences, 100101 Beijing, China
zhangx@irsa.ac.cn
[2] State Key Laboratory of Marine Environmental Science (Xiamen University), 361005 Xiamen, China

Summary. Geographic information sharing and service plays an important role in the implement of Digital City. The traditional development model that is based on common geographic information system can not accommodate for the geographic information sharing and service in distributed web environment. In this article, we firstly introduce the requirement of the distributed geographic information sharing and service and the character of service-oriented software architecture. Then the principle of the service-oriented architecture and the service-oriented geographic information sharing and service software architecture for digital city is put forward which includes GIS service, called VGS (Virtual GIS Services) and AGS(Application GIS Services) which reflects the operation service upper the VGS. At last, the web service based model integration in SOA is introduced. And an application example is stated to demonstrate the advantage and application model of the service-oriented geographic information sharing and service software architecture, which is the ocean current numerical simulation model call based on web service. The development and deployment of the model services example was used in Xiamen City, China.

1 Introduction

With the broadly use of the Internet and Intranet, the increasing amount of web users need more and more geographic information. But the distributed nature of the geographic data, the complex of the geographic data analysis, the slow speed and high delay problem of network are the bottleneck to the development of the usage of geographic information on web. It is respected that the software construction model will solve the problem mentioned above.

In the past 40 years, GIS has been implemented in different computer architecture including micro computer, computing workstation and personal computer [1]. Now GIS is introduced into a distributed network environment [2]. And the distributed industrial geographic information service system is the new trend of the GIS development.

Nowadays, as an important area of geographic information application, Digital City has been being paid more and more attentions. It is claimed that more than 80 % of the public sector information has a geographic dimension in that it is

K.M. Węgrzyn-Wolska and P.S. Szczepaniak (Eds.): Adv. in Intel. Web, ASC 43, pp. 381–386, 2007.
springerlink.com   © Springer-Verlag Berlin Heidelberg 2007

referenced by address or location (which refers to geographic data). Building Geographic Information Service System is of great importance to the realization of Digital City [3-6].

2 Necessary

In a city, the geographic information is heterogeneous, which is produced, maintained and renewed by different department such as Surviving and Mapping Department, Environment Department and so on. So the Geographic information Service System must be built as distributed software on the Internet. The amount of the geographic information in a city is huge, so the transmission speed of the geographic information on the web is a key factor affecting the quality of the Geographic information Service System. As the network bandwidth is limited, the network must be overburdened if the geographic information is transmitted on the network. The spatial analysis upon the geographic information is complex. It is common that a data processing consists of more than one procedure. For example, the overlay processing may demand many procedures. In the past, for the reason of the technology development the GIS software architecture is in the personal computer model or in single department or company. This kind of usage module is not suitable for the nature of the geographic information stated above. A new geographic information technology is respected to solve the problem of geographic information sharing and service for Digital City.

The new software architecture should satisfy the requirement as follow.

Firstly, the new software architecture should realize the maintenance and update of the geographic information and software in a distributed web environment.

Secondly, the new software architecture should be scalability. The required hardware and software must be not expensive. And the new hardware and software resource must be usable continually.

Thirdly, the new software architecture should have the standard geographic information accessing and software programming interface.

Fourthly, the new software architecture should support key GIS data format and software functions. This kind of character can ensure that the new software architecture is compatible for the older systems.

The Service-Oriented Architecture i.e. SOA is suitable for the requirement stated above.

3 Principle of the Service-Oriented Architecture

The basic unit of SOA is "Service". These "Services" are a group of software modules that can execute certain operation flow. Generally, SOA consists of a series of "services" that based on the internet. These "Services" can communicate with each other. Moreover they have loose coupling, and can be reused. According to SOA, service user does not need to care about the specific service he is communicating with. Because the bottom establishment or the service "bus" will make the right choice on behalf of the user, the bottom establishment has hidden technique from requesters as much as possible. Especially the technology standard comes from different application technology (for example J2EE or .NET) cannot affect the SOA users. If already had a service realization, we may reconsider a "better" service realization to replace, since the new service realization has better service quality. SOA can supply a platform to construct an application service that has loose coupling, a transparent position and is independent of protocols. Thus it can reduce the problems caused by different architectures' mutually operation and ceaseless changing product.

SOA evolves from the traditional tight coupling application architectures such as CORBA. SOA does not necessarily need the web service, and the web service also may carry on the development separated from SOA. But it's an extremely ideal method to construct SOA by using the web service technology; because the web service can be the loose coupling relations. The tight coupling paradigm makes it hard for the application system to adapt the service demand changes neatly, because both of the two ends of the distributed computation must follow the same API restraint. The change of one application system can possibly affect many other application systems. Moreover, SOA is also different from the object-oriented technology since the granularity of the object seemed be much smaller, and it base on the software code development stratification plane. But, in the SOA paradigm, "service" is much closer to the application. It may be a service flow such as a GIS model computation process and so on.

There are three categories in the SOA role: service provider, service requestor and service broker (in figure 1). The roles of SOA interact on each other by three basic operations: publishing, searching, and binding. The service provider publishes service to the service broker. The service requester searches the required services by the service broker and is bound to these services. Those are the same as the web service in some cases.

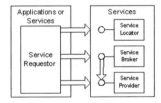

Fig. 1. The roles in the SOA

4 The Design of the SOA for Geographic Information Sharing and Service of Digital City

Based on the analysis of the requirement of the SOA for geographic information sharing and service of Digital City, we think that the services can be divided into two layers. One is GIS service, called VGS (Virtual GIS Services) and the

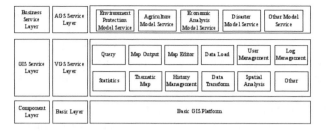

Fig. 2. The Layer architecture of the SOA for geographic information sharing and service of Digital City

other is AGS (Application GIS Services), which reflects the operation service, upper the VGS. It's easy to see that VGS is the abstract service for the bottom GIS module's function and the AGS is the knowledge deposit oriented industry.

The purpose to bring forward VGS is to reuse the GIS services, functions and the corresponding function management logic which have been implemented by every system. VGS has divided the GIS service from different levels and different points of view, thus the similar GIS system developers simply need to pay attention to the unimplemented service, construct system fleetly by using the existing resource, and bring forward an effective resolution decision-making for the ceaselessly variation requests of clients, so implements the demand of replying the change of every requests. The purpose of designing is to decompose, split, thin and classify the GIS service demands of the application system so that the development of the blurry, variable application will be easy and explicit. It liberates application developers from the multifarious coding work and helps them pay attention to fit the application and satisfy users by its set of complete services with more expressive force. According to this purpose, the internal module demarcation of VGS strives for the equilibrium of maturity and expressive force. Because maturity decides whether VGS can provide complete support for users, and expressive force determines the complexity of the function module logical organization structure when carrying on application development by VGS. The services, abstracted to different granularities according to this technique thinking way, can define a new service kit. Thus application system can construct different application system by using these services. And in case the kit falls together with certain model, a kind of study ability will be formed thus the decision-making support will be implemented. The difference of AGS and VGS lies in the function layer. AGS orients the industries and operations, not GIS. The purpose of AGS is to build an effective industry knowledge deposit mechanism, so it's similar with VGS. Currently, the interfaces provided by the grid services are still limited. The SOA for geographic information sharing and service of Digital City are still in the continuous developing process. It will take the management enlargement and the safe etc into consideration on the next move.

5 Web Service Based Model Integration in SOA

The open computing environment of Web causes the Spatial Analysis Model and the Applied Mathematics Model very good combination. Using the distributed computing environment, the acquirement of data becomes more extensive and the interaction model computation becomes stronger. The multi-models computation can synthetically compare, analyze and provide the ability of on-line real-time processing. Based on GeoServer, it can provide the function of GIS analysis and expression. It can also establish a kind of model reuse mechanism based on the web service model in this foundation.

In the virtual environment made up of web service, users can use any client software to call the web service. Taking SOAP as the communication protocol, web services is based on the opening and standard specification. SOAP calls the long-distance service by the XML standard encapsulation and web service can be transmitted through the entity transmission level. This is the foundation of software reuse under isomerism conditions. The web service demonstrates the service provided by itself through WSDL to the client and makes model reuse realization more convenient.

Decision Support System (DSS) communication structure is based on the web service technology (Fig.2). Firstly, service management machine provides the service directory according to the model service content. The client submits the the request of model call service (the SOAP request), based on the directory content and description. This request is handed to the XML switch by the service directory. The XML switch converts the XML structure requested by SOAP to the language structure which can be distinguished by the model service, then proposes call request to model service according to specific request parameter. The client can inquire the service it calls whether exists in model service directory. If the service exists, then it activate the corresponding model service and run the model to get a result, then give the model output result to the XML conversion program. The XML conversion program converts the model output to the XML structure and sends the response based on HTTP protocol to the client by service management machine.

6 Application Example

Xiamen City of China is on the seashore Taiwan Strait. The ocean dynamical environment real-time solid monitoring system of Taiwan Strait and adjacent maritime region is composed of the oceanic shore/platform monitoring net, the high frequency terra wave radar monitoring net, the buoys monitoring net, boat based monitoring net and remote sensing monitoring net. The monitoring nets form the three dimension solid monitoring system which monitors the oceanic dynamical environment and ecological environment from underwater, ocean surface and air. The observing data is located in a distributed web environment. The web service based ocean current numerical simulation computation model integration can acquire the distributed observing data and be called by "Shipwreck Salvation decision-making support system" to forecast the floating trail of the shipwreck objects.

7   Conclusions

In this article, we firstly introduce the requirement of the distributed geographic information sharing and service and the character service-oriented software architecture. Then the principle of the service-oriented architecture and the service-oriented geographic information sharing and service software architecture for digital city is put forward which includes GIS service, called VGS (Virtual GIS Services) and AGS(Application GIS Services) , which reflects the operation service upper the VGS. At last, an application example is stated to demonstrate the advantage and application model of the service-oriented geographic information sharing and service software architecture.

Acknowledgments. This work was sponsored by Open fund of State Key Laboratory of Marine Environmental Science (Xiamen University) (No. MEL0507).

References

1. Coleman D. Geographical Information Systems in Networked Environments in Longley, Goodchild, Maguire and Rhind. Geographical Information Systems:Principles,Techniques, Applications and Management[M]. New York:John Wiley & Sons,317-29,1999.
2. Peng Zhong-Ren, Ming-Hsiang Tsou. Internet GIS:Distributed Geographic Information Services for the Internet and Wireless Networks[M]. Hoboken, New Jersey:John Wiley & Sons,2003.
3. T. Ishida, Understanding Digital Cities, Digital Cities, Technologies and Future Perspectives, 1765 (1997).
4. J. Dale and D. Deroure, A Mobile Agent Architecture for Distributed Information Management, 98 (1997).
5. M.H Tsou and B.P. Buttenfield, Agent-based Mechanisms for Distributing Geographic Information Services on the Internet, First International Conference on Geographic Information Science, 223(2000).
6. Y.X. Huang, Web-Based Geographic Information Services, 63(2002).

Graded Reasoning in n-Valued Łukasiewicz Propositional Logic*

Hongjun Zhou and Guojun Wang

College of Mathematics and Information Science,
Shaanxi Normal University, Xi'an 710062, China
sdzhjun@stu.snnu.edu.cn,
gjwang@snnu.edu.cn

Summary. In the present paper we deal with graded reasoning in n-valued Łukasiewicz propositional logic $\mathbf{Ł}_n$. Firstly we propose an approach to measure the extent to which a theory over $\mathbf{Ł}_n$ is consistent. Secondly, with the concept of consistency degrees of theories, we give several methods of graded reasoning in n-valued Łukasiewicz propositional logic $\mathbf{Ł}_n$.

1 Introduction

Deciding whether a theory (i.e., a set of formulas) is consistent or not is one of the crucial questions in any logic systems. The reason is that in classical logic, a contradictory theory (i.e., a theory which is not consistent) turns into a useless theory in which every thing is provable. Quite surprising is that the same result holds also in fuzzy and many-valued logic systems. Moreover, how to measure the extent to which a theory is consistent is also one of the crucial questions in logic systems. For trying to grade the extent of consistency of different theories, many authors have proposed different methods in fuzzy (continuous valued) logic systems and have obtained many good results [5, 2, 10, 14, 11, 12]. Especially in [12], the authors, from logical point of view and based on deduction theorems, completeness theorems and the concept of satisfiability degrees of formulas, introduced, in classical and fuzzy propositional logic systems, a more natural and reasonable definition of consistency degrees of theories. In other words, we have studied successfully the consistency of theories where the set of truth values jumped from $\{0, 1\}$ to $[0, 1]$. A natural question then arises: how to harmoniously fill in the gap of consistency of theories between classical and fuzzy logic systems? That is to say, how to establish the concept of consistency degrees of theories in n-valued logic systems such that it approximates the consistency of theories in fuzzy logic system when n turns to infinity, and takes the classical case as a special case when $n = 2$?

In [13], we have adapted the results of [12] to the n-valued R_0-logic (more precisely, the n-valued NM-logic) and have partially answered the above question.

* Projected by the Natural Science Foundation of China under Grant 10331010, and the Superior Dissertation Foundation of Shaanxi Normal University.

K.M. Węgrzyn-Wolska and P.S. Szczepaniak (Eds.): Adv. in Intel. Web, ASC 43, pp. 387–391, 2007.
springerlink.com
© Springer-Verlag Berlin Heidelberg 2007

Following the train of thought from special case to general, the present paper has two purposes. The first one is to consider the consistency degrees of theories and to adapt the results of [12] to L_n. The second is to consider the general reasoning, to propose several other methods of graded reasoning, and further to compare these results. As we will see, the methods of graded reasoning given in the present paper in which the satisfiability degrees of formulas is a crucial tool are different and far from that of [5, 2].

2   Preliminaries

Consider the length of the paper, for the representation of Łukasiewicz propositional n-valued logic system L_n, we refer to [1, 8, 6, 3, 7].

Definition 2.1. *(Halmos [4]).* Suppose that $(X_k, \mathcal{A}_k, \mu_k)(k = 1, 2, \cdots)$ are probability measure spaces, let $X = \prod\limits_{k=1}^{\infty} X_k$, then $\prod\limits_{k=1}^{\infty} \mathcal{A}_k$ generates on X a σ-algebra $\mathcal{A}$. There exists on X a measure μ satisfying the following conditions: (i) $\mathcal{A}$ is the set of consisting of all μ-measurable sets, (ii) For any measurable subset E of $\prod\limits_{k=1}^{m} X_k, E \times \prod\limits_{k=m+1}^{\infty} X_k$ is μ-measurable and μ is called the infinite product measure of $\mu_1, \mu_2, \cdots$, and $(X, \mathcal{A}, \mu)$ is called the infinite product of $\{(X_k, \mathcal{A}_k, \mu_k)\}_{k=1}^{\infty}$. $(X, \mathcal{A}, \mu)$ can be abbreviated as X if no confusion arises.

Definition 2.2. Suppose that $n \geq 2$ is a fixed natural number, and $(Y, \mathcal{B}, \eta)$ is an evenly distributed probability measure space where $Y = \{y_1, \cdots, y_n\}$, i.e., $\eta(\emptyset) = 0, \eta(Y) = 1$ and $\eta(y_i) = \frac{1}{n}(i = 1, \cdots, n)$. Let $(X_k, \mathcal{A}_k, \mu_k) = (Y, \mathcal{B}, \eta)(k = 1, 2, \cdots)$, and $(X, \mathcal{A}, \mu)$ be the infinite product of $\{(X_k, \mathcal{A}_k, \mu_k)\}$, then $(X, \mathcal{A}, \mu)$ is called an n-valued logic measure space.

Definition 2.3. *(Wang and Li [9]).* Suppose that $A \in F(S)$ in $L_n, n \geq 2$, define

$$\tau_n(A) = \sum_{i=0}^{n-1} \frac{i}{n-1}\mu([A]_{\frac{i}{n-1}}) = \sum_{i=1}^{n-1} \frac{i}{n-1}\mu([A]_{\frac{i}{n-1}}),$$

where $[A]_{\frac{i}{n-1}}$ is the class of $\frac{i}{n-1}$-models of A, i.e.,

$$[A]_{\frac{i}{n-1}} = \{\vec{v} \in L_n^{\infty} \mid \vec{v} = \varphi(v), v(A) = \frac{i}{n-1}, v \in \Omega_n\}, \quad i = 0, 1, \cdots, n-1.$$

Then $\tau_n(A)$ is called the n-valued satisfiability degree of A in L_n.

Lemma 2.4. Suppose that A and B are formulas in L_n, $n \geq 2$, then

(i) A is a tautology iff $\tau_n(A) = 1$,
(ii) A is a contradiction iff $\tau_n(A) = 0$,
(iii) $\tau_n(\neg A) = 1 - \tau_n(A)$,

(iv) If $\vdash_n A \to B$ holds, then $\tau_n(A) \le \tau_n(B)$,

(v) If A and B are logically equivalent then $\tau_n(A) = \tau_n(B)$.

(vi) Define

$$\rho_n(A, B) = 1 - \tau_n((A \to B) \wedge (B \to A)), A, B \in F(S).$$

Then it is easy to check that $\rho_n(A, B)$ is a pseudo-metric on $F(S)$.

3 Consistency Degrees of Theories Based on Deduction Theorems in $\mathbf{L}_n$

As space is limited, we only list the main theorems in the following. Some explanations and the proofs are also omitted here.

Definition 3.1. Suppose that Γ is a theory of $\mathbf{L}_n, n \ge 2, 2^{(\Gamma)}$ is the set of all finite subsets of $\Gamma, \Sigma = \{A_1, \cdots, A_m\} \in 2^{(\Gamma)}$. Let

$$\Sigma^n \to \bar{0} = \begin{cases} A_1^{n-1} \& \cdots \& A_m^{n-1} \to \bar{0}, & m > 0, \\ \bar{0}, & m = 0, \end{cases}$$

and define

$$\mu_n(\Gamma) = \sup\{\tau_n(\Sigma^n \to \bar{0}) \mid \Sigma \in 2^{(\Gamma)}\}.$$

Then $\mu_n(\Gamma)$ is called the degree of entailment of $\bar{0}$ from Γ, or say, $\bar{0}$ is a consequence of Γ in the degree $\mu_n(\Gamma)$.

The calculation of $\mu_n(\Gamma)$ can be simplified as follows:

Theorem 3.2. Suppose that Γ is a theory of $\mathbf{L}_n, n \ge 2$, then

$$\mu_n(\Gamma) = 1 - \inf\{\tau_n(A_1^{n-1} \& \cdots \& A_m^{n-1}) \mid A_1, \cdots, A_m \in \Gamma, m \in \mathbf{N}\}.$$

Example 3.3. Calculate $\mu_n(\Gamma)$ for (i) $\Gamma = \emptyset$, (ii) $\Gamma = \{p\}$, where p is a propositional variable, (iii) $\Gamma = S = \{p_1, p_2, \cdots\}$.

Solution. (i) $\mu_n(\Gamma) = 0$; (ii) $\mu_n(\Gamma) = 1 - \frac{1}{n}$; (iii) $\mu_n(S) = 1$.

Theorem 3.4. Let Γ be a theory of $\mathbf{L}_n, n \ge 2$. If Γ is inconsistent then $\mu_n(\Gamma) = 1$, but not vice visa.

Definition 3.5. Suppose that Γ is a theory of $\mathbf{L}_n, n \ge 2$, define

$$i_n(\Gamma) = \max\{[\tau_n(\Sigma^n \to \bar{0})] \mid \Sigma \in 2^{[\Gamma]}\},$$

and $i_n(\Gamma)$ is called the polar index of Γ in $\mathbf{L}_n$, where $\Sigma^n \to \bar{0}$ is defined as in Definition 3.1.

Theorem 3.6. Suppose that Γ is a theory of $L_n, n \geq 2$, then

(i) Γ is consistent iff $i_n(\Gamma) = 0$,
(ii) Γ is inconsistent iff $i_n(\Gamma) = 1$.

Definition 3.7. Suppose that Γ is a theory of $L_n, n \geq 2$, define

$$consist_n(\Gamma) = 1 - \frac{1}{2}\mu_n(\Gamma)(1 + i_n(\Gamma))$$

and call $consist_n(\Gamma)$ the consistency degree of Γ in L_n.

Theorem 3.8. Suppose that Γ is a theory of $L_n, n \geq 2$, then

(i) Γ is completely consistent, i.e., all members of $D(\Gamma)$ are tautologies, iff $consist_n(\Gamma) = 1$,
(ii) Γ is consistent iff $\frac{1}{2} \leq consist_n(\Gamma) \leq 1$,
(iii) Γ is consistent and $\mu_n(\Gamma) = 1$ iff $consist_n(\Gamma) = \frac{1}{2}$,
(iv) Γ is inconsistent iff $consist_n(\Gamma) = 0$.

4   Methods of Graded Reasoning in L_n

Definition 3.1 indeed offers a method to measure the extent to which the contradiction $\overline{0}$ is a consequence of a theory Γ. If we replace $\overline{0}$ with a general formula A then we get a method to measure the extent to which the formula A is a consequence of a given theory Γ. Moreover, the pseudo-metric ρ_n defined in Lemma 2.4 induces naturally another method of graded reasoning in L_n by evaluating the distance of A to the deductive closure of Γ. It is of interest to prove that these two seeming completely different methods are equivalent. Based on the concept of Hausdorff distance on a nonempty set, we can introduce a third method of graded reasoning in this logic. As space is limited, all these issues are omitted here. We will add this part when presenting our talk at the conference.

References

1. S. Gottwald, A Treatise on Many-Valued Logics, Studies in Logic and Computation, Research Studies Press. Baldock, 2001.
2. S. Gottwald and V. Novák, On the consistency of fuzzy theories, Proc. of 7th IFSA World Congress, 1997, Academi, Prague, 168-171.
3. P. Hájek, Metamathematics of Fuzzy logic, Kluwer Academic Publishers, Dordrecht, 1998.
4. P.R. Halmos, Masure Theory, Springer-Verlag, New York, 1974.
5. V. Novák, I. Perfilieva and J. Močkoř, Mathematical Principles of Fuzzy Logic, Kluwer Academic Publishers, Boston, 1999.
6. W.A. Pogorzelski, The deduction theorem for Lukasiewicz many-valued propositional calculi, Studia Logica 15(1964)7-23.
7. E. Turunen, Mathematics Behind Fuzzy Logic, Advances in Soft Computing, Physica-Verlag, Heidelbery, 1999.

8. R. Tuziak, An axiomatization of the finite valued Lukasiewicz Calculus, Studia Logica 47(1988)49-55.
9. G.J. Wang, B.J. Li, Theory of truth degrees of formulas in Lukasiewicz n-valued propositional logic and a limit theorem, Sci. China, Information Science, E 35(6)(2005)561-569 in Chinese.
10. G.J. Wang, W.X. Zhang, Consistency degrees of finite theories in Lukasiewicz propositional fuzzy logic, Fuzzy Sets and Systems 149(2005)275-284.
11. H.J. Zhou, G.J. Wang, A new theory consistency index based on deduction theorems in several logic systems, Fuzzy Sets and Systems 157(2006)427-443.
12. H.J. Zhou, G.J. Wang, Generalized consistency degrees of theories w.r.t. formulas in several standard complete logic systems, Fuzzy Sets and Systems, in press.
13. H.J. Zhou, G.J. Wang, W. Zhou, Consistency degrees of theories and methods of graded reasoning in n-valued R_0-logic(NM-logic), International Journal of Approximate Reasoning 43(2006)117-132.
14. X.N. Zhou, G.J. Wang, Consistency degrees of theories in some systems of propositional fuzzy logic, Fuzzy Sets and Systems 152(2005) 321-331.

Measurement Centred Evaluation in Adaptive Web-Based Learning Systems Based on CBR Approach

Houda Zouari Ounaies[1], Yassine Jamoussi[2], and Henda Ben Ghezala[3]

[1] ENSI, National School of Computer Science
houda.zouari@isetr.rnu.tn
[2] ENSI, National School of Computer Science
yassine.jamoussi@ensi.rnu.tn
[3] ENSI, National School of Computer Science
henda.benghezala@riadi.rnu.tn

Summary. Several Web-based learning systems are currently available. Nevertheless, most existing e-learning platforms lack efficient adaptivity to learner needs. Moreover, adaptive system development is quite costly. Evaluation of these systems and assessment of adaptation components may enable identifying failures and improving them, in one hand, and exchanging successful adaptive components, in the other hand.

Adaptive behaviour is strongly determined by functional requirements available to the application domain. In Web-based system, granular functions are increasingly provided by Web services. First, in this paper we focus on measurement criteria related to both adaptive functions and related Web services. Second, Case-Based Reasoning are adopted in order to capitalize evaluation results, learn from past experiences and reuse successful patterns.

Keywords: Evaluation,adaptive learning systems, measurement, functional requirements, Web services, CBR.

1 Introduction

Web-based learning is an emergent area. Space and time flexibility encourage strongly stakeholders to adopt this technology. Currently, learners are increasingly interested to educational systems that fit their changing and disparate needs. Nevertheless, most e-learning platforms lack adaptivity [4]. Many researches advocate the high cost of adaptive systems development as a critical rationale of adaptivity deficiency [6][7]. Consequently, interoperability between adaptive Web-based educational system and reuse of successful adaptive functional components is widely necessary.

A measurement centred-evaluation of these systems is a key issue for identifying failures and improving them, in one hand, and reusing successful adaptive patterns, in the other hand. Service Oriented Approaches (SOA) enhance e-learning flexibility. Accordingly, they are recommended by numerous e-learning organizations such as DEST (Australia), JISC-CETIS (UK), IC (Canada), ADL

K.M. Węgrzyn-Wolska and P.S. Szczepaniak (Eds.): Adv. in Intel. Web, ASC 43, pp. 392–397, 2007.
springerlink.com © Springer-Verlag Berlin Heidelberg 2007

(US) [8] and they are mainly based on Web Services. Adaptive educational system can be decomposed to a set of autonomous functional components delivered by Web Services.

Evaluating the adaptation performance from a functional dimension necessitate to consider specific measurement criteria. Expert and real users are involved in the evaluation process. In order to deal with judgment uncertainty in both measurement and interpretation steps, we propose to capitalize from past evaluation results. Data-mining techniques such as Case Based Reasoning (CBR) can bring much benefit to supports measurement assessment and enhances decision making.

In this paper we propose a measurement centred evaluation approach. This research is twofold goal. First, we focus on adaptive functional dimension measures based on learning Web Services. Second, on the basis of these measurements, past evaluation experiences are characterized and capitalized by means of CBR approach. In the following section, we describe the inter-relation between learning Web Services and adaptation components.

2   Learning Web Services in Adaptive Context

SOA are being increasingly integrated into e-learning field. Several e-learning systems based on Web services technology are currently available such as UbiLearn, dotLRN, Moodle, Blackboard, WebCT, etc. Behaviours contained in adaptive learning systems are fully or partially exposed as services. In 2004, JISC proposes the ELF framework which is extended afterwards by IC, DEST, and LSAL. ELF supplies a list of functions that can be provided as services, in e-learning applications. Nevertheless, learners are asking for more and more personalized services. Consequently, adaptation issue is becoming essential in Web-based learning systems. Evaluation of adaptive learning systems measure adaptation performance and assess at which extent functionalities delivered to the users fits their needs.

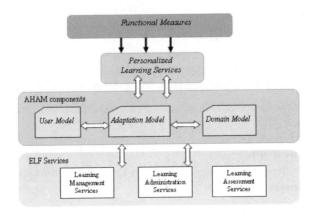

Fig. 1. Components interaction in Service based adaptive learning system

Most of recent adaptive systems are based on AHAM model (Adaptive Hyper-media Application Model) [5]. In our work we focus on adaptive Web based educational system adopting AHAM principles. Inter-relations between different adaptive learning systems components based on ELF framework are illustrated in 1.

Adaptive functionalities measures are needed to assess adaptation behaviour provided by personalized learning services. AHAM components generate adaptive presentation specifications. In the following section, we focus on the measurement dimension related to the evaluation of adaptive Web-based learning systems.

3   Measurement Focus in Adaptive Web-Based Learning Systems

Measurement-centered evaluation framework provides an objective basis to assess adaptivity in Web based learning system, indicating their strengths and shortcomings. Measures of software internal attributes have been extensively used to help software managers, customers and users to characterize, assess, and improve the quality of software products [2]. One of our current concerns is to measure adaptive functional requirement in order to help evaluators to understand at which extent internal adaptation attributes fit user needs.

David Chin proposes the following measures to be considered in adaptive system evaluation [3]:

- Frequency of certain behaviors,
- Qualities of a behavior in a particular situation,
- Number of errors,
- Error rate,
- Time to complete a task,
- Proportion/qualities of tasks achieved,
- Interaction patterns,
- Learning time/rate.

We propose to relate these measures to Web services that are behind adaptive functionalities delivered to the learner. Moreover, we consider the following measurement criteria as fundamental in our measured evaluation framework:

1. Responsiveness: describes timeliness of proposed functions and Web services behind them.
2. Appropriateness & acceptance: deal with the quality of experience of learner with the delivered function.
3. Availability: describes the accessibility to the needed functionalities and the related Web services.

Decisions are taken to face possible failures expressed by measurement criteria. Experts and users are both implicated in the evaluation process. Nevertheless, in many cases the decision makers could be uncertain, due to the complexity of the decision environment and to information or knowledge lacking. In order to reinforce decision making, we suggest to consider past evaluation experiences and to capitalize from past evaluation results. Case-Based Reasoning (CBR) is being widely adopted in various domains. In our research, we aim to use this approach to support assessment process, in one side, and to estimate failures origins, in the other side.

4 Adaptive Learning System Evaluation Based on CBR Approach

CBR is based on knowledge reuse and past experience capitalization to solve problems. As described in [1], a new problem is solved by finding a similar past case, and reusing it in the new problem situation. Past experiences are stored in a Case Base. One case is composed by both problem situation and past experienced solution. A new solution is obtained by comparison between new situations and past cases. In the following section, we describe a learning situation in an adaptive Web-based context.

Measures characterize the learning situation. One situation is related to one session launched by a learner. It is described first by user model state which reflects the learner knowledge about concepts of the application domain. Adaptive Web-based system presents functionalities fitting user needs. In SOA context, each function corresponds to a set of orchestrated Web Services. 2 illustrates UML class diagram describing adaptive learning situation.

Different learning scenarios are possible. Measurement criteria characterize each specific situation and reflect the validity of delivered functionalities adaptation. Solutions to resolve possible failures are proposed. Adjustments results are stored as a new case in the CBR system.

Currently, we are implementing an automated case creation tool. Information from the current state of the user model, the delivered functionalities and related Web services are stored automatically in the Case Base. Nevertheless,

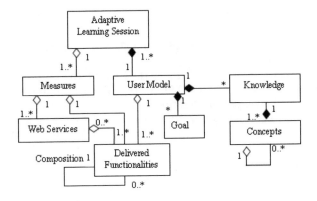

Fig. 2. Adaptive learning situation

measurement criteria which deal with user quality of experience are assigned by both learner and expert. This CBR project aims to develop a system that is able to create cases using a mostly automated process, and to develop a Java-based case-based reasoning engine that can be integrated to any adaptive web-based educational system. The cases capture the adaptation performance expertise of learning system and best adjustment practices that can be reused in similar situation.

Nevertheless, a set of factors can compromise the validity of measurement adopted in the evaluation such as data contamination or a non-sufficient number of users. Currently, we are testing this tool by our lab students. Measures and evaluation results are stored in the case base. An important number of cases are needed to allow an effective knowledge capitalization.

5   Conclusion and Future Work

This paper has introduced a measurement centred-evaluation approach for adaptive web-based learning system. It focuses on functional adaptation measurement issue. Currently Web services are increasingly adopted in e-learning applications. This technology enables using functions that are distributed around the Web and facilitate interoperability across learning systems. Evaluation of these systems aims to identify failures and to reuse successful adaptive patterns.

Evaluation is mostly based on learner and expert decisions. In order to deal with decision uncertainty, we propose to reinforce the evaluation process by adopting Case-Based Reasoning. Adaptive learning situations are stored in the Case Base. Measures related to delivered functionalities and to Web services behind them, characterize each situation. CBR allows learning from past evaluation experiences and enhances assessment decision making in the current evaluation. Currently, we are conducting several experiments using manually or semi-manually case stored. Finally, we are studying the integration of automated and real-time case creation in adaptive e-learning environment.

References

1. Aamodt, A., Plaza, E.: AICom - Artificial Intelligence Communications, IOS Press, Vol. 7: 1, (1994) 39-59
2. Brito, F., Poels, G., Sahraoui, H., Zuse, H.: Quantitative Approaches in Object-Oriented Software Engineering, ECOOP workshop, Nice, Lecture notes in computer science, Springer-Verlag, Berlin Heidelberg New York (2000)
3. Chin, D.: Empirical Evaluation of User Models and User-Adapted Systems, User Modeling and User-Adapted Interaction Vol. 11, No. 1/2, Kluwer Academic Publishers, (2001) 181-194
4. Graf, S., List, B.: An Evaluation of Open Source Platforms Stress adaptation Issues. Proceedings of the International Conference on Advanced Learning Technologies. Kaohsiung, Taiwan (2005) 163-165
5. Muntean, C.: Quality of Experience Aware Adaptive Hypermedia System, Thesis dissertation, Dublin (2005)

6. F. Mödritscher, C. Gütl, V. M. García-Barrios, H. Maurer: Why do Standards in the Field of E-Learning not fully support Learner-centred Aspects of Adaptivity?, proceedings of ED-Media conference, Lugano (2004)
7. Stewart C., I. Celik, A. Cristea, Ashman, H.: Interoperability between AEH User Models, ACM conference APS'06, Odense, Denmark (2006)
8. Wilson, S, Blinco, K, Rehak, D.: Service-Oriented Frameworks: Modelling the infrastructure for the next generation of e-Learning Systems. Available at http://www.jisc.ac.uk/uploaded_documents/AltilabServiceOrientedFrameworks.pdf, (2004)

Author Index

Printing: Mercedes-Druck, Berlin
Binding: Stein+Lehmann, Berlin